A LEVEL & AS

HUMAN GEOGRAPHY

AQA Geography

Series editors Simon Ross   Alice Griffiths

Tim Bayliss   Lawrence Collins

Catherine Hurst   Bob Digby

Andy Slater

OXFORD
UNIVERSITY PRESS

OXFORD
UNIVERSITY PRESS

Great Clarendon Street, Oxford, OX2 6DP, United Kingdom

Oxford University Press is a department of the University of Oxford. It furthers the University's objective of excellence in research, scholarship, and education by publishing worldwide. Oxford is a registered trade mark of Oxford University Press in the UK and in certain other countries

© Oxford University Press 2016

Series editor: Simon Ross, Alice Griffiths

Authors: Tim Bayliss, Lawrence Collins, Catherine Hurst, Bob Digby, Andy Slater

The moral rights of the authors have been asserted

Database right of Oxford University Press (maker) 2016

First published in 2016

All rights reserved. No part of this publication may be reproduced, stored in a retrieval system, or transmitted, in any form or by any means, without the prior permission in writing of Oxford University Press, or as expressly permitted by law, by licence or under terms agreed with the appropriate reprographics rights organization. Enquiries concerning reproduction outside the scope of the above should be sent to the Rights Department, Oxford University Press, at the address above.

You must not circulate this work in any other form and you must impose this same condition on any acquirer

British Library Cataloguing in Publication Data
Data available

ISBN 978-0-19-836654-6

10 9 8 7 6 5 4 3 2

MIX
Paper from responsible sources
FSC® C007785

Paper used in the production of this book is a natural, recyclable product made from wood grown in sustainable forests. The manufacturing process conforms to the environmental regulations of the country of origin.

Printed in Great Britain by Bell and Bain Ltd., Glasgow

Acknowledgements

The publisher and authors would like to thank the following for permission to use photographs and other copyright material:

Cover: © Patrick Bingham-Hall; **Cover:** Earth Imaging/Getty Images; **p6:** Susan Schulman; **p11:** The Grameen Foundation; **p13:** GEORGE OSODI/AP/Press Association Images; **p16:** JHumbrella.com; **p19:** Getty Images / Mike Kemp / Contributor; **p20:** Arnos Design Ltd; **p21:** mauritius images GmbH/Alamy Stock Photo; **p21:** Vanderwolf Images/Shutterstock; **p27:** Tom Oldham/REX/Shutterstock; **p28:** Jeff Morgan 16/Alamy Stock Photo; **p29:** epa european pressphoto agency b.v. / Alamy Stock Photo; **p29:** The Coca Cola Case ©2009 Argus Film/National Film Board of Canada. All rights reserved; **p30:** Shutterstock/Thinglass; **p33:** Eric St Pierre/Equal Exchange Coop; **p34:** Joki Desnommée-Gauthier.; **p35:** Paulo Whitaker/Reuters; **p36:** Marc Andersen/Alamy Stock Photo; **p37(r):** GreenPalm; **p38:** Nature Picture Library/Alamy Stock Photo; **p41(t):** ZUMA Press, Inc./Alamy Stock Photo; **p42:** Ingram Pinn/Financial Times 2008; **p45:** Xinhua/Alamy Stock Photo; **p46(b):** www.wordle.net; **p46(t):** Image created by Reto Stockli with the help of Alan Nelson, under the leadership of Fritz Hasler/VISIBLE EARTH/NASA; **p48(t):** Archive Pics/Alamy Stock Photo; **p51:** Nigel McCall/Alamy Stock Photo; **p54(b):** Wolfgang Kaehler/Getty Images; **p54(t):** R PLACZKIEWICZ/REX/Shutterstock; **p55:** Anonymous/AP/Press Association Images; **p56:** AF archive/Alamy Stock Photo; **p58:** blickwinkel/Alamy Stock Photo; **p59:** Professor Michael Ashley/University of New South Wales; **p60(b):** Prasit Photo/Getty Images; **p60(t):** popa/Toonpool; **p61(b):** Pavel Constantin/Cartoon Movement; **p61(tl):** Sean Gallup/Getty Images; **p61(tr):** Michael Kemp/Alamy Stock Photo; **p62:** International Association of Antarctica Tour Operators (IAATO); **p64:** Pool/Getty Images; **p66:** Gregory Davies/Alamy Stock Photo; **p67:** Rudi Van Starrex/Getty Images; **p69:** Peter Jordan_NE/Alamy Stock Photo; **p69:** ZUMA Press, Inc./Alamy Stock Photo; **p70:** Guy Corbishley/Alamy Stock Photo; **p71:** Ivy Close Images/Alamy Stock Photo; **p71:** YHA (England and Wales); **p72:** ton koene/Alamy Stock Photo; **p73:** Paisan changhirun/Alamy Stock Photo; **p74:** Greek Photsnews/Alamy Stock Photo; **p74:** Robert Harding/Alamy Stock Photo; Copyright © David Nicholls 2014, reproduced by Hodder & Stoughton Limited; **p76(l):** Allan Wright/Alamy Stock Photo; **p76(r):** Jack Sullivan/Alamy Stock Photo; **p77:** eye35 images/Alamy Stock Photo; **p78:** paul weston/Alamy Stock Photo; **p79(l):** Source: Department for Education Crown copyright -HR to get license www.clickanduse.hmso.gov.uk; **p80(b):** Colin Palmer Photography/Alamy Stock Photo; **p80(t):** 1987: REUTERS/Stringer, 2013: REUTERS/Carlos Barria; **p81(b):** International Photobank/Alamy Stock Photo; **p81(t):** Dorset County Museum; **p82(b):** Martin Bond/Alamy Stock Photo; **p82(t):** Jon Arnold Images Ltd/Alamy Stock Photo; **p83(r):** Guy Corbishley/Alamy Stock Photo; **p85(b):** Raymond Long/Alamy Stock Photo; **p85(t):** Anthony Woodhouse; **p86(b):** Robert Harding/Alamy Stock Photo; **p86(m):** The Photolibrary Wales/Alamy Stock Photo; **p86(t):** Courtesy of Alice Liddell Innovative Community Enterprise Ltd; **p87(t):** VisitGreatBritain.com; **p88(b):** Jeff Morgan12/Alamy Stock Photo; **p88(t):** Heritage Image Partnership Ltd/Alamy Stock Photo; **p91(b):** Tagxedo/Creative Commons Attribution-Noncommercial-ShareAlike License 3.0; **p92(l):** EDB Image Archive/Alamy Stock Photo; **p92(r):** Anthony Palmer/Alamy Stock Photo; **p93(b):** Department for Communities and Local Government (OGL); **p93(t):** Datashine/James Cheshire and Oliver O Brien/UCL Geography. Place names and buildings: Ordinance Survey © Crown copyright and Database rights 2014-15.; **p94:** Department for Communities and Local Government (OGL); **p95(b):** © Consumer Data Research Centre 2016/© Crown copyright & database right 2014-5; **p95(t):** Anthony Palmer/Alamy Stock Photo; **p96:** Paul Felix Photography/Alamy Stock Photo; **p97(b):** Courtesy of the British Film Institute; **p97:** Steven Sidhu; **p98:** Maurice Savage/Alamy Stock Photo; **p99(b):** www.wordle.net; **p99(m):** Alice Griffiths; **p99(t):** Chilterns Conservation Board; **p101(l):** Mark King; **p101(r):** Alice Griffiths; **p101(t):** Crown Copyright (2016)100043706/Chilterns Conservation Board; **p102:** Roy Porter; **p106(b):** Wikimedia Commons/Public Domain; **p106(t):** Kristoffer Tripplaar/Alamy Stock Photo; **p107(b):** Jim West/Alamy Stock Photo; **p107(t):** Everett Collection Historical/Alamy Stock Photo; **p108(b):** Jim West/Alamy Stock Photo; **p108(t):** BRIAN HARRIS/Alamy Stock Photo; **p110(b):** AF archive/Alamy Stock Photo; **p110(t):** Pictorial Press Ltd/Alamy Stock Photo; **p111(b):** Bill Rankin/www.radicalcartography.net; **p111(tr):** Eric Fischer/© OpenStreetMap, CC-BY-SA; **p112:** © 2014 (Alana Semuels/Los Angeles Times/MCT) All rights reserved. Distributed by Tribune Content Agency; **p116:** Bloomberg/Getty Images; **p119:** From "World Urbanisation Prospects" by The Dept of Ecomomic and Social Affairs, © 1970 and 2014, Unite Nations. Reprinted with the permission of the United Nations; **p123:** Paul Chesley/Getty Images; **p124(r):** By Keepscases (Own work) [CC BY-SA 3.0 (http://creativecommons.org/licenses/by-sa/3.0 or GFDL (http://www.gnu.org/copyleft/fdl.html)], via Wikimedia Commons; **p126:** Bloomberg/Gett Images; **p127:** Huw Jones/Getty Images; **p129:** Alistair Laming/Alamy Stock Photo; **p131:** Bluesky International Limited; **p133:** Lawrence Collins; **p134:** Bay Ismoyo/Getty Images; **p135:** Tim Bayliss; **p139:** John Michaels/Alamy Stock Photo; **p140(b):** Trevor Christopher/Shutterstock; **p141:** Tim Bayliss; **p142:** Ladd Company/Warner Bros/REX/Shutterstock; **p143:** littleny/Shutterstock; **p144:** "Changing Tastes in Britain stamps" designed by Catell Ronca and Rose Design © Royal Mail Group Limited, 2005; **p145(l):** Mark Phillips/Alamy Stock Photo; **p145(r):** Matthew Lewis/Getty Images; **p149:** Arup/University College London; **p150(t):** Jodie Griggs/Getty Images; **p152:** Hulton Archive/Getty Images; **p153(b):** Tim Bayliss; **p153(t):** Walter Bibikow/Getty Images; **p155:** Bruce McGowan/Alamy Stock Photo; **p157(b):** photosilta/Alamy Stock Photo; **p157(t):** Oxford University Press; **p158:** Royal Haskoning; **p160:** Airfotos/Northumbrian Water; **p161:** Tim Bayliss; **p163:** Huguette Roe/Shutterstock; **p164:** Aurora Photos/Alamy Stock Photo; **p165:** Mark Edwards/Still Pictures; **p167:** Tim Bayliss; **p168(b):** © Bill & Melinda Gates Foundation/Michael Hanson; **p168(t):** Gui Yongnian/123RF; **p170(b):** Jeff Morgan09/Alamy Stock Photo; **p170(t):** Tim Bayliss; **p171:** Chris Cooper-Smith/Alamy Stock Photo; **p172(b):** San Francisco County Transportation Authority (SFCTA); **p173:** Raf Makda/View Pictures/REX/Shutterstock; **p174:** Tim Bayliss; **p175:** ImageBROKER/Alamy Stock Photo; **p176(b):** LaiQuocAnh/Shutterstock; **p176(t):** CP DC Press/Shutterstock; **p178:** Sean Caffrey/Getty Images; **p179(b):** By Halley Pacheco de Oliveira (Own work) [CC BY-SA 3.0 (http://creativecommons.org/licenses/by-sa/3.0)], via Wikimedia Commons; **p179(t):** Peter M. Wilson/Alamy Stock Photo; **p181:** Tim Bayliss; **p183:** Transport for London; **p191:** Tim Bayliss; **p192(bl):** age fotostock/Alamy Stock Photo; **p192(br):** Nick Turner/Alamy Stock Photo; **p192(tl):** Adrian Sherratt/Alamy Stock Photo; **p192(tl):** All Canada Photos/Alamy Stock Photo; **p192(tr):** Celtic Collection - Homer Sykes/Alamy Stock Photo; **p193(b):** Tim Bayliss; **p193(tr):** Jacques Jangoux/Alamy Stock Photo; **p194:** CRSHELARE/Shutterstock; **p198:** AP/Press Association Images; **p199(b):** Maximilian Buzun/Alamy Stock Photo; **p199(m):** REUTERS/Alamy Stock Photo; **p199(t):** Martin Hughes-Jones/Alamy Stock Photo; **p200:** daij/Fotolia; **p201(b):** Thornton Cohen/Alamy Stock Photo; **p201(m):** Philip Quirk/Alamy Stock Photo; **p201(t):** Art Directors & TRIP/Alamy Stock Photo; **p203:** journalturk/iStockphoto; **p204:** Bloomberg/Getty Images; **p205(b):** Oli Scarff/Getty Images; **p205(t):** Wolfgang Flamisch/Getty Images; **p206:** Gapminder; **p207:** © Crown copyright/Contains public sector information licensed under the Open Government Licence v3.0.; **p209:** © A. L. Hansell, L. A. Beale, R E. Ghosh, L. Fortunato, D. Fecht, L. Järup and P. Elliot 2014/Oxford University Press; **p211(b):** wlablack/Shutterstock; **p211(t):** Clara Molden/PA Archive/Press Association Images; **p212:** Iain McGillivray/Alamy Stock Photo; **p213:** epa european pressphoto agency b.v./Alamy Stock Photo; **p214(b):** Lawrence Berkeley National Lab/Roy Kaltschmidt; **p214(t):** Wikimedia Commons/Public Domain; **p215(t):** izf/Shutterstock; **p217:** www.worldmapper.org; **p218:** Jake Lyell/Alamy Stock Phot **p220:** Xinhua/Alamy Stock Photo; **p221(b):** lembi/Shutterstock; **p221(t):** Ella Ling & Malaria No Mor UK; **p226:** Courtesy of Global Initiative for Asthma www.ginasthma.org ; **p227:** TonyV3112/Shutterstock; **p230:** Kai-Otto Melau/Alamy Stock Photo; **p234:** www.worldmapper.org; **p235(b):** Ti Graham/Alamy Stock Photo; **p235(t):** Barry Lewis/Alamy Stock Photo; **p236:** Dmitry Chulov/Shutterstock; **p237:** epa european pressphoto agency b.v./Alamy Stock Photo; **p239:** U.S. Census Bureau/www.census.gov; **p240:** Anadolu Agency/Getty Images; **p242:** TMAX/Fotolia; **p244(b):** Use with kind permission of Michael Crisafulli/www.VernianEra.com; **p245:** Kevin Ebi/Alamy Stock Photo; **p248:** SCIENCE PHOTO LIBRARY; **p249:** BJERREGAARD JYTTE Polfoto/Press Association Images; **p250(t):** Chronicle/Alamy Stock Photo; **p251(b):** EcoHealth, Climate Change and Global Health: Quantifying a Growing Ethical Crisis, Volume 4, 2007, p397-205, Jonathan A. Patz, Holly K. Gibbs, Jonathan A. Foley, Jamesine V. Rogers, Kirk R. Smith, With permission of Springer.; **p251(t):** Courtesy of GRID Arendal/www.grida.no/graphicslib/detail/number-of-extra-skin-cancer-cases-related-to-uv radiation_1456; **p258:** ATTA KENARE/Getty Images; **p261(b):** Tim Bayliss; **p261(t):** OS data © Crov copyright and database right 2016; **p262:** Lawrence Collins; **p263:** Tommy (Louth)/Alamy Stock Phot **p266:** Per-Anders Pettersson/Getty Images; **p268(b):** Tim Bayliss; **p268(t):** Bloomberg/Getty Image **p271:** Julia Waterlow/Eye Ubiquitous/Alamy Stock Photo; **p272:** WhisperToMe/Wikimedia Commons/Public Domain; **p274:** Bloomberg/Getty Images; **p278:** Courtesy of the Library of Congress; **p282:** Courtesy of UNICEF; **p284:** John Walters/Daily Mail/REX/Shutterstock; **p287:** Tim Bayliss; **p288(t):** Roger Bacon/REUTERS/Alamy Stock Photo; **p289(b):** Courtesy of the Pacific Institute; **p289(t):** STR/Getty Images; **p289(t):** Tribune Content Agency LLC/Alamy Stock Photo; **p297:** Courtesy of WaterAid www.wateraid.org/uk; **p298:** REUTERS/Alamy Stock Photo; **p299:** Youssef Boudlal/Reuters; **p304:** jerdad/Fotolia; **p306:** Pavel L Photo and Video/Shutterstock; **p307:** epa european pressphoto agency b.v./Alamy Stock Photo; **p308:** pdm/Fotolia; **p309:** Gerd Ludwig/National Geographic Creative; **p310:** Paul Rookes/Shutterstock; **p311(b):** Gamma Rapho/Getty Images; **p311(t):** TEPCO/Alamy Stock Photo; **p313(m):** Environment Images/UIG/Getty Images; **p313(t):** Ashley Cooper pics/Alamy Stock Photo; **p314(b):** © Khongkitwiriyachan/Dreamstime.com; **p316:** Rodrigo Baleia/Greenpeace; **p319:** Jonathan Player/Alamy Stock Photo; **p322(t):** Ravell Call/AP/Press Association Images; **p324:** iofoto/Fotolia; **p327(b):** RESOLUTION/Balan Madhavan/Alamy Stock Photo; **p327(t):** Lawrence Collins; **p331(b):** David McNew/Getty Images; **p331(t):** Lucy Nicholson/REUTERS; **p333:** Danita Delimont/Alamy Stock Photo; **p336:** Tim Bayliss; **p337:** Simon Ross; **p339:** Kevin Wheal/Alamy Stock Photo; **p340:** Simon Ross; **p343(b):** Tim Bayliss; **p343(m):** Alice Griffiths; **p343(t):** © catrinhelen/Stockimo/Alamy Stock Photo; **p344(b):** Alice Griffiths; **p344(t):** © Bubbles Photolibrary/Alamy Stock Photo; **p345:** Alice Griffiths. All other photographs: Shutterstock.

Page design/layout: Kamae Design. Artwork by Kamae Design and Ian West.

Every effort has been made to contact copyright holders of material reproduced in this book. Any omissions will be rectified in subsequent printings if notice is given to the publisher.

Third party website addresses referred to in this publication are provided by Oxford University Press in good faith and for information only and Oxford University Press disclaims any responsibility for the material contained therein.

Approval message from AQA

This textbook has been approved by AQA for use with our qualification. This means that we have checked that it broadly covers the specification and we are satisfied with the overall quality. Full details of our approval process can be found on our website.

We approve textbooks because we know how important it is for teachers and students to have the right resources to support their teaching and learning. However, the publisher is ultimately responsible for the editorial control and quality of this book.

Please note that when teaching the AQA A Level and AS Geography courses, you must refer to AQA's specifications as your definitive source of information. While this book has been written to match the specifications, it cannot provide complete coverage of every aspect of the courses.

A wide range of other useful resources can be found on the relevant subject pages of our website: www.aqa.org.uk.

Contents

Contents

How to use this book

This is one of two books in this series, written for the AQA GCE in Geography. This particular book (Human geography) has been written to meet the content requirements of the A Level course, but can equally well be used for the separate AS course.

Skills questions indicated by the **S** icon are aimed at meeting the geographical and statistical skills requirements for both AS and A Level.

Practice questions have been included for both AS and A Level, with marks allocated. Please note that the Practice questions used in this book allow students a genuine attempt at practising exam skills, but are not intended to replicate the exact nature of final exam questions.

At appropriate points, chapters focus on providing fieldwork opportunities. These, plus associated questions, will help to prepare you for the fieldwork requirements for both AS and A Level.

In April 2015, the residents of Numbi watched the erection of a mobile phone mast – the first in their region. Numbi is a remote and poor town in the DRC, one of the world's least-developed countries. How do you think being connected to the rest of the world could change things for Numbi and its people?

Your exam

 'Global systems and global governance' is a core topic. You must answer all questions in Section A of Component 2: Human geography. Component 2 makes up 40% of your A Level.

Your key skills in this chapter

 In order to become a good geographer you need to develop key geographical skills. In this chapter, whenever you see the skills icon you will practise a range of quantitative and relevant qualitative skills, within the theme of 'global systems and global governance'. Examples of these skills are:

- Using different types of data to develop critical perspectives on data categories and approaches 1.15
- Use and analysis of text and discursive/creative material 1.18
- Using atlases and other map sources 1.7, 1.14
- Presenting quantitative data and interpreting graphs 1.3, 1.4, 1.10
- Analysing data, including applying statistical skills 1.14

Fieldwork opportunities

'Global systems' and 'global governance' might sound like far off, nebulous concepts that you might struggle to investigate in the real world. However, globalisation touches everyone's life in some way today. Evidence for this is, in fact, everywhere – in your home, school and local high street, from food and fashion to the media you consume.

Finding out about the operation of a TNC in your local area and people's views on its activities would be a good start for fieldwork on this topic. You needn't go to Antarctica!

1 Industrial change and local opinion

Companies are bought and sold every day. Some are taken over by a global player who might rebrand its operations in the TNC's own image, and/or 'rightsize' the workforce. But what do local customers or employees think? What impact will this change of ownership have on their life and, in their opinion, is the development a good thing for the wider community? Design a range of interview questions and consider who you might like to interview and how you might get access to these people. Consider whether the approach of a vox pop or a quick question posed to a local social media group result in a greater response for you to analyse?

2 Investigating branding and 'glocalisation'

To what extent do TNCs market their products to suit their UK customers? Using online resources, find out about how far TNCs tailor the presentation of their operations, goods and services to suit an audience. Often you can access different versions of the same website by clicking on the appropriate flag in the top right-hand corner. In the case of some global brands there may be little difference, but this might prompt you to think about which national culture its ethos and imagery originally come from. And, if the same branding is used worldwide, might this be seen as 'cultural imperialism'?

3 Production and consumption: workers' rights and the environment

Find out if initiatives such as Fairtrade, #whomademyclothes or GreenPalm have any impact in other European countries or North America. What effects have the activities of the EU or other international agencies had on the homogenisation of production standards to protect (or reduce) the rights of workers and/or the environment (including animal rights) along with consumer awareness of such issues?

In this section you will learn about dimensions of, and factors in globalisation

Some words or phrases are so commonly used that we rarely think about the complex processes that the term describes. For example, sustainability, climate change – even geography! Whilst globalisation is not new, most geographers would argue it is a lot more than just 'the world becoming global'.

Globalisation is a term used frequently within trade and economics and describes a process of opening up world trade and markets to **transnational companies (TNCs)** and an increasingly interconnected world (Figure **1**). However, a definition of globalisation should also involve the associated effects of this process on people, culture, political systems, environment and the quality of life of every human on the planet!

▶ **Figure 1** *Globalisation describes the often complex processes associated with an increasingly interconnected world*

Dimensions of globalisation

Globalisation has many interconnected aspects. Some of these are outlined below, with some examples.

Flows of information, technology and capital

- Cheap, reliable and near instantaneous communication between virtually all parts of the world allows for information and capital to be shared at unprecedented levels.
- Money flows electronically around the world. **Highy developed economies (HDEs)** invest in **less developed economies (LDEs)** to take advantage of cheaper production costs.
- Technology, for example the internet and associated mobile technologies, largely ignores political boundaries when connecting people and places (see 1.3).
- Countries such as India provide a range of financial and IT services for higher income countries (so-called outsourcing).

Flows of products and labour

- Global transport systems have never been cheaper or more efficient in moving both people and goods.
- High speed rail networks (such as HS1 in the UK), international airport hubs (Dubai has now overtaken Heathrow as the world's busiest airport) and containerisation, for example, have revolutionised travel.

- People move around the world for employment. This includes specialised workers, for example, who move between different units/companies of a TNC on a short-term basis and unskilled migrant workers using a range of transport modes.
- Tourists now travel increasing distances to more remote and exotic locations, encouraged by global marketing and low-cost flights.

Flows of services and global marketing

- Services, such as global marketing, follow the flows of capital, information, people and products.
- Marketing is now globalised and uses international strategies to deliver inter-continental imagery/ messages
- TNCs use the same adverts to advertise their products in different parts of the world.
- **Global products**, such as Coca-Cola or Nike, rely on a common global brand with the same identity the world over.

Patterns of production, distribution and consumption

- TNCs dictate where their products are made – generally where labour costs are cheaper in LDEs.
- Products are distributed around the world to meet the demands of consumers in HDEs.

ANOTHER VIEW

Colonialism

In reality, Christopher Columbus followed the Russians, Chinese and even Vikings when he accidentally 'discovered' the Americas (somewhere near the Bahamas). But the half-truth persists and he set out a model for the 450 years of European **colonialism** that followed. Historians might argue as to the long-term effects of such globalisation, but most would agree that colonialism sought to extract as much wealth as possible from the land, and also the people. In the 1890s, Cecil Rhodes put the case for British colonialism as 'We must find new lands from which we can easily obtain raw materials and at the same time exploit the cheap labour that is available from the natives of the colonies. The colonies (will) also provide a dumping ground for the surplus goods in our factories'.

Factors in globalisation

Globalisation is influenced by a number of factors. Some of these are outlined below.

New technologies, communications and information systems

- Information can now be shared easily and cheaply with an audience of billions at the click of a button.
- Mobile phones are one of the most important technologies for LDEs as they connect different people, markets and so trade, in ways that were previously not possible (see 1.2).

Global financial systems

- Banks and financial services operate across the world. They are linked together by vital transmission systems that allow lending and flows of money.
- The 2007 collapse of US house prices led to a credit squeeze (when banks no longer wished to lend money) and then a global banking crisis in 2008.

Transport systems

- The world has never been more accessible. A global transport network allows the movement of people and goods across vast distances.
- Without the friction of time and space, there are both new opportunities as well as new threats (such as the spread of disease).

Security

- As national boundaries have become less of a barrier to more mobile and better informed populations, traditional security measures have reduced in relative significance.
- High profile leaks of sensitive information have brought the issue of cybersecurity to greater prominence due to our reliance on information systems in all walks of life. For example, in 2016 the leak of 11.5 million financial and legal records exposed wrongdoing on a global scale.
- In the UK, the average cost of the most severe online security breaches for big corporations now starts at nearly £1.5 million.

Trade agreements

- Without a system of global trading rules, countries would resist some foreign imports whilst possibly favouring others.
- The **World Trade Organisation** (1994) oversees over 97 per cent of world trade. It provides a forum for negotiations and ensures that trade agreements are followed. The WTO agreement is over 26 000 pages, which hints at the complexity of world trade today.

ACTIVITIES

1 Why is the term 'globalisation' so widely used yet still so difficult to define?

2 Search the internet for a list of the top TNCs in the world. How many have you heard of? Evaluate the information you find (look at the sectors of industry and the countries of origin) and draw conclusions about why they have been so successful.

STRETCH YOURSELF

In groups or in pairs, discuss whether you think that globalisation is something that we should be worried about.
Think of advantages and disadvantages and compare your findings with the other groups.

In this section you will learn about issues associated with unequal flows of people, money, ideas and technology in Uganda

You may have played the game shown in Figure **1**. It involves removing a brick from a tall stack and then placing it at the top of the pile. A wise move will allow the stack to grow taller whilst keeping solid foundations below. As the game continues and the stack of bricks grows taller, more risks have to be taken and eventually the blocks tumble. Global systems are certainly no game, but there are winners and losers. Look again at Figure **1**, 1.1 – such is the interdependence of global systems that any 'tumble' might have an impact globally.

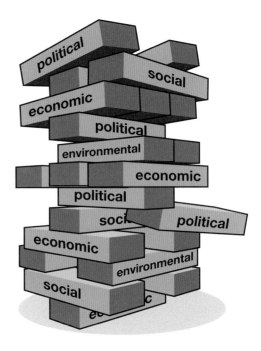

Figure 1 As represented by this brick-stacking game, expanding global systems are part of the same whole. Economic, political, social and environmental 'bricks' are all dependent on each other for support – a weakness or wobble in one may lead to the collapse of others.

Think about

The British Empire

The origins of the British Empire can be traced as far back as the fifteenth century when English seamen sailed and traded all over the world. English colonies were subsequently formed in America (late 1500s), West Indies (1620s) and Canada (1760s). Between 1815 and 1914, 10 million square miles of territory and 400 million people were added to the British Empire, including India and colonies across Africa (Uganda was made a colony in 1894). The British way of life – government, laws and religion – was imposed, sometimes by force, and national economies 'served' the needs of the Empire. It was not until after the Second World War that the British Empire was dismantled and replaced by a voluntary organisation of former colonies called the Commonwealth.

Uganda and global systems

Uganda is a land-locked country in East Africa that lies within the Nile basin (Figure **2**). Uganda shouldn't be a poor country – it is green and fertile, and has plenty of resources such as copper and cobalt.

However, civil war, corruption and HIV/AIDS have all acted as checks to development. Nearly one-third of the 35 million population live below the national poverty line and, of the many indicators of development, life expectancy stands out – it is only 59 years.

In this context, have global systems worked for or against the interests of Uganda and its people?

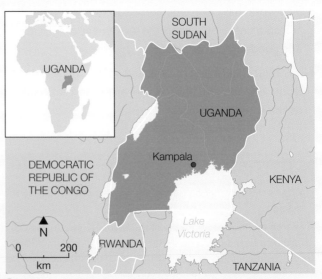

Figure 2 Uganda – a developing country affected by global systems

Inequalities and injustices

Poverty in Uganda is greatest in rural areas, particularly in the north and north-east where the majority of the population are smallholder subsistence farmers. When Uganda was part of the British Empire, the British East Africa Company strongly influenced the country's exports. Low-valued primary products such as the cash crops of coffee, tea and cotton were exported around the Empire and they continue to dominate Ugandan exports.

However, fish is now a traditional food staple for Ugandans living on the shore of Lake Victoria, and is one of the country's most profitable exports. This trade is unsustainable (Figure **3**). Overfishing and the predatory Nile Perch (introduced by the British) have resulted in stocks of indigenous fish being reduced to extinction levels. This in turn has caused fish factories to close and an associated knock-on effect to the local economy.

⊘ Figure 3 *Overfishing in Lake Victoria is causing significant environmental and economic damage*

Stability, growth and development

Shambas, the smallholdings owned by subsistence farmers, are an unlikely location to see the positive effects of global systems. Against a traditional backdrop of coffee bushes and banana and yam plants, occasional mobile phone antennae on the tops of farm buildings are evidence of recent technological changes.

The installation of fixed cables in Uganda is cost prohibitive, but cheap wireless technology now means that telephony and particularly access to the internet is possible even in remote rural areas. The so-called 'Village Phone' model offers loans to people wishing to start a mobile phone business (Figure **4**). The loan allows the purchase of a mobile (and increasingly smart) phone, a car battery to charge it and a booster antenna that can pick up signals from 25 km away. There is a rapidly growing market of users who are willing to pay for this mobile phone service – for example, farmers pay to access the internet and to gain information about the price they might pay for seeds at market or information on new farming techniques.

⊘ Figure 4 *The Village Phone model*

Did you know?

There are around 700 million mobile phone subscriptions in sub-Saharan Africa.

ACTIVITIES

1 Evaluate the effects of unequal flows of people, money, ideas and technology on global systems on Uganda. (Hint: consider stability, growth, development, inequalities, conflicts and injustices.)

2 Making reference to a named country, focus on two aspects and assess whether there were benefits of being part of the British Empire for that country. (Aspects could be, for example, infrastructure, investment, public health.)

STRETCH YOURSELF

Uganda gained independence from Great Britain in 1962. Assess whether independence brought greater political stability. You could compare the three periods: 1962–71, 1971–79 and 1979 to the present day. You may find online resources to be useful (such as www.nationsencyclopedia.com).

In this section you will learn about:
- the use of the internet to influence geopolitical events
- the negative impacts of a single-product economy

'When America sneezes we all catch a cold.' This is a statement as much about our interconnectedness with America (and whether this is a positive) as it is about an unequal relationship. The USA is an example of a country, a world superpower, that uses global systems to drive home its advantages over the rest of the world. Other countries, where resources are more limited, are only able to respond to global events in a more limited way. Rather than being leaders, they are followers or recipients of change.

1	Twitter	6	Baidu.com
2	Taobao	7	Yahoo
3	Linkedin	8	YouTube
4	Tencent QQ	9	Facebook
5	Wikipedia	10	Google

Figure 1 *Top 10 most visited websites in the world*

China and the internet

What are the world's most popular websites? You might be surprised by the results (Figure **1**). When you consider that China has the largest number of internet users in the world – 650 million in 2015, nearly 25 per cent of the world total – then the number of Chinese websites in the top ten should provide some context.

The internet is perhaps the embodiment of globalisation – allowing flows and sharing of money, ideas and technology around the world. Most countries struggle to manage the effects of this huge giant, yet China controls the internet at source and uses it to influence geopolitical events and, in turn, its citizens.

The Chinese central government controls what its citizens see on the web using two methods.

- The 'Great Firewall' is a system of online censorship started in the late 1990s. It works by blocking access to foreign websites, filtering key words and **bandwidth throttling**. Attempts to bypass this, for example using virtual private networks to access US sites such as Google and Facebook, have been of limited success.

- The 'Golden Shield' is a system of domestic surveillance set up in 1998 by the Ministry of Public Security. This uses more traditional methods such as fines, arrests, libel lawsuits and dismissals to enforce censorship.

Ironically, China's connection to the outside world matters – particularly trade links to HDEs. However, the communist government has largely succeeded in ensuring that its population sees a filtered view of the online world. In 2010, Chinese activist Liu Xiaobo was awarded the Nobel Peace Prize, but this news never reached the majority of Chinese. Liu was already in jail and all news about his imprisonment was also censored.

Open sesame!

Alibaba.com is a highly successful, Chinese e-commerce company. Its web portals have helped boost the success of China's manufacturing sector by connecting exporters to businesses in over 190 countries worldwide. Since the establishment of its first web marketplace in 1999, the Alibaba Group has grown to become a leading facilitator of international trade. It is also influential amongst domestic shoppers. The Alibaba Group's websites include Taobao (an equivalent of eBay, see Figure **1**), AliPay (an online payment service with 400 million users) and Tmall.com that enables Chinese consumers to buy branded products from international as well as Chinese companies, listing some 70 000 brands.

Think about

The internet in numbers

Every 60 seconds ... 204 million emails are sent, 75 hours of video uploaded, 4 million Google searches and 100 000 Facebook friend requests are made, and the big technology companies, the majority based in the US, earn US$142 000! Yet today only around 40 per cent of the world's population have an internet connection – so these numbers are only set to increase.

Nigeria: negative impacts of a single-product economy

Look at Figure **2**. Nigeria is a country that should have done well from globalisation. It has proven oil reserves of around 36 billion barrels and natural gas reserves of over 2800 billion cubic metres! It is an example of a **single-product economy** – oil and gas accounts for more than 80 per cent of its national income. Without the global trade in oil and Nigeria's membership of **OPEC**, the country's development would surely have been limited … or would it?

Global demand for oil has fuelled Nigeria's economy but at significant cost (Figure **3**). The focus on oil alone has resulted in a dramatic decline in the traditional industries of agriculture and manufacturing. Rural–urban migration has increased, resulting in increased levels of rural poverty and overcrowding in cities such as Lagos and Abuja. As Nigeria had neither the technology nor the skills to exploit the oil, the world's major oil companies were encouraged to develop these reserves. These global giants have been criticised for having scant regard to the local environment and indigenous local people. For example, oil spills are commonplace in the Niger Delta and land rights of local people are reported to have been abused.

The high income that is usually generated from oil results in the Nigerian currency being significantly overvalued, making imported consumer goods cheap. However, this results in domestically manufactured goods being too expensive and unable to be exported – process known as **Dutch disease** and common in other resource-rich **emerging major economies (EMEs)**. Deindustrialisation is a consequence of this, which drives more people into the oil and gas industries exacerbating the problems identified earlier. A greater emphasis on exports of oil and gas makes Nigeria less internationally competitive in manufactured goods and increases its reliance on foreign imports.

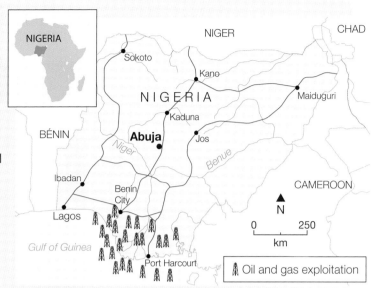

▲ *Figure 2 Nigeria – an economy dominated by oil*

▲ *Figure 3 Environmental costs of oil in Nigeria*

ACTIVITIES

1 Suggest how the internet is capable of influencing geopolitical events.

Ⓢ
2 A country's Human Development Index (HDI) is a ranking of human development. Look at the data of the components of Nigeria's HDI in Figure **4** and present the data appropriately. Evaluate and interpret the trends in the data.

3 Research the current global market conditions for oil prices. What effect could they have on Nigeria's opportunities for long-term growth?

4 Research similar data for an emerging country in south-east Asia. What similarities and differences can you infer from your findings?

	Life expectancy at birth	Expected years of schooling	Mean years of schooling	GNI per capita (2011 PPP$)	HDI value
2005	48.7	9.0	5.2	3606	0.467
2010	51.3	9.0	5.2	4825	0.493
2011	51.7	9.0	5.5	4926	0.499
2012	52.1	9.0	5.7	5018	0.505
2013	52.4	9.0	5.9	5166	0.510
2014	52.8	9.0	5.9	5341	0.514

▲ *Figure 4 Nigeria's component indices and HDI since 2005*

In this section you will learn about international trade and investment

Anyone who uses an online auction site will understand the feeling of a 'good deal' (Figure **1**). Either as a buyer or seller, there is much satisfaction to be had from selling an unwanted item or for purchasing an in-demand product that is cheaper than on the high street.

Trends in international trade

Most economists and geographers would agree that nations are better off when they buy and sell from one another. Indeed, since no single country has everything it needs and materials and resources are unevenly distributed, international trade is inevitable. So, the foreign producer is able to sell more and make increased profits; the consumer has access to products that might not be available domestically or better meet their specific needs. Yet international trade remains highly contentious – foreign products might be bought in cheaply but the (more costly) domestic seller loses a sale. Countries exert their political and economic power globally to ensure that they gain from international trade whilst other countries lose out.

The value and volume of trade has increased dramatically since the Second World War (Figure **2**). International trade (exports) is expanding faster than the world's economic output (as measured by GDP). This trade is seen as one of the main 'engines' of economic growth. However, it is also clear that the idyll of free trade breaking down barriers between countries at different stages of development is not yet a reality (Figure **3**).

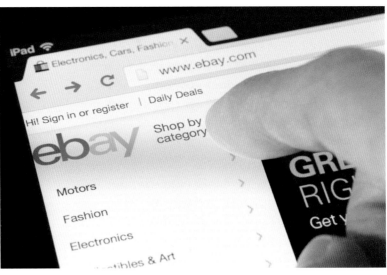

▲ **Figure 1** *eBay was launched in 1995; if all international trade was free of barriers such as trade agreements, legal systems and corruption (as it is on eBay) then the world's real income could be 30 per cent higher, according to the European Trade Study Group*

▼ **Figure 2** *Value of world merchandise exports (1948–2013)*

	1948	1953	1963	1973	1983	1993	2003	2013
	Value (US$ billion)							
World exports	59	84	157	579	1838	3684	7380	18 301

ANOTHER VIEW

Does international trade and investment make the world a safer place?

Deaths from war have fallen since the end of the Second World War while, over the same period, the volume of global trade has soared. Global systems may provide greater opportunities for international conflict (e.g. the global trade in arms sales or the increasing use of social media to influence others), but might also offer security and support the *status quo*. In a global war, increased risk and cost make international trade difficult. Components and raw materials may become in short supply – either because they are imported or because materials are diverted to make military hardware. Far more devastating would be the effects on the economies of individual countries. Such is the reliance of firms and businesses on international trade and investment that many would simply not survive. Would Chinese, Russian or American industries support a regional or international war and risk losing multi-billion or trillion dollars of investment? But then again, the causes of conflict do not always follow logic nor reason ...

Advantages of international trade	Disadvantages of international trade
Comparative advantage – a country specialises in producing only those goods that can be produced efficiently and at the lowest opportunity cost (see 'The Umbrella City' below).	*Over-specialisation* – If demand falls or if the same goods can be produced more cheaply overseas, then production needs to shift to other products. Specialised production centres tend to be less flexible and less able to diversify (see 1.3, Nigeria).
Economies of scale – producing a narrower range of goods and services means that a country can produce in higher volumes and at a cheaper cost per item/unit.	*'Product dumping'* – profit lines can be dangerously tight, even to the point where goods are sold at a potential loss. Dumping refers to exporting at a price that is lower in the foreign market than the price charged domestically.
Purchasing power – increasing trade results in increased competition that lowers prices and allows consumers to be able to buy more for their money.	*Stunted growth or decline of local and emerging industries* – new home-grown industries may find it difficult to grow and become established when faced with existing foreign competition, where costs are lower.
Fewer domestic monopolies – domestic prices may be kept high when a single firm controls a large proportion of the domestic market (25% or greater), as there is less competition. Imports from overseas competitors help to lessen this effect.	*Protectionism and tariffs* – a country/government may protect important domestic industries by imposing additional taxes and tariffs on imported goods and/or encouraging exports.
Transfer of technology – application of new technologies is incentivised as this may lead to design improvements and cost savings as well as supporting innovation and enterprise (see 1.2, Uganda).	*De-skilling* – traditional skills and crafts may be lost when production technology replaces manpower. So-called 'screwdriver jobs' may dominate.
Increased employment – increased production for export is likely to result in increased employment. In turn, as a result of the multiplier effect, more jobs will be created across the whole economy.	*Exploitive and labour-intensive industries* – the biggest cost for most industries is labour; this is particularly true for consumer manufacturing industries. By squeezing this cost, even if working conditions are compromised, profits can be maximised.

Figure 3 *Advantages and disadvantages of international trade*

The Umbrella City

Umbrellas have been around for more than 3000 years. It is likely that the Chinese were the first to create a 'collapsible fabric dome' to protect from rain. The umbrella was subsequently exported via the Silk Road to Asia and then across Europe, gaining particular popularity within the Roman Empire. Umbrellas are now recognised worldwide, a global product that comes in many shapes and sizes (a staggering 5000 models are on sale on *Amazon*). Today, around 70 per cent of the world's umbrellas are still made in China. At the centre of this production phenomenon is Songxia, in the city of Shaoxing (Figures **4** and **5**). Described as the umbrella capital of the world, around half a billion umbrellas are made here annually in more than 1200 factories – a single worker makes 300 umbrellas a day. Songxia retains its prominent position as a result of comparative advantages:

◆ *Specialisation* – all kinds of umbrellas are manufactured, from traditional rain umbrellas to golf, mini, children's, wedding, parasol and fashion umbrellas.

◆ *Access to domestic and international markets* – good road networks connect Songxia to the large population centres of Hangzhou city (70 km) and Shanghai (260 km), and also Ningbo Port (80 km), from where umbrellas are exported.

Did you know?
Britain buys more items per capita on eBay than any other nation.

Figure 4 *Location of Songxia, China; centre of the world's umbrella production*

- *Cheap production costs* – as a labour-intensive industry, umbrella manufacture in China benefits significantly from low labour costs and a flexible, local and efficient 40 000 workforce (mainly female) prepared to work long and sometimes unsociable hours.

- *Government support* – local government support for this single-product city includes tax incentives for producers, as well as preferential policies for all parts of the supply chain (such as fabric weaving, dyeing and printing, manufacturing of ribs, poles, handles).

- *Songxia Umbrella Industrial Park* – established with local government support to strengthen the competitiveness of the many local manufacturers and to raise the brand awareness of Songxia Umbrellas in domestic and global markets.

Figure 5 *Songxia Umbrella Industrial Park; if you have an umbrella there is a high probability it was manufactured here!*

Patterns of international trade and investment

Although the ratio of exports to GDP has risen for most countries, even in LDEs, the 49 poorest countries only account for 0.6 per cent of global trade (compared to 37 per cent of the top five exporting countries). Such differences are exaggerated further by the type and limited range of exports from LDEs, which tend to be dominated by a limited number of low value **primary products** such as crops or raw materials. Such single product economies are vulnerable to market price fluctuations and may have limited options for generating other sources of foreign income should natural disaster strike (e.g. drought or flood) or even if tastes or fashions change.

CIS comprises Armenia, Azerbaijan, Belarus, Kazakhstan, Kyrgystan, Moldova, Russia, Tajikistan, Uzbekistan, Georgia (until 2008) Six East Asian Traders are: Hong Kong, Malaysia, Republic of Korea, Singapore, Chinese Taipei, Thailand

	1948	1953	1963	1973	1983	1993	2003	2013
	Value (US$ billion)							
World	59	84	157	579	1838	3684	7380	18301
	Share (%)							
World	100.0	100.0	100.0	100.0	100.0	100.0	100.0	100.0
North America	28.1	24.8	19.9	17.3	16.8	18.0	15.8	13.2
United States	21.7	18.8	14.9	12.3	11.2	12.6	9.8	8.6
Canada	5.5	5.2	4.3	4.6	4.2	3.9	3.7	2.5
Mexico	0.9	0.7	0.6	0.4	1.4	1.4	2.2	2.1
South and Central America	11.3	9.7	6.4	4.3	4.5	3.0	3.0	4.0
Brazil	2.0	1.8	0.9	1.1	1.2	1.0	1.0	1.3
Argentina	2.8	1.3	0.9	0.6	0.4	0.4	0.4	0.4
Europe	35.1	39.4	47.8	50.9	43.5	45.3	45.9	36.3
Germany	1.4	5.3	9.3	11.7	9.2	10.3	10.2	7.9
France	3.4	4.8	5.2	6.3	5.2	6.0	5.3	3.2
Italy	1.8	1.8	3.2	3.8	4.0	4.6	4.1	2.8
United Kingdom	11.3	9.0	7.8	5.1	5.0	4.9	4.1	3.0
Commonwealth of Independant States (CIS)	–	–	–	–	–	1.5	2.6	4.3
Africa	7.3	6.5	5.7	4.8	4.5	2.5	2.4	3.3
South Africa	2.0	1.6	1.5	1.0	1.0	0.7	0.5	0.5
Asia & Oceania	14.0	13.4	12.5	14.9	19.1	26.0	26.1	31.5
China	0.9	1.2	1.3	1.0	1.2	2.5	5.9	12.1
Japan	0.4	1.5	3.5	6.4	8.0	9.8	6.4	3.9
India	2.2	1.3	1.0	0.5	0.5	0.6	0.8	1.7
Australia and New Zealand	3.7	3.2	2.4	2.1	1.4	1.4	1.2	1.6
Six East Asian Traders	3.4	3.0	2.5	3.6	5.8	9.6	9.6	9.6

Figure 6 *World merchandise exports by region and selected economy, 1948–2013*

In contrast, world merchandise exports, particularly manufactured consumer goods, have traditionally been dominated by North America, Europe and East Asia (Figure **6**). China overtook Japan as the leading Asian exporter in 2004, just three years after it joined the WTO. It then overtook the USA in 2007 and Germany in 2009 to become the world's leading exporter. However, this pattern is beginning to change. Between 1993 and 2013, the share of developed economies' merchandise declined significantly from over 70 per cent to just over 60 per cent and exports from the so-called NIC4 (Hong Kong China, Republic of Korea, Singapore and Chinese Taipei) also fell. In contrast, over the same period, the share of developing economies' exports in world trade and that of the so-called BRICS (Brazil, Russia, India, China and South Africa) increased markedly.

Did you know?

Japan is the world's fourth largest consumer of oil, despite having no oil reserves at all!

Foreign Direct Investment (FDI), the amount of capital invested in foreign countries, has largely mirrored the changes in world trade (Figure **7**). For example, while China (perhaps unsurprisingly) was the world's largest recipient of FDI in 2014, Singapore and Brazil were ranked 4th and 5th respectively. FDI flows to developed countries as a whole dropped by 14 per cent compared to 2013 also mirroring the small but significant shift in world trade from HDEs to LDEs. (A *transition economy* is one that is changing from a centrally planned economy to one driven by market forces.)

Region / Economy	2012	2013	2014	Growth rate 2013–14 (%)
World	**1324**	**1363**	**1260**	**−8**
Developed economies	**590**	**594**	**511**	**−14**
Europe	310	225	305	36
European Union	282	235	267	13
North America	213	302	139	−54
Developing economies	**650**	**677**	**704**	**4**
Africa	55	56	55	−3
North Africa	17	15	13	−17
Other Africa	39	41	42	2
Latin America and the Caribbean	178	190	153	−19
South America	145	133	118	−11
Cental America	27	53	31	−41
Caribbean	6	5	4	−16
Developing Asia	414	427	492	15
West Asia	48	46	44	−4
East Asia	217	220	254	16
South Asia	32	35	43	23
South-east Asia	116	127	151	19
Transition economies	**84**	**92**	**45**	**−51**

▶ **Figure 7** *FDI inflows (US$ billion) by major region, 2012–14 (data ©UNCTAD)*

ACTIVITIES

S

1 Create a line graph using the data in Figure **2**. Use the *x*-axis for the year and think carefully about the scale you use for the *y*-axis. What do you notice about the *rate* of the growth of global exports?

2 Study Figure **3**. List factors that might influence the volume and patterns of international trade and investment.

3 Study Figure **6**. With reference to specific figures and countries/and or regions, analyse changes in world exports. Write your answer as concise evidence-based bullet points.

4 You have designed a new wind resistant umbrella and already have a large order book of international sales! In which country would you choose to manufacture your product? Explain your answer.

STRETCH YOURSELF

One viewpoint is that markets work most efficiently when the buyer and seller both receive a fair price. How does globalisation influence this position of equality?

In this section you will learn about terms of trade and the impacts of metal extraction

What are terms of trade?

When players are picked for a team game, it is likely that the best players will be picked first. Those chosen last are likely to be given marginal roles where their influence on the game is slight. Many argue that world trade is a game played out between international powers who pick so-called **terms of trade** at the expense of developing countries.

The phrase 'terms of trade' refers to the cost of goods that a country has to import, compared with the price at which they can sell the goods they export. HDEs tend to import primary products from LDEs and subsequently turn these into manufactured goods for export to world markets – in other words, the value of the product increases as it passes through the hands of HDEs (Figure **1**). In general, the prices of manufactured goods have continually increased over the last few decades, whereas the prices of primary products have fluctuated (Figure **2**). This means that LDEs need to export increased volumes of primary products to purchase the manufactured goods that they require. So the terms of trade for LDEs are often uncertain and less favourable. The rapid industrialisation of China has offered opportunities for some regions, particularly Africa, to improve terms of trade but this is not an equal power relationship.

Impacts of metal extraction

In the last 20 years, there has been a dramatic increase in demand for metal from the emerging economies of east Asia, particularly China (Figure **3**). To meet this growth in demand, metal supplies that were previously uneconomic to exploit have been developed. In short, the frontiers of metal exploitation have shifted from north to south – from the advanced markets and developed economies to those that are emerging and developing. This global shift, despite political risks, is particularly apparent in Africa.

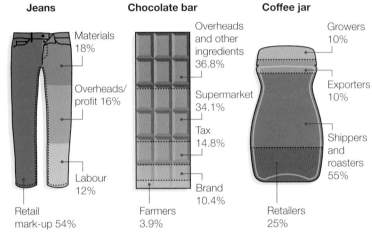

Figure 1 *Free trade, where goods are traded globally without barriers, does not necessarily mean that the trade is fair or the profits evenly distributed*

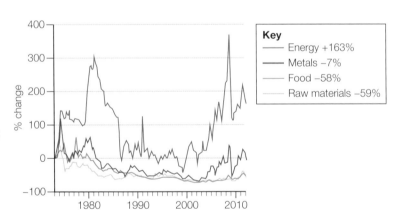

Figure 2 *Fluctuating commodity prices, 1973–2012 (index 1973 = 0); adjusted for inflation, energy prices have fallen since 2014 but are still well above 1973 levels*

Figure 3 *Percentage growth in China's consumption of metals (2002–14)*

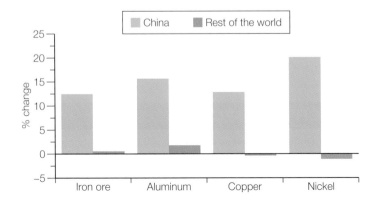

New mines in Africa result in more investment and jobs and, in turn, increased government revenues. New trade routes have been established and the geopolitical implications of increased African–Asian cooperation are also significant. China in particular argues that its economic policies offer a 'no strings attached' route to development. However, critics point to unfulfilled Chinese promises of development assistance, flooding of cheap Chinese manufactured goods into African local markets and to working practices that are unsafe and unethical.

Collum Coal Mining Industries, Zambia

Located 200 km south-west of Lusaka, Collum Coal Mine supplies fuel to Zambia's copper and cobalt mines.

In 2011, police were called to the mine to investigate an incident in which 11 African workers were shot by the operation's Chinese managers during a protest against poor working conditions. The following year, a Chinese manager was killed during a pay dispute and the year after, the mine was seized by the Zambian Government, in light of the company's poor 'safety, health and environmental record'. Collum Coal Mining Industries were also accused by ministers of failing to accurately declare the total amount of coal produced, thereby avoiding tax payments.

Look at Figure **4**. In a globalised economy, the consequences of decision-making at continental level impacts at every other level. When leaders of the premier economic group, the G20, met in November 2015, many shared an economic concern – steel plant closures across Europe, the US and Mexico. The common cause was the competition from cheap Chinese steel. As the Chinese construction boom faltered in 2015, the demand for steel in the country fell and Chinese steel producers looked to international markets to absorb the overproduction. Chinese steel is state-subsidised and that makes its price very competitive – too competitive, the leaders of the G20 argued. Chinese 'steel dumping' – where steel is sold at a rate even below the cost to produce it – created an oversupply of steel. This has caused other steelmakers to lose jobs or to close altogether.

Tata Steel, UK

In 2016, when Tata Steel announced its intention to sell its UK business, risking the jobs of its thousands of employees, there was a call for the British Government to renationalise the steel industry (see 2.2). Instead the Government said it would work with Tata to find a buyer, to protect steel jobs. Such an arm's length intervention, which stopped short of direct conflict with the Chinese Government, saved face for the UK's ongoing negotiations to encourage Chinese investment, not least in the new £18 billion nuclear plant, Hinkley Point C.

In 2015, the EU had imposed anti-dumping duties for six months on selected steel imports from China and Taiwan. Britain's Brexit vote in 2016 raised questions amongst potential buyers of Tata Steel's UK operations, about whether or not a UK outside the EU would have a freer hand to impose more protective import tariffs.

⊘ **Figure 4** *Half of the 1.6bn tonnes of steel made globally each year comes from China, giving it huge pricing power. In 2015, following 18 months of steel dumping and a rapid fall in the price of global steel, Britain's steel industry was unable to compete and jobs were lost.*

ACTIVITIES

1 Is free trade the same as fair trade (see 1.9)? Explain your answer with reference to trading relationships between large developed economies, emerging economies and less-developed economies. Does Figure **1** help to give the whole picture?

2 Who does Chinese 'steel dumping' really benefit? Consider the different groups of people who gain from cheaper steel being made available within the UK and the EU.

STRETCH YOURSELF

The West has held back from investing in parts of Africa, including signing trade agreements, where doubt exists over governance (such as fears of corruption and abuse of human rights). Following trade and aid agreements with these same African governments, should China be seen as a saviour or a sinner? For example, many millions of Africans have been lifted out of poverty while others accuse China as acting like yet another colonial power (see Figure **5**, 1.18) The annual Global Attitudes survey may help to structure and to inform your argument (www.pewglobal.org – search for 'China in Africa').

In this section you will learn about:
- trade blocs
- trading agreements and their impact on economies and societies
- the role of the EU in trade agreements and governance

Major trade blocs

The boundaries of Europe might, at times, seem rather confusing. Some countries, such as Russia and Turkey, are part European and part Asian, Israel plays in European football competitions, and the Eurovision Song Contest even included Australia in 2015! Today many would think of Europe in the context of the European Union (EU), an economic and political union that was created by the Maastricht Treaty in 1992. However, the birth of the EU dates back to 1957 when the Treaty of Rome created the Common Market, an example of an early trading association or **trade bloc**.

Look at Figure **1**. One consequence of global systems has been to bring countries together into trading blocs, which support free trade between member countries without incurring tariffs or charges. Countries outside the bloc have to pay an additional tariff to trade within the bloc. Opponents of trade blocs argue that by offering unfair advantages to member countries, they restrict the development of a global economy. The advantages of the largest trade blocs are increased further as any similar trade agreements between LDEs are weaker and achieve limited advantages.

Did you know?

The euro is the second most traded currency (behind the US dollar) and is used daily by more than 320 million people.

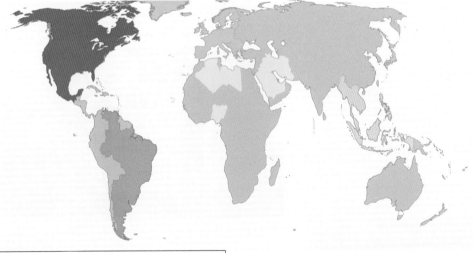

Key
- ASEAN (Association of Southeast Asian Nations)
- OPEC (Organization of the Petroleum Exporting Countries)
- NAFTA (North American Free Trade Agreement)
- MERCOSUR (Mercado Común del Sur)
- EU (European Union)

◀ **Figure 1** *Main trade blocs of the world*

◀ **Figure 2** *In the 2016 UK referendum, more than 17.4 million people voted to leave the EU and 16.1 million voted to remain. Facebook 'remain' friends shared this image following the referendum.*

Trading organisations

As well as trade blocs, several other organisations help to encourage trade of different types and from different countries. They attempt to govern and set rules of trade and include:

◆ *World Trade Organisation (WTO)*. Formed in 1993, it aims to cut trade barriers (subsidies, tariffs and quotas) that stop countries trading freely, so that goods can flow more easily.

◆ *Organisation for Economic Cooperation and Development (OECD)*. A global 'think tank' for 30 of the world's wealthiest nations.

◆ *Organisation of Petroleum Exporting Countries (OPEC)*. Consists of 11 states who supply 40 per cent of the world's oil. OPEC tries to regulate the global oil market to ensure a good fair price.

◆ *G8*. The Group of 8 (Canada, France, Germany, Italy, Japan, Russia, UK, USA) represents 65 per cent of the world's trade and meets annually to discuss economic development. In 2005, G8+5 was formed to include China, India, Brazil, Mexico and South Africa.

◆ *G20*. Includes all of the (finance) leaders of the G8+5 countries plus South Korea, Australia, Turkey, Saudi Arabia, Argentina, Indonesia and also the leader of the EU. It also includes representatives from the IMF and World Bank. Like the G8, the G20 discusses the global economy and methods to encourage economic growth.

◆ *World Bank*. Promotes investment globally and provides loans for countries under certain conditions.

◆ *International Monetary Fund (IMF)*. Standardises global financial relations and aims to promote global monetary and exchange stability by monitoring the global economy and encouraging the growth of international trade. It can force countries to privatise (or sell off) government assets, which are then bought by large TNCs, and open up trade in return for refinancing debt.

An example of a trade bloc and regional governance – the European Union

Look at Figure **3**. The EU has a unique institutional set-up that allows for regional governance as well as economic cooperation. Founded in the years following the Second World War, the origins of the EU were based on a simple premise – that countries who trade with each other (and thereby become economically interdependent) are less likely to be in conflict. The huge trade bloc (or single market) of the EU is now only one aspect of its work. Legally binding treaties, agreed by all member countries, govern life for all 500 million EU citizens, 175 million of whom have a common currency, the euro. EU policies cover every aspect of our lives, from human rights to the environment. Only by exiting the EU via a referendum (as in the case of the UK, in 2016, see Figure **2**), do the benefits (and drawbacks) of membership no longer apply.

European Council
Sets the EU's overall political direction and priorities. It is made up of the heads of state or government of EU member countries.

European Commission
An executive body that is responsible for proposing and implementing EU laws, monitoring treaties and the day-to-day running of the EU. Members are appointed by EU national governments.

European Parliament
Represents the 500 million EU citizens and is directly elected by them. Adopts the laws proposed by the Commission. Shares power over EU budget and legislation with the Council of the European Union.

Council of the European Union
Represents the governments of member countries and promotes/defends national interests. The government ministers share power over EU budget and legislation with the European parliament.

Member countries
Implement the laws passed by the EU. The Commission ensures that the laws are properly applied and implemented.

⬆ **Figure 3** *The unique, albeit sometimes complex and bureaucratic, institutions of the EU have delivered half a century of relative stability, peace and prosperity for its member states.*

The Transatlantic Trade and Investment Partnership (TTIP)

A new trade deal is being negotiated behind closed doors between the US and EU, with the aim of liberalising 'one third of global trade'. In 2014, the European Commission stated that TTIP has the potential to boost the EU's economy by 120 billion euros, the US economy by 90 billion euros and the rest of the world by 100 billion euros. However, the lack of transparency around such wide-ranging trade talks has been criticised and their content has been leaked to the media on several occasions. TTIP has been criticised by unions, NGOs and environmental pressure groups, all of whom fear that access to markets for TNCs will be promoted over existing EU laws that protect health, the environment and minimum labour rights. In 2015, the BBC reported that food safety had become a stumbling block in the talks.

Once negotiated, all of the members of the EU will have to vote on the agreement in order to ratify it; however, there is concern over whether individual national governments will have any power to veto a deal. TTIP negotiations are so complex they are not expected to conclude until 2019–20.

Greece and the European Union

Look at Figure **4**. With debts of over 4 billion euros, including over 3 billion to the European Central Bank, Greece came close to leaving the EU in the summer of 2015 – a so-called *Grexit*. But how did Greece get into such an economic meltdown and would a Grexit have been a bad thing for the Greek economy and people?

	2010	2011	2012	2013	2014
Population (million)	11.2	11.1	11.1	11.1	11.0
GDP per capita (EUR)	20226	18678	17459	16491	16290
Economic growth (GDP, annual variation in %)	−5.5	−8.9	−6.6	−3.9	0.8
Investment (annual variation in %)	−20.9	−16.8	−28.7	−9.5	2.7
Exports (goods and services, annual variation in %)	4.6	0.0	1.2	2.1	9.0
Imports (goods and services, annual variation in %)	−5.5	−9.0	−9.1	−1.6	7.4
Industrial production (annual variation in %)	−5.9	−5.7	−2.0	−3.2	−2.2
Retail sales (annual variation in %)	−6.3	−10.2	−12.2	−8.1	−0.4
Unemployment rate (%)	12.7	17.9	24.6	27.5	26.6
Fiscal balance (% of GDP)	−11.1	−10.2	−8.7	−12.3	−3.5
Public debt (% of GDP)	146	171	157	175	177
Trade balance (EUR billion)	−29.8	−24.3	−19.8	−15.8	−17.3

Figure 4 *Greece economic data – the economy returned to recession in 2015*

Why is Greece so much in debt?

It is a simple truism – you cannot indefinitely live beyond your means. Money can be borrowed to buy a lifestyle but even this will need to be repaid, usually with interest, at some point in the future.

Greece first applied to join the EU in 1975 but its application was rejected – the reasoning being that membership would 'pose serious problems for both Greece and the [EU] community … and a substantial economic programme [is needed] to enable Greece to accelerate the necessary structural reforms'. It was perhaps surprising that five years later Greece joined the EU in a so-called 'Mediterranean expansion'.

Since gaining membership, increasing amounts of money were spent by a government funding a massive, inefficient and unsustainable public sector (in which public sector wages doubled in a decade and accounted for 40 per cent of the total economy). This spending spree was paid for partly by increasing taxes (even though widespread tax evasion remained) and also government borrowing. However, the situation was suddenly made a lot more challenging after the global financial downturn of 2008. By 2010, the country had run out of money and was granted euro bailouts by the IMF and the EU of 110 billion euros in 2010 followed by a further 109 billion in 2011.

If Greece still had its own currency, it could have been devalued to make its exports more competitive and its imports less so. This is not possible with the euro. Greece only stood a chance of being more competitive by increasing efficiency – reducing the wage bill by increasing unemployment (peaking at almost 28 per cent in 2013). Other countries, such as Germany, have benefited financially during this time due to increased exports to Greece, even though it has also been heavily involved in the financial bailouts.

The influx of more than one million migrants in 2015 from the Middle East has also done little to help Greece's recovery, although an EU directive in April 2016 aimed to stem the flow of migrants, despite protestations from the UN High Commissioner for Refugees (UNHCR).

Was it right for Greece to stay in the EU?

For:
Greece imports nearly 50 per cent of its food and 80 per cent of its energy from abroad and consequently benefits from being part of the European free market. Any so-called Grexit would drastically change Greece's trade balance and might cause bankruptcies and high inflation with an associated huge knock-on social cost. Greece would find it very difficult to borrow further and would therefore have to pass on increased living costs to the population (everything from pharmaceuticals to a loaf of bread would be more expensive).

Against:
An independent Greece might be able to trade more freely and to take advantage of its location and geography. Greece would be free of EU legislation and control, and might be able to newly position itself as a regional trading hub and gateway into the Middle East, Balkans or to Russia. National needs might be met in a more sustainable and local way. A Grexit would see the return of the drachma as currency and, with it, flexibility of exchange rates. Imports may become more expensive but may, in turn, spur job growth and economic growth.

ACTIVITIES

1. How far do trade blocs encourage globalisation? (Hint: consider the effects of 'regionalism' versus the concept(s) of free trade.)
2. Study Figure **3**. To what extent do you feel part of Europe?
3. Study Figure **4**. Was Greece right to stay in the EU? Use evidence in support of your argument.

STRETCH YOURSELF

NGOs such as Oxfam criticise global trading organisations because they operate in the interests of their members and not for the benefit of the entire global community. Instead, many NGOs advocate ethical or fair trade agreements even if this may involve a higher price for the consumer. What is your view of global trading organisations? Compare evidence from one of the websites of the trading organisations on page 21 to the ethical trade pages of Oxfam or other NGOs.

In this section you will learn about the nature and role of TNCs – their spatial organisation, linkages, production, impacts, trading and marketing patterns

What is a transnational corporation (TNC)?

Think of your favourite fast food or designer clothing label. They may have one thing in common – they are examples of transnational companies (TNCs) or companies that operate in more than one country. As the world's economy has become globalised, TNCs have grown in size and number. Many TNCs are now truly global companies producing global products – mainly consumer goods with a strong, recognisable brand that are distributed, marketed and sold throughout a large number of countries (e.g. Coca-Cola or Nike). There are also many TNCs in the primary and tertiary sectors, such as banking, telecommunications and mining (see 5.2).

Spatial organisation

Look at Figure **1**. In general, TNCs usually adopt a hierarchical model of organisation and impose top-down decision-making. In other words, branch plants tend to be recipients of change rather than initiating change or making decisions. This may result in branch plants being vulnerable to sudden closure or job losses as efficiency savings or spending reviews ripple down and along the hierarchy, for example, UK steelwork closures by Tata Steel in 2015 (see 1.5).

Did you know?

The ten largest TNCs in their field control around 86 per cent of the telecommunications sector, 85 per cent of pesticides industry, 70 per cent of the computer industry and 35 per cent of pharmaceutical industries.

Figure 1 *BP spatial organisation*

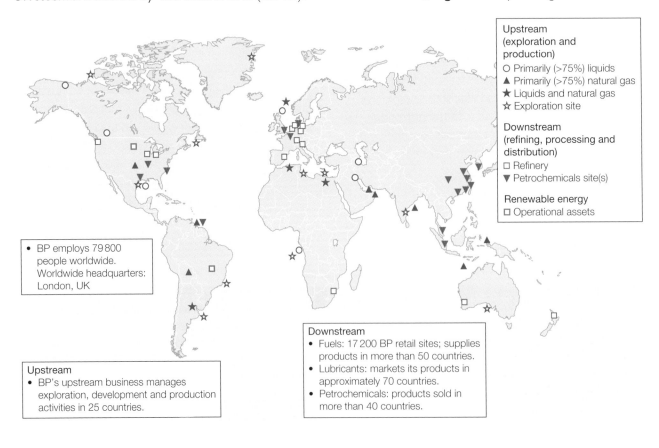

Upstream (exploration and production)
○ Primarily (>75%) liquids
▲ Primarily (>75%) natural gas
★ Liquids and natural gas
☆ Exploration site

Downstream (refining, processing and distribution)
□ Refinery
▼ Petrochemicals site(s)

Renewable energy
☐ Operational assets

- BP employs 79 800 people worldwide. Worldwide headquarters: London, UK

Upstream
- BP's upstream business manages exploration, development and production activities in 25 countries.

Downstream
- Fuels: 17 200 BP retail sites; supplies products in more than 50 countries.
- Lubricants: markets its products in approximately 70 countries.
- Petrochemicals: products sold in more than 40 countries.

Linkages and production

TNCs maintain strong links between all parts of their organisation. This allows TNCs to control and coordinate economic activities in different countries and also between units of the same corporation in more than one country. In this way, the TNC can lessen the impacts of trade restrictions, such as quotas, and negotiate more favourable terms of trade. **Horizontal integration** involves improving links between different firms in the same stage of production. This usually occurs when a TNC acquires competitors in the same industry and uses common structures in the hope of making cost savings. It results in a smaller number of firms controlling a given section of the market.

Vertical integration describes an industry where one company either owns or controls multiple stages in the production and distribution chain. See Figure **1** for the example of BP. This type of organisation gives the company significant economic advantages over competitors – for example, increased bargaining power with end purchasers or nation states.

Conversely, in *No Logo* (2000), Naomi Klein writes of the truncation or vertical *disintegration* of production in which big brand name corporations such as Nike and Levi Strauss 'bypass production completely' by 'sourcing' products from suppliers, rather than directly employing factory workers themselves. The advantages of this method of organisation include lower wage bills, the avoidance of costly bills for pension schemes and less stringent health and safety policies for workers. Klein suggests such an approach gives a TNC 'greater flexibility to allocate resources and capital to its brands'.

Either way, by not having all production units in 'the same basket' or region, TNCs are able to take advantage of spatial differences in factors of production and government policies across all the countries in which they operate. For example, TNCs can exploit differences in the costs of labour, raw materials, land and buildings as well as the availability of capital, tax incentives, subsidies, and more favourable government policies. Location of production units thus becomes flexible and allows TNCs to choose least-cost locations on a global scale. In this way, trade barriers may be negated by locating production in the host economy, resources can be shifted as local conditions change and specific local factors can be taken advantage of, such as access to skilled workers.

The impact of TNCs

The impact of TNC activities on host countries is described in some detail in Figure **2**. Whether formally or informally integrated across the world, TNCs bring both positives and negatives for their host nations. The use of new technology involved (transferring skills from HDEs to their hosts) is a longer term benefit by seeding the next generation of designers and manufacturers that will undercut their premium, branded goods. However, the dominant flow of capital is from the centres of production. Profits are transferred to another country, either an HDE where the company is headquartered or, perhaps, a remote tax haven in LDEs that supply an essential workforce in the chain of production (which reduces the benefits of a TNC's operation for the public purse).

Crickhowell, coffee and tax-avoiding TNCs

TNCs often have elaborate company structures, which sit beneath the overarching umbrella of their organisation, to funnel their funds around the world so that paying tax is avoided. The governments of some countries encourage such activities by giving tax breaks that attract large corporations to rent empty offices in places thousands of miles away from their employees. In 2014, a BBC investigation followed the attempts of shopkeepers of Crickhowell, a small market town in Wales, to move their profits at first to the Isle of Man and then to the Netherlands, mirroring the legal structure of large corporations, like Caffè Nero and Starbucks.

	Favourable	Unfavourable
Host country	Increase employment and thereby raise living standards	Many jobs are of low skill in LDEs
	Improve levels of skill and expertise	Managerial positions tend to be brought in rather than developed locally
	Foreign currency brought in, improving the balance of payments	Majority of profits are sent back to home country
	Socio-economic multiplier effect. This includes increased purchasing power (particularly in LEDCs) that, in turn, leads to demand for consumer goods and further economic growth	Multiplier effects can also be negative, for example, on the environment; political muscle of TNCs can be too aggressive and corners can be cut in terms of health and safety and employee rights
	Encourage a transfer of technology into the country, for example, the growth of telecommunications	Investment may only be short-term and TNC may pull out at short notice
Country of origin	Development of higher-order jobs such as in research and development or management	Workforce may need to relocate or make increased visits to operations overseas
	Overseas investment adds to income for the whole nation (via tax and other multiplier effects)	As a result of loopholes, corporation (business) tax is not paid fully by all TNCs
	Wider share ownership – individuals and companies more willing to become involved in foreign investments	Speculative investments in TNCs for quick returns helped contribute to the global 2008 financial downturn

▲ **Figure 2** *Impacts on countries in which TNCs operate*

Reasons for growth of TNCs

Working around the world, TNCs are able to take advantage of locational factors and minimise exposure to risks. The **flexibility of production** and **spreading of locational risks** means that should, for example, the local economic climate change or a natural hazard occur, global operations are not seriously affected.

Cheap labour	Lower wage demands from LDEs and also from unemployed in HDEs e.g. the Chinese have been accused of exploiting cheap African labour in their pursuit of cheap raw materials (see 1.5)
Mergers and takeovers	Allow big businesses to buy out smaller competitors or increase their market share (leading to an eventual **monopoly**) e.g. UK-based EMI's US$20 billion merger with Time–Warner created the world's largest music firm
Flexible workforce	Willingness to travel to jobs overseas or to be retrained in situ e.g. training of call centre employees in India for UK-based companies
Availability of finance to fund expansion	TNCs from emerging economies have heavily invested overseas to increase their portfolios e.g. the Indian TNC, Tata Steel bought the Anglo–Dutch company Corus Steel in 2006 and Tata Motors acquired the high-end Jaguar Land Rover and Daimler car brands from Ford in 2008
Fewer environmental restrictions	Environmental controls may either be less or weakly enforced e.g. oil exploration in the Niger delta (see 1.3)
Globalised transport network	e.g. containerisation has revolutionised the movement of goods by creating a standard transport product that can be handled anywhere in the world – ships. trucks, barges and rail wagons are all designed to handle containers of the same shape and size
Technological developments	e.g. refrigeration or freeze drying allows perishable fruit and vegetables from LDEs to be transported to markets in HDEs (see 1.9)
Governmental encouragement	e.g. financial incentives such as tax breaks
Cheap land	e.g. areas suffering from effects of deindustrialisation or changes in land use – Honda (Swindon) and Nissan (Washington, Tyne and Wear) both built car plants on sites of former airfields

▲ **Figure 3** *Further reasons that encourage growth of a TNC*

Trading and marketing patterns

Mapping the geography of a TNC is challenging because the largest firms might have offices, branch plants and retail outlets in almost every country of the world! However, as a general rule production takes place in LDEs or emerging economies (particularly China), design, research and development centres are in HDEs and the headquarters are in the country of origin (Figure **1**). Whilst TNCs may operate all over the world, most promote common patterns of consumption – in other words, they sell and market the same or similar product everywhere. In so-called *glocalisation* (see extension), such global products might, somewhat ironically, be tweaked to meet local needs. For example, Cadbury make their chocolate sweeter in China and McDonalds do not serve beef or pork at its India locations (the McAloo Tikki, a spiced potato croquette, is a unique menu item not found anywhere else).

Sport and TNCs

Sport is a global business and the most popular attract large investors or sponsors. Such sponsorship regularly comes from the big TNCs – they have deep pockets where marketing their brand is concerned. Some of the greatest sportsmen and women alive are paid huge sums to advertise global brands, becoming millionaire 'brand ambassadors' (Figure **4**). But is this always appropriate? For example, the official Olympic sponsors include fast food chains and confectionery companies and do not fit well alongside the healthy message of sport. (Neither does overt sponsorship and branding sit well with education. Klein, in *No Logo* questioned the sponsorship activities of TNCs in US schools – in classrooms, dining halls and sports facilities.)

Perhaps more sinister is the deliberate manipulation of sport to suit a particular self-interest. In summer 2015, the US indictment brought against FIFA officials included allegations of bribes paid indirectly by Nike to secure sponsorship rights to the Brazilian national football team. The newly appointed head of world athletics, Sebastian Coe, was forced to relinquish his ambassadorial role with Nike worth £100000 a year following controversy of the award of the 2021 World Athletics Championships to Eugene, USA (the home city of Nike).

◀ **Figure 4** *Olympic sprint champion Usain Bolt and Richard Branson promoting Virgin Media's fibre optic broadband and the Virgin Galactic spaceflights*

1.8 World trade: Coca-Cola

In this section you will learn about the impact of Coca-Cola on those countries in which it operates

Coca-Cola: a global brand

Every day, 1.9 billion products made by the Coca-Cola brand are consumed globally. As a global product, it is arguably one of the world's most recognisable brands. Indeed, many people believe that it was a result of the 30-year Coca-Cola advertising campaign that Father Christmas now wears the traditional red and white robes! (Figure **1**).

The world's most valuable brand?

Look at Figure **3**. In 2015 Coca-Cola was rated by Interbrand as one of the most valuable brands in 2015 (behind Google and Apple). It has a long history of success – shareholders have received 53 years of consecutive dividend increases. The brand has grown in strength as a result of astute marketing, investment and strategic worldwide acquisitions. The company has been adept in both spreading locational risks as well as manoeuvring into new market areas. For example, Coca-Cola sometimes uses existing bottling companies rather than building new factories. Where there is a shortage of bottling capacity, such as in the expanding markets of eastern Europe, India and Africa, Coca-Cola has been quick to seize the initiative and commit to billions of dollars of investment. Coca-Cola now has 20 main brands (from tea to fruit juices) that generate over US$45 billion a year in revenue and sales in nearly 200 countries.

Figure 1 *The Coca-Cola Santa; some people believe the soft drinks firm established the red and white colour scheme – a reflection of just how much influence the largest TNCs have, or are perceived to have, on our established cultures and traditions*

	Social	Economic	Environmental
Positive	Training and education programmes (e.g. global commitment to enable **empowerment** of 5 million female entrepreneurs by 2020 – the 5by20 programme) The Coca-Cola Foundation (Figure **4**) awards grants to companies throughout the world (e.g. Anandana in India focuses on social development as well as environmental sustainability) Massive employment opportunities both directly and indirectly in related industries	Franchise operation means that many local bottlers profit from sale; this supports the local economy directly and indirectly Investment in new plants in expanding markets such as Asia and North Africa (e.g. US$90 million R & D centre in Shanghai) Investment in new markets drives economic growth (US$2 billion investment in India since 2011)	Uses marketing network to increase awareness of recycling and distribution network for disaster relief Initiates sustainable agricultural schemes (e.g. constructing rainwater harvesting system at tea suppliers in China) Replenishes the water it uses (e.g. by funding local projects to protect watersheds)
Negative	Harsh working conditions in some bottling plants Millions spent countering the links to obesity; Coca-Cola spent £4.86 million setting up the European Hydration Institute (a research foundation promoting hydration) which controversially recommended that people consume sports and soft drinks Workers encouraged to abandon union membership in some LDEs (e.g. Guatemala 2010)	Long hours for little pay (in poor working conditions in some bottling plants) Majority of profits are returned to shareholders back in USA Vulnerability of bottling plants to the effects of top-down decision-making in 'remote' Atlanta (the headquarters of Coca-Cola)	Exhaustion of local water supplies (in 2012 Coca-Cola used more water than around 25% of the world's population) Water pollution (e.g. Kerala, India 2004) Pesticide residue identified in some Coca-Cola products (e.g. the Centre for Science and Environment found that Coke contained 30 times the amount of pesticide residue considered acceptable by the EU)

Figure 2 *Impact of Coca-Cola on host countries*

▼ **Figure 3** *Coca-Cola in numbers (2015)*

Global operating revenue	US$46 billion
Market share in soft drinks market	25.9%
Number of Facebook fans	93.42 million
Brand value	US$79.21 billion
Number of products	3500 beverages
World population who recognise the red and white logo	94%
Average number of servings per day	1.9 billion

▶ **Figure 4** *Ukrainian former heavyweight boxing world champion, Vitali Klitschko and a boy play basketball on a playground in Cherkasy, Ukraine, in 2012, funded by the Klitschko Brothers Foundation and the Coca-Cola Foundation*

Is Coca-Cola a bitter drink to swallow in Uraba, Colombia?

Long under the control of paramilitaries, a dangerous civil war has raged around the Uraba region for decades. It might, therefore, seem to be a strange place for Coca-Cola to establish a bottling plant. Far away from order and control, it could be said that it might be easier to make your own rules and working practices, and maximise revenue. In the 1990s, it was alleged that the owner of the Coca-Cola plant formed a coalition with the paramilitaries and used his fearsome partners to 'persuade' workers to leave trade unions and accept poorer working conditions. Some have even suggested that trade union members have been murdered! Coca-Cola were acquitted of these allegations in the US courts but this has not stopped film-makers and social media persisting in their argument (Figure **5**). In 2005 Coca-Cola commissioned a report into its South American plants by a private company, Cal Safety Compliance Corporation, which stated 'that workers in Coca-Cola plants enjoy freedom of association, collective bargaining rights, and a work atmosphere free of anti-union intimidation'.

Did you know?
The average Mexican drinks the most Coca-Cola products – 728 servings per year!

THE COCA-COLA CASE

A FILM BY GERMÁN GUTIÉRREZ AND CARMEN GARCIA

THE TRUTH THAT REFRESHES!

▶ **Figure 5** *The Coca-Cola Case documentary film presents a chilling series of allegations against the TNC*

ACTIVITIES

1 Look again at section 1.7. Suggest reasons why Coca-Cola is one of the world's most successful global products?

2 Summarise attitudes to Coca-Cola. Try to identify specific groups of people in your answer and aim for a balanced viewpoint.

3 Why do you think the Coca-Cola Foundation was created?

4 Discuss the social and economic impacts of TNCs on the countries in which they operate.

STRETCH YOURSELF

On their website, Coca-Cola's 2020 Vision starts with their mission to 'refresh the world … inspire moments of optimism and happiness [and] … to create value and make a difference'. With specific reference to the impacts of Coca-Cola around the world, evaluate this mission statement.

In this section you will learn about world trade in bananas and Fairtrade

Bananas and Fairtrade

Question: *Why are you eating a banana with the skin on?* Answer: *Oh, it's all right. I know what's inside!*

Joking aside, how many people are likely to fully understand the unequal world trade in bananas? The banana trade is big business – it comes fifth (behind cereals, sugar, coffee and cocoa) in terms of world trade in agricultural produce. But is this a fair trade?

The banana trade and UK consumption

Look at Figure **2**. Bananas are the world's most popular fruit and tend to be grown in the lower latitudes with the tropical conditions necessary for growth. Many HDEs are entirely reliant on imports so without global systems, including the use of refrigerated ships, the bananas would never find their way from banana plant to supermarket shelf.

It may therefore be surprising that in the UK, bananas are, on average, cheaper than apples. Supermarkets may use them as *loss leaders* – selling so cheaply that no profit is made and using them as an incentive to lure in shoppers (bananas are often near the front of a supermarket). Other factors keeping the price of bananas artificially low include the use of bananas on price comparison websites (showing where the cheapest food items are sold), price match promises and the relatively short shelf life. The huge buying power of supermarkets means that they are able to negotiate the best price possible from plantations and farmers. All these factors have combined to keep prices low and to squeeze profits from those lower down the supply chain.

However, **Fairtrade** bananas have an increasing representation on supermarket shelves – in 2013, one in three bananas sold in the UK carried the Fairtrade certification mark. Many of those are grown on small-scale farms and sold via cooperatives. For example, 85 per cent of the Windward Island banana crops are Fairtrade. Further Fairtrade bananas are grown on certified plantations across the Caribbean and Central America.

In recent years, sales of certified Fairtrade products have continued to soar in Britain, despite being several pence more expensive and a tough economic climate for shoppers, demonstrating the appetite for food traded on fairer terms. This is seen as a shift in consumer awareness and behaviour. Fairtrade's broad appeal is reflected in (and has been bolstered by) the growing participation of large food companies such as Cadbury's, Nestlé and Ben & Jerry's.

⊙ *Figure 1 The first Fairtrade label was launched in 1988 in the Netherlands. The purpose of Fairtrade is to create opportunities, through sustainable development, for producers and workers who have been economically disadvantaged or marginalised by the conventional global trading system.*

Did you know?

In 2015, a kilo of bananas cost around 68p in a UK supermarket. In 2002, the same bunch would have cost £1.08 – 59 per cent more!

What is 'Fairtrade'?

Originally a grass roots movement of NGOs, Fairtrade products are now commonplace on supermarket shelves across the country with the UK Fairtrade market doubling every two years (Figure **1**).

Over the last 40 years, the price paid to farmers has halved for many products. Fairtrade aims to pay farmers a guaranteed minimum price, offer fair terms of trade and pay an additional development premium for reinvestment in the local community. For example, Fairtrade cooperatives (organised groups of farmers) are able to develop local infrastructure, build schools and health clinics, and provide training and some financial services such as bank credit and insurance services. Today there are 1210 Fairtrade-certified producer organisations in 74 countries benefiting around 6 million people.

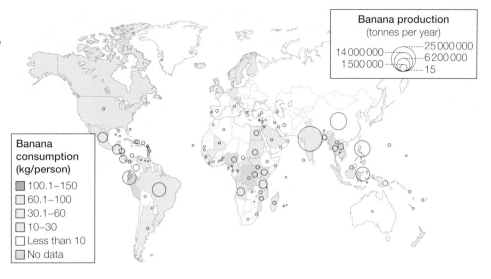

▶ **Figure 2** *Areas of banana production and consumption. Bananas are grown in tropical regions where the average temperature is 27 °C and the annual rainfall is 2000– 2500 mm.*

Banana TNCs

The unreasonable pressures of UK supermarkets on suppliers to keep prices low has helped to create unethical methods of banana cultivation. Bananas sold in the UK are typically grown in Latin American plantations owned by fruit-exporting TNCs. The landscape is likely to have had natural vegetation removed to create the vast plantations, which are heavily treated with pesticides (in Latin America the banana is referred to as 'the chemical fruit'). They are farmed by workers who are paid little and live and work in poor conditions. It is worth noting, however, that the industry's poor environmental and socio-economic record has its roots in the nineteenth century.

TNCs have always dominated plantation monoculture. Originally, this was because of the huge capital investment needed to develop a reliable harvest from bushes or trees, in tropical and sub-tropical climates where there was no tradition of export prior to colonial times. (Of course in the colonial era, indentured or slave labour kept wage bills low on, for example, sugar plantations.)

Also, strong vertical and horizontal integration enabled economies of scale and secured market access. Transport was always key in the profitable sale of fresh fruit. Even in the twenty-first century, the same TNCs own the banana plantations and the linked companies that chemically treat, ship and process the fruit, often crossing several national boundaries. Consequently, fruit companies such as Dole and Chiquita Brands International have the marketing power to bring bananas from tropical plantations to consumers all over the world.

What is a 'banana republic'?

In the nineteenth century, shipping magnates and railroad entrepreneurs discovered how lucrative the banana trade could be, with a growing North American taste for the fruit. These profiteers worked, notoriously, to consolidate their monopolies, manipulating governments in Central America to ensure the cheap purchase of huge swathes of land for banana plantations. Large areas were required to keep costs low. Spare land was needed to replace areas where the soil has become exhausted by such monoculture. Their method of operating in these countries, being almost entirely export-orientated, gave little benefit to host nations and gave rise to the term 'banana republic'. The term refers to a country whose economy is dependent on one or few commodities (pay for workers is, consequently, kept low as people have few alternatives). They are often run by an incompetent or corrupt government 'in hock' to large international corporations. Politically, such countries were intrinsically unstable (consider Guatemala or Honduras in the twentieth century). At the turn of the Millennium, five companies still controlled 70 per cent of the banana export market worldwide.

However, trends in production are shifting away from TNCs towards smaller and independent producers (Figure **3**) and new buyers are entering the market despite its highly competitive nature. This, in part, has been helped by the growth in Fairtrade organisations, as well as consumer pressure that has pushed supermarkets to purchase either from smaller wholesalers or directly from producers, and not from TNCs.

The growth of Fairtrade

When and how did Fairtrade get so popular and, apparently, so successful?

The Fairtrade movement came into being:
'to raise awareness of trade injustices and imbalances of power in the conventional trade structures and to advocate for change in policies to favour equitable trade.'
(World Fairtrade Organisation)

How do small-scale producers compete with large TNCs and sell their products all over the world? In part, the secret of Fairtrade's success is teamwork. Across its history, this movement of people that includes many NGOs has sought to link organisations in the economic 'North' and 'South'. For example, by enabling charities in HDEs to help develop the capabilities of cooperative and marketing organisations in LDEs, that in turn advised, organised and in some cases lent seed money to disadvantaged producers. What began as regular international conferences of similarly minded organisations (guided by a moral or religious compass), has subsequently been formalised as four international trade networks: Fairtrade Labelling Organisations International (FLO), World Fairtrade (WFTO), Network of European Worldshops (NEWS!) and European Fair Trade Associaton (EFTA).

Secondly, it is also worth remembering the movement's *longevity* – the first Fairtrade products were sold in the US in the late 1940s and in Europe in the late 1950s. At first, its focus was on handicrafts, an industry with strong links to missionaries (the origin of many international NGOs). Handicrafts supplemented family income and, importantly, boosted women's participation in the economy, often a key aim of participating charities. The first Fairtrade coffee was sold in the Netherlands in 1973 and Fairtrade sugar has been sold from the very beginning. Many further Fairtrade foodstuffs followed, gaining a foothold in European and US markets. These products were sold via so-called 'world shops' like Oxfam, in the UK. The first Dutch 'world shop' opened as early as 1969. Across a number of HDEs, these shopfronts were used very effectively to raise awareness of related trade issues at the point of sale.

Thirdly, and in parallel with the growth of development politics on the high street, was the important role of the governments of LDEs, advocating 'trade not aid' in international forums, including the UN Conference on Trade and Development, in 1968 in Delhi.

From the late 1980s onwards, the labelling of Fairtrade products as such has boosted consumer awareness, and the influence of TNCs has significantly decreased. This new brand's credibility has been strengthened by the creation of an international monitoring system. FLO-CERT is the independent company with the job of inspecting and certifying producer organisations and auditing traders.

▼ **Figure 3** *Market shares of selected companies in global banana exports by volume*

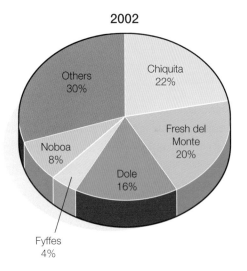

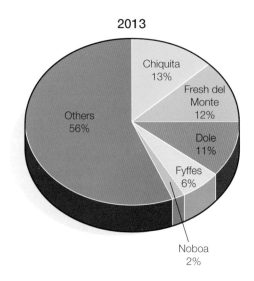

The El Guabo Association of Small Banana Producers

The El Guabo Association of Small Banana Producers was formed in 1997 in south-west Ecuador (Figure **4**). Today, it is one of the world's largest producers of Fairtrade bananas and exports around 30 000 boxes a week to the USA and Europe. Before Fairtrade, the 339 family farms sold their bananas through intermediaries at a price that was too low to cover basic costs. Fairtrade has resulted in a range of individual and community benefits:

Economic

◆ Stabilised incomes and improvements to standards of living

◆ Guaranteed fair wage and long-term supply contract, including direct access to new and international markets

◆ Producers able to raise additional capital for reinvestment, for example, tanks to wash bananas

◆ Migrant labourers are helped, for example, assistance to buy their own land

Social

◆ Health care benefits to families of cooperatives, for example, free use of El Guabo clinic

◆ Provision of educational and medical supplies

◆ Affiliation with a social security system, for example, payment of retirement benefits

◆ Support for the poorest groups, for example, food baskets

◆ Improved education provision, for example, new school for children with special needs

◆ Marginalised groups helped to find employment, for example, HIV/AIDs sufferers.

▼ **Figure 4** *Members of El Guabo, a fair trade banana cooperative from Ecuador*

ACTIVITIES

1 Study Figure **2**.
   a Describe the world trade in bananas?
   b To what extent is this trade sustainable? You may find it useful to refer to www.fairtrade.org.uk or http://sustainablefoodtrust.org. Search for 'bananas'.

2 Study Figure **3**. Analyse the data and comment on changes in the market share of global banana exports.

3 'If banana price inflation had kept up with the pace of Mars bars then the fruit would cost £2.60 a kilo now.'
   a Explain this statement.
   b Who or what ultimately pays the cost of cheap bananas?
   c Suggest problems facing the small independent banana grower in an LDE (consider the comparative advantages of TNCs presented earlier in this section and in 1.7).

STRETCH YOURSELF

'Trade is an important development tool. Trade is not, however, an end itself.' Using a named example, discuss this statement.

In this section you will learn about:
- geographical consequences of global food systems
- impacts of palm oil trade on lives across the globe

Global food systems

So – are global systems good or bad? There is, quite literally, food for thought. Global food TNCs are arguably some of the most powerful on the planet – they operate over a vast area, have more than one billion employees and even have their own armed forces! They directly or indirectly control what we eat and even how we live. In fact, they impact on our lives and those of just about everyone else on planet Earth like no other global system.

Who controls what I eat?

Do we really have a choice over what we eat? This may seem a silly question – after all, there is plenty of choice in the supermarkets and, as a result of good competition, prices in UK supermarkets are comparatively cheap. However, look again at the different brands on offer and you might be surprised to learn that many are owned by the same TNC (Figure **1**). Ten firms control 28 per cent of global food production (Figure **2**) and less than ten TNCs control more than 50 per cent of the food on sale in a typical supermarket. Since these TNCs operate to maximise their profits, the food they sell is manufactured as cheaply as possible whilst appealing to our taste buds.

Figure 1 *A food choice? Many of the food products that we buy from the supermarket are owned by a handful of very large food TNCs.*

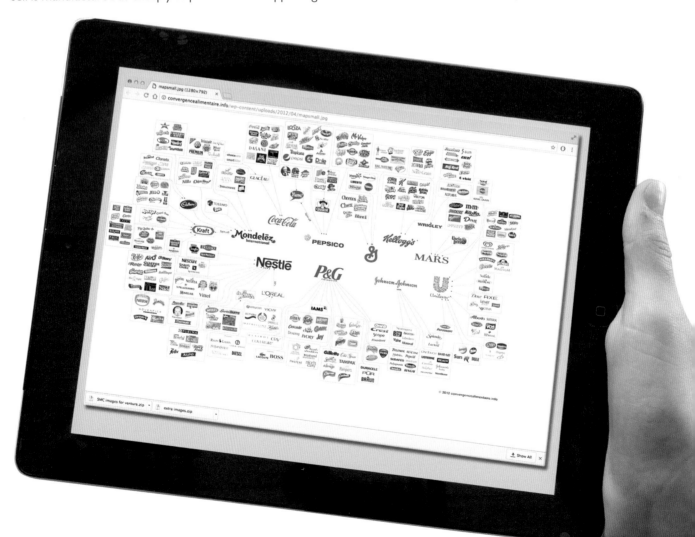

Following EU legislation in January 2015, GM crops have begun to enter the UK food chain, mainly via animal feed. However, it is likely to be some time before UK public opinion supports the widespread planting of GM crops, such as seen in other parts of the world (85 per cent of all corn crops in the US are now GM). Other examples of cost savings within the food manufacturing process include replacing natural ingredients in food and drink with cheaper synthetic ones or adding chemicals.

Nevertheless, artificial additives are important in extending shelf lives of food and drink and help to present food attractively for potential customers. Unsold supermarket food is simply thrown away – over 40 per cent of bakery items and over two-thirds of salad sold in bags are thrown away by supermarkets!

Who controls how much I eat?

Obesity is a global problem – but so too is hunger. Are the food TNCs responsible? Around 1.9 billion of the world's population are overweight, potentially adding to the profits of the TNCs, while the more than 800 million who are undernourished are largely invisible to these companies. It is illogical to blame the food TNCs for making us 'fat' since it is our choice to purchase the more convenient ready-meal or burger. Yet it is also evident that cheap processed foods can not only affect our health (see 4.16) but also keep profits of TNCs large.

Look at Figure **3**. As the global population has rocketed, so food TNCs have found more inventive ways of meeting the increased demand, thereby avoiding any global Malthusian crisis (see 4.25). There is plenty of food for the hungry 800 million plus – indeed, more food is being grown than ever before and increasing use of **agrotechnologies** is likely to continue this upwards trend, at least in the short-term. So where is all the food going? Almost half of all the grain harvested worldwide in 2014 did not go into food processing but into animal feed, biofuels or raw materials for industry.

◈ **Figure 2** *Just ten firms control 28 per cent of all food production.*

1 Associated British Foods
2 Coca-Cola
3 Danone
4 General Mills
5 Kellogg
6 Mars
7 Mondelez International
8 Nestlé
9 Pepsi
10 Unilever

◈ **Figure 3** *Bom Futuro farm in Brazil is the biggest soya producer in the world; it is a staggering 5400 km² in area*

The trade in palm oil

What do Cadbury chocolate, Findus fish fingers and Ginsters Cornish pasties have in common? They are all top-selling products that contain palm oil. Palm oil is highly versatile – it has a high melting point and so makes products smooth, creamy and easy to spread. It is also cheap, has the highest yield of any oil crop and is widely available. Consequently, as our demand for processed food products has increased, so has worldwide demand for palm oil – it is the most-used vegetable oil worldwide. In the UK, we consume around 320 000 tonnes of palm oil every year and it is found in many products on the supermarket shelf – from processed foods to candles, hair and cleaning products.

Figure 4 *Red fruits of the oil palm, harvested from a plantation in Pahang, Malaysia*

So what...?

Palm oil needs consistent high humidity and temperatures, and a lot of land to grow – in other words, an equatorial or tropical rainforest climate. This has placed enormous pressures on tropical rainforest biomes, particularly in south-east Asia – 90 per cent of the world's palm oil is grown in Malaysia (Figure **4**) and Indonesia. Rainforest has been cleared to allow plantation monoculture on a massive scale. In Indonesia alone, 13 million hectares of rainforest (an area three times the size of Switzerland) have been transformed into oil palm plantations.

Aside from the environmental impacts, land appropriation by the TNCs has often forced local inhabitants off the land, sometimes violently. Deforestation results in a lifestyle change for local inhabitants. The chemicals used in palm oil production pollute the water and soil making other forms of agriculture impossible. Consequently, locals are obliged to work for low wages cultivating crops that they will most likely never taste.

Sustainable palm oil

Palm oil plantations have tripled in the last decade. The environmental damage and unfair labour practices of the nineteenth century that were associated with the production of commodity crops on plantations in colonial Latin America, are now practised in the twenty-first century in palm oil production in south-east Asia. So is the answer to stop using palm oil as an ingredient in the things we buy? The Roundtable on Sustainable Palm Oil (RSPO), which represents 2000 members from over 75 countries (including producers, processing companies, retailers, banks and NGOs), argues not. Instead, the RSPO campaigns to raise awareness of the production of palm oil using more sustainable methods, among both producers and consumers. RSPO-certified palm oil must be produced where there is:

◆ environmental responsibility and conservation of natural resources and biodiversity

◆ responsible consideration of employees, and of individuals and communities affected by growers and mills

◆ responsible development of new plantings.

(From *8 Principles for growers to be RSPO Certified*, RSPO)

Between January 2015 and January 2016, sales of RSPO certified sustainable palm oil have increased three-fold. However, critics of the mark suggest that the fact that RSPO members can still clear pristine rainforest calls into question the 'sustainable credentials' of the organisation, as do ongoing investigations into claims of forced labour and child labour on certified plantations.

Green Palm certificates: connecting small-scale producers to the global market

Manufacturers under increasing consumer pressure to go green with their palm oil ingredients need some guarantees on traceability. However, the cost of segregating RSPO-certified oil from all the rest is high, having a knock-on effect on take up of RSPO oil products (Figure **6**). These costs can be reduced via the issuing of Green Palm certificates to producers and plantations that meet the standards of the RSPO. Producers then sell their certificates to TNCs who can demonstrate their commitment to the greener, fairer production of palm oil, even if it isn't actually in their products. Unilever are among a number of TNCs backing RSPO and Green Palm.

However, NGOs who criticize the environmental credentials of RSPO and Green Palm certificates argue that orangutan populations remain under threat from large-scale producers that still drain peatlands and clear pristine forest that form essential green corridors for this endangered animal (Figure **5**). In 2016, RSPO suspended palm oil giant IOI after allegations of deforestation in Indonesia.

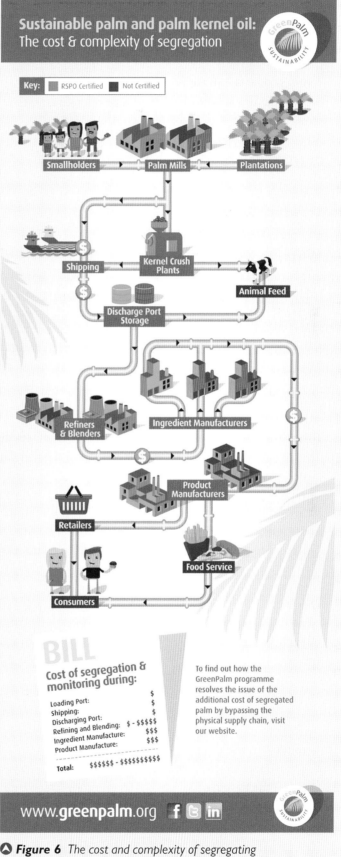

Sustainable palm and palm kernel oil: The cost & complexity of segregation

Key: RSPO Certified / Not Certified

BILL
Cost of segregation & monitoring during:

Loading Port: $
Shipping: $
Discharging Port: $ - $$$$$
Refining and Blending: $$$
Ingredient Manufacture: $$$
Product Manufacture:

Total: $$$$$$ - $$$$$$$$$$

To find out how the GreenPalm programme resolves the issue of the additional cost of segregated palm by bypassing the physical supply chain, visit our website.

www.**greenpalm**.org

▲ **Figure 6** *The cost and complexity of segregating RSPO-certified oil from non-RSPO certified palm oil*

▲ **Figure 5** *Orangutan in Sarawak, Malaysia*

Agumil: A palm oil company operating in Palawan, Philippines

'AGUMIL is taking over our indigenous people's lands through forced and fraudulent land sales. It is quite contrary to national laws. We are losing our lands and our livelihoods. We are calling on the Philippines Government to uphold our rights.'

(Motalib Kimel, Chairman of Coalition Against Land Grabbing (CALG), Nov 2015)

Source: www.grain.org

⬢ **Figure 7** *A typical palm oil plantation on deforested land, Borneo, Malaysia*

What can we do?

Despite being established in 2004, only around 20 per cent of palm oil produced globally is certified by the Roundtable on Sustainable Palm Oil (RSPO). For the remaining 80 per cent, it's business as usual. While certified sustainable palm oil has, so far, been industry-led, could a tipping point come from heightened consumer understanding and action?

Source: www.theguardian.com

Respecting land and forest rights: a guide for companies

The Interlaken Group is made up of many stakeholders, including TNCs such as Nestle, Coca-Cola and Unilever. The Group developed a guide to help companies do their part to respect local land and forest tenure rights. It supports the United Nations endorsed document, 'Voluntary Guidelines on the Responsible Governance of Tenure of Land, Fisheries and Forestry in the Context of National Food Security' (VGGT). VGGT emphasizes the following priorities:

- ◆ Respect for legitimate land tenure rights
- ◆ Do no harm
- ◆ Support for smallholders
- ◆ Broad-based consultation
- ◆ Land and forest tenure impact assessments
- ◆ Accountability, monitoring and enforcement.

Source: www.ifc.org

Impacts on local people

It is as if people in the impacted oil palm communities are dying little by little because they no longer have the plants needed to cure themselves. Before, they only walked half an hour to get the raw material for building their houses, for their artefacts and medicinal plants. Now they have to walk half a day to the other side of the mountain before they can find the plants they need.

Source: www.forestpeoples.org

◼ Indonesia 51%
◼ Malaysia 42%
◻ Papua New Guinea 5%
◼ Brazil 1%
◼ Columbia 1%

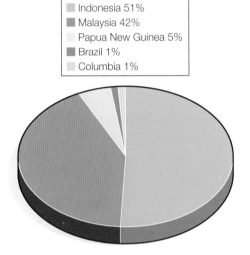

⬢ **Figure 8** *Sources of RSPO certified palm oil*

	India	China	EU (28 countries)
1997	1400	1700	1900
2002	3600	2600	3400
2007	3800	5500	4500
2012	7600	6100	6100

⬢ **Figure 9** *Consumption of palm oil (thousand tonnes)*

	UK (thousand mt)	World (thousand mt)
2009	155	344
2010	179	1281
2011	262	2491
2012	278	3479
2013	322	4500
2014	396	5400

⬢ **Figure 10** *Certified sustainable palm oil sales, supported by RSPO supply chain (thousand tonnes)*

Roundtable on Sustainable Palm Oil (RSPO)

The RSPO has developed a set of environmental and social criteria which companies must comply with in order to produce certified sustainable palm oil. When they are properly applied, these criteria can help to minimize the negative impact of palm oil cultivation on the environment and communities in palm oil-producing regions.

It has more than 1700 members worldwide who represent all links along the palm oil supply chain. They have committed to produce, source and/or use sustainable palm oil certified by the RSPO.

Source: www.rspo.org

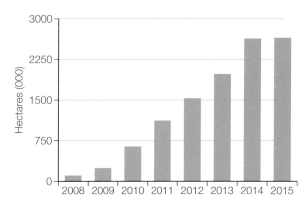

▲ **Figure 11** *Total area of sustainable palm oil production*

Kraft foods – a food TNC using palm oil

Following a BBC Panorama investigation, Kraft, a food TNC, were asked to comment on their use of sustainable palm oil.

'We buy our palm oil from Roundtable on Sustainable Palm Oil (RSPO) members, but more needs to be done to enforce guidelines and address deforestation. The RSPO needs to reach broad consensus on its certification standards, in particular with regard to climate change impacts of palm oil production.'

Source: www.bbc.co.uk

Getting 'it' right

'Land rights are a common issue across West Africa. One of the challenges across the region is that there is no real land-use planning, there is no regulation of what is a protected area and what is for housing and farming.' (David Hoyle, ProForest)

Source: www.proforest.net

Invisible palm oil?

It's an invisible ingredient, really, palm oil. You won't find it listed on your margarine, your bread, your biscuits or your KitKat. It's there though, under 'vegetable oil'. And its impact, 7000 miles away, is very visible indeed.

What, then, is 'unsustainable' palm oil? Step one: log a forest and remove the most valuable species for furniture. Step two: chainsaw or burn the remaining wood releasing huge quantities of greenhouse gas. Step three: plant a palm oil plantation. Step four: make oil from the fruit and kernels. Step five: add it to biscuits, chocolate, margarine, soaps, moisturisers and washing powder. At breakfast, when millions of us are munching toast, we're eating a small slice of the rainforest.

Source: www.independent.co.uk

ACTIVITIES

1 What influences your choice when buying a food product? Rank the following in order of importance: price, quality, brand. Justify your choice.

2 Does it really matter if most of our food only comes from a few global food TNCs? Consider your answer to activity **1** before responding.

3 Sort the impacts of palm oil cultivation into primary and secondary impacts.

4 With reference to Figure **9**, comment on changes in palm oil consumption.

5 Study Figure **10**. Compare UK sustainable palm oil sales with the global trend? (Hint: a comparative line graph or statistical manipulation of the data would help to support any answer.)

6 Suggest reasons why supply chains (the global systems that move palm oil from the tropical plantation into the products on the supermarket shelf) do not always meet principles of sustainability.

7 To what extent is it possible to achieve sustainable global systems? Your answer might refer to any global system but should be illustrated with specific and appropriate examples.

In this section you will learn about global governance, and norms, laws and institutions

What do governments do?

Governments have always governed, regulated, organised and, of course, controlled. It has long been recognised that TNCs also wield power in locations across the globe (1.7). Our increasingly interconnected 'globalised' world faces unprecedented challenges related to the environment, resource dependency, energy, economic inequalities, communications, conflict, health, human rights, trade and so on – all of which have profound implications on our material and social welfare.

Some even suggest that today's global challenges threaten humanity's very survival, but there is no single 'world government' to manage these challenges. Furthermore, global governance isn't just a problem for national governments and international institutions to ponder. The term 'global governance' is bigger than that and affects all our everyday lives.

What is global governance?

Alas there is not yet a widely agreed definition of global governance. Many geographers describe it as a movement rather than a simple entity. Others refer to it as a concept. In short, although the phrase first emerged in the late 1980s, both its exact meaning and priorities are still contested today.

The Commission on Global Governance describes it as the:

'... *sum of the many ways individuals and institutions, public and private, manage their global affairs* ...'

Note how this definition goes beyond the obvious regulatory institutions. The range of so-called **actors** managing global affairs includes nation states, profit-making companies, international and regional organisations (such as the EU) and their participation in this process may be reactive or proactive (Figure **1**). Global governance is essentially, therefore, an international process of consensus-forming which, in turn, generates guidelines and agreements 'governing' the actions of those same players.

Actor	Action	For example...
National governments	**Reactive** Legislate and invest, to implement international initiatives, laws, targets or agreements	The UK Government's recent support for renewable energy sources, to reduce dependence on fossil fuels and climate-changing emissions is, in part a response to new EU laws on dirty power stations.
	Proactive Lobby for and contribute to international discussions, votes and decisions	Some national governments have more influence than others on the UN Security Council. China, France, Russia, the UK and the USA are all permanent members of the UN Security Council and often take leadership roles in conflict resolution.
TNCs	**Reactive** Compensate and adopt new codes of conduct or working practices imposed by legislation or socio-economic pressure	In the wake of collapse of the eight-storey Rana Plaza building near Dhaka, in Bangladesh in 2013, clothing company Primark (part of multinational company Associated British Foods) paid compensation and provided emergency aid to the victims of the disaster (Figure **2**).
	Proactive Sponsor and support the work of NGOs and international organisations	Tobacco companies such as British American Tobacco and others sponsor anti-smoking NGOs within LDEs, for example in south-east Asia, to publicise the health risks of smoking, thereby protecting themselves from the sort of law suits seen in HDEs in recent decades.
International organisations	**Reactive** Respond to global events to offer advice to national governments and publicise the work of NGOs already on the ground	The WHO was late to offer guidance on how to contain the Ebola virus in the early stages of the recent epidemic in West Africa.
	Proactive Sponsor, facilitate and publicise international issues and agreements to address them	In 2013 the UNI Global Union, in alliance with leading NGOs, sponsored and created the legally binding *Bangladesh Accord on Fire and Building Safety*, to protect working conditions in the ready-made garment industry. The accord has been signed by 200 clothing companies (from over 20 countries), leading to the inspection of 1500 factories in the first two years.

Figure 1 *The role of different actors in global governance may be proactive or reactive*

Norms and laws

Global governance is central to the idea that norms, rules and laws make, and remake, global systems. As we have seen, these systems have a geography – they create consequences for citizens, ecosystems, human and physical environments in different places and at different scales.

Norms are the values, traditions and customs that govern individuals' behaviour in any particular society – some may be unspoken and never written down, but they're widely understood – 'It's the done thing.'

Norms often become enshrined in laws that reflect the acceptable standards associated with a specific cultural background. Laws are obligatory and normally protect the rights and interests of all who live under them. For example, it is a norm in British society to keep pet animals, to love them and protect them from cruelty. Specific laws, such as the Animal Welfare Act, reflect these norms and, when enforced, provide protection from abuse and negligence (Figure **3**).

But not all societies agree as to what is normal and reasonable. If you have ever been on holiday to a country where scrawny stray cats loiter around the legs of evening diners you can testify to this! It is this that makes global governance so complex.

Moreover, even those signatories to international agreements and treaties that are committed to systems of global governance differ on how they interpret and enforce norms and laws. This can have important consequences in areas such as conservation management and in the protection of human rights.

WHO will save us?

In a globalised world, there are many transnational problems that need addressing, for example, climate change, which was discussed again at the UN Climate Change Conference in Paris in 2015. Likewise, international and civil wars, humanitarian emergencies, unprecedented flows of refugees, outbreaks of infectious diseases, financial market instability, sovereign debt crises, **trade protectionism**, and the development of poorer countries all need regulation, monitoring and control, on a global scale.

Perhaps only international organisations such as the UN, the World Health Organisation (WHO), World Bank and the International Criminal Court are in a position to do this. But their role in world politics is controversial. Although some see them as legitimate and effective, others seek and find faults – not least questioning their strong links to the most powerful nations that provide the largest amount of funding for them. The functioning, power and effectiveness of these institutions, inevitably, differ widely and these realities are examined in the following pages.

△ **Figure 2** *Relatives of the victims of the collapse of the Rana Plaza factory building in 2013, near Dhaka; they were supported by local and international NGOs and, eventually, received compensation from TNCs whose suppliers were located in the Rana Plaza*

△ **Figure 3** *Animal cruelty is an uncomfortable reality; the RSPCA is England and Wales' oldest welfare charity, with a long history of protecting animals and representing their rights.*

STRETCH YOURSELF

How can it be law if it cannot be enforced? To experienced international lawyers this is an old and rather tiresome question, not least because it assumes that international law cannot be enforced! Investigate the legitimacy of this question in the context of global governance. Pick an international law, such as EU competition law or the International Bill of Human Rights, and find out how and why it has or has not been enforced.

ACTIVITIES

1 **a** Make a list of economic, social and environmental challenges that may be termed 'global'.

   b For each challenge, suggest three or four actors that could have a role in helping to address it.

2 State and justify your own definition of the term 'global governance'.

3 What does 'building consensus' look like in the context of global governance?

4 Why are the roles of international institutions like the WHO, UN or World Bank controversial?

In this section you will learn about:
- issues associated with attempts at global governance
- inequalities and injustices in global governance

Sovereignty and global governance

Central to any understanding of global governance is the fixed reality of nearly 200 individual sovereign states and their national territories. Some might argue that their claims to exclusive sovereignty and security are watertight, but in an increasingly interconnected, globalised world this is not realistic. Just as rivers do, people, money, ideas and information leak out of individual nations and flow across borders into other countries.

In reality, nation states exercise sovereign authority over their territories and interests beyond national boundaries – governments 'pool sovereignty' when it is in everyone's best interests. For example, EU countries adopt a regional-based approach to economic planning and migration. Nation states also collaborate in areas such as military intelligence, training, border security and joint missions. The world is transnational and highly interdependent even for the largest military powers (e.g. the USA working within NATO).

◀ **Figure 1** *Why can they not tear down the barriers? The Doha Development Round showed the failure of the US and many European nations to recognise that globalisation no longer belongs just to the West.*

ANOTHER VIEW

We need to bang heads together

Institutions like the UN, IMF, G20 and WTO appear too often to be slow and ineffectual in tackling the world's most pressing problems. The guiding principles of these organisations are clear and they could have far more impact in promoting global economic growth, resolving conflicts and so on. However, this can happen only if nation states allow these principles to be implemented and these institutions are empowered to produce concrete results rather than simply issuing declarations.

For example, trade is a major engine of economic growth and job creation; it is also important as a deterrent to reckless political or military actions that would otherwise threaten the benefits of these economic ties – hence the importance of promoting **trade liberalisation**. However, since commencing in late 2001, the **Doha Development Round** talks, aimed at lowering trade barriers globally, have repeatedly stalled (Figure **1**). The most significant differences have been between HDEs (including EU countries, the USA and Japan) and major developing economies (such as India, Brazil and South Africa). Since 2014, the Doha Round's future has remained uncertain. Yet few would disagree that the best way forward is to tackle trade agreements, government subsidies and so on at the global rather than regional or **bilateral** level.

Issues associated with attempts at global governance

Deepening economic globalisation as well as increasing migration, trade and capital flows make individual states more and more susceptible to policies adopted by others. Likewise climate change and increased activities in the global commons (see 1.13) do not fall within the jurisdiction of any one particular country. This is why established organisations such as the WTO, WHO and the **United Nations Environment Programme (UNEP)** are championed by many as ideal to lead in trade, health and environmental issues.

Inequalities and injustices in global governance

The Centre for International Governance Innovation (CIGI) has identified a number of critical inequalities and injustices in almost every sector of global governance. These so-called 'gaps' are believed to be impeding progress in the global economy, security, development and the environment. The CIGI observations cover three key themes – jurisdiction, incentive and participation (Figure **2**).

Examples of their concerns include:

◆ environmental governance involving so many agencies and agreements that duplication and incoherence is restricting progress.

◆ the need for collaboration to improve among the IMF, G20 and OECD if international jurisdiction and regulation of finance and capital flows are to improve

◆ that there are 'no rules of the game' existing to deal with the unsustainable debt burdens of some sovereign states

◆ that no agency is tasked with thinking about new and emerging long-term trends in agriculture and food security – and their implications for current policies and practices

◆ the need for better coordination to deal with epidemics and vaccine stockpiling.

The jurisdictional gap	The gap between the increasing need for global governance in many areas (such as health and water security) and the lack of an authority with the power, or jurisdiction, to take action
The incentive gap	The gap between the need for international cooperation and the motivation to undertake it. This is closing as globalisation provides increasing impetus for countries to cooperate, but there are concerns that, as Africa lags further behind economically, its influence on global governance processes will diminish
The participation gap	This refers to the fact that international cooperation remains primarily the affair of governments, leaving civil society groups on the fringes of policy-making. On the other hand, globalisation of communication is facilitating the development of global civil society movements

Figure 2 *Key gaps in global governance*

Their report also detailed the need for an international law to ensure and enforce water security, and feared that the International Atomic Energy Agency (IAEA) is not empowered to impose an effective regulatory framework on the nuclear industry.

Furthermore, troubling gaps were highlighted in the global governance of science and technology, urbanisation, migration, energy and even transnational crime. For example, cybersecurity is, in some ways, the newest international frontier. Given the speed of technological change, therefore, it comes as no surprise that there is little in the way of governance, but there is also the danger that some forms of regulation could be worse than none. So the international challenge will be how best to maintain a free flow of information while limiting various forms of 'cyber-aggression' without giving national governments license to curb the flow of information for political purposes.

Figure 3 *A summary of the work of the United Nations*

Purpose	Main organisations or bodies	Examples of achievements	Examples of weaknesses
Maintain world peace and security	• Security Council (1945) • Office of Peacekeeping Operations (1992) – formerly the UN Office of Special Political Affairs • Office for Disarmament Affairs (1998)	• Developed and strengthened international relations – since its inception, there have been no further world conflicts and over 170 peaceful UN settlements have ended regional conflicts. • Positive peace-keeping missions in Africa that have largely managed to maintain a fragile peace. • Supported nuclear disarmament.	• Powerful military states have ignored UN opinions and mandates – notably the Soviet Union (now Russia) and China. • The UN has sometimes only had a minor influence on the outcome (e.g. 1950s–70s Vietnam war, 1962 Cuban Missile Crisis, or present-day Chinese expansion into the South China Sea). • Numbers of nuclear powers in the world has kept on increasing – most recently, North Korea.
Provide long-term humanitarian and development assistance and uphold human rights	• United Nations Children's Fund (1946) • UN Women (2010) • UN Population Fund (1969)	• Improved the quality of life for millions of the most vulnerable, particularly children and women. • The ambitious Millennium Development Goals achieved notable successes.	• Failures in fully addressing security situations in Afghanistan and Middle East have hampered efforts at development and rebuilding.
Eradicate and prevent global hunger	• World Food Programme (1963) • Food and Agriculture Organisation (1945) • UN Conference on Trade and Development (1972)	• Negotiated fairer trade agreements between HDEs and LDEs. • Applied agro-technologies to increase food yields in LDEs.	• Globalisation and capitalism support an unequal movement of goods between LDEs and HDEs.
Protect refugees worldwide	• United Nations High Commissioner for Refugees (1950)	• First responder to human and environmental disasters around the world.	• Resettlement of refugees has not always been achievable (e.g. Darfur refugee camps set up in 2003 are now semi-permanent towns).
Eradicate and combat the spread of global disease	• World Health Organisation (WHO) (1948)	• Reduced global mortality rates, including eradication of smallpox (1979) and near eradication of polio.	• AIDS pandemic continues to cross countries and continents.
Develop friendly relations	• General Assembly (1945)	• Independent arbitrator and upholds a strong moral code	• Financial dependence on HDEs leaves UN open to criticism of impartiality (e.g. US-led invasion of Iraq in 2003 in the search for 'weapons of mass destruction' that were never found by UN weapons inspectors)
Settle legal disputes (within international law)	• General Assembly (1945) • International Court of Justice (1945)	• International laws established within a world legal framework.	• Limited repercussions for countries that do not follow mandates (laws). • Unipolarity, where one state exercises most of the cultural, economic and military influence, exists, e.g. US geopolitical influence in the Middle East may have shielded Israel from wider UN criticism.
Promote and support wise use and sustainable development of the global environment	• United Nations Environment Programme (1972)	• Established global environmental norms underpinned by legally binding agreements.	• Truly global and ambitious agreements on climate change are slowly being realised.
Poverty reduction and improvement of living standards worldwide within a sustainable framework	• International Monetary Fund (1945) • United Nations Development Programme (1966) • World Bank (1945) • Economic and Social Council (1946) • United Nations Educational, Scientific and Cultural Organisation (1946)	• Promoted better understanding of world development issues and inequalities.	• Economic powers in practice are overshadowed by more powerful players such as the WTO or the wishes of individual and influential HDEs.

With great power comes great responsibility

It has been said that in the UN Security Council chamber and other forums, China is increasingly willing to take the lead and behave like the great power it has become (Figure **4**). Other writers suggest that China is largely preoccupied with protecting its interests and those of its allies rather than working to strengthen the international system. China has changed the way it does business and, as its wealth and power grow, so also will its interests expand. While the USA and the UK say they want rising powers to be more responsible and active, how might the West feel, for example, about China intervening in the Middle East peace process? How will the contribution of this new superpower be received then?

⊙ **Figure 4** *Chinese President Xi Jinping addressing the UN General Assembly in September 2015*

ACTIVITIES

1   Look at Figure **1**. Describe and explain the viewpoint taken by the cartoonist.

2   The key gaps in global governance cover jurisdiction, incentive and participation. For each theme:

   a   explain the term

   b   illustrate it with a specific example.

3   For **either** cybersecurity **or** migration investigate and comment on contemporary concerns in global governance.

4   To what extent has the United Nations succeeded at global governance?

EXTENSION

A global government?

Read again about the work of the EU, WTO and other international organisations that regulate and reproduce global systems in section 1.6. Now study Figure **3**. Arguably the most globally significant sets of norms, laws and institutions arise from the work of the United Nations. Established in 1945 immediately after the Second World War and initially to maintain world peace, the UN has since developed and subsequently enforced international laws and policies that impact directly or indirectly on every one of world's citizens. If the world had a global government, then it would be called the United Nations.

Extract from the Charter of the United Nations (Article 1)

The purposes of the United Nations are:

◆   to maintain international peace and security, and to that end: to take effective collective measures for the prevention and removal of threats to the peace, and for the suppression of acts of aggression or other breaches of the peace, and to bring about by peaceful means, and in conformity with the principles of justice and international law, adjustment or settlement of international disputes or situations which might lead to a breach of the peace

◆   to develop friendly relations among nations based on respect for the principle of equal rights and self-determination of peoples, and to take other appropriate measures to strengthen universal peace

◆   to achieve international co-operation in solving international problems of an economic, social, cultural, or humanitarian character, and in promoting and encouraging respect for human rights and for fundamental freedoms for all without distinction as to race, sex, language, or religion

◆   to be a centre for harmonizing the actions of nations in the attainment of these common ends.

STRETCH YOURSELF

'Strengthening existing international institutions is essential to implementing true global governance.' Discuss.

In this section you will learn about global commons – the concept, rights and benefits

Look at Figure **1**. Since the Apollo missions of the 1960s and 1970s, pictures of Earth from space have become the most published images of all time – and understandably so! Breathtakingly beautiful, they provoked deep thought on global environmental and political issues. The inspiration for so many was a simple realisation that this 'sparkling blue and white jewel' was all we had and so should be nurtured, protected and even managed as a whole. As the United States astronaut Edgar Mitchell continued to describe '… this is Earth … home.'

The environmental movement has embraced such sentiments, but politically – with the individual sovereign rights of nearly 200 states and their national territories to consider – managing the Earth as a cohesive whole is not possible. The world community faces profound challenges. Climate change, economic inequalities, resource consumption, fossil fuel dependency, water and food security, and 'extreme events' such as natural disasters, recessions, mass migrations and wars all demand an international perspective. Political interdependence is a necessary reality and illustrated particularly well in the governance of the global commons.

△ **Figure 1** *Earth from space*

Global commons – the concept

The global commons are defined as those parts of the planet that fall outside national jurisdictions and to which all nations have access.

Four global commons are identified. These are:

◆ the high seas and deep oceans

◆ the atmosphere

◆ the northern and southern polar regions – Antarctica in particular

◆ outer space.

Notice how these territories are guided by the principle of a common heritage of humankind. This has led to a widening of the concept to include resources of interest or value to the welfare of all nations – such as biodiversity and tropical rainforests (*AQA Geography A Level and AS Physical Geography*, 6.1, 6.7). Others define the set of global commons even more broadly – to include science, education, information and peace. Some go as far to suggest that they are 'the collective heritage of humanity — the shared resources of nature and society that we inherit, create and use' (Figure **2**).

> **Word clouds**
> A word cloud is a visual representation of text. The size of each word indicates its frequency of use. Word clouds allow analysis of large volumes of text (e.g. for word association or bias/opinion). By comparing different word clouds, key words might be revealed, which may say as much about the author as the content (Figure **2**).

△ **Figure 2** *Global commons – the collective heritage of humanity?*

Global commons – rights and benefits

Rights and benefits associated with global commons can only be assured through coherent, coordinated and collective decision-making at the global level – in short, their effective management on a global scale.

International law already identifies four global commons – specifically the high seas, the atmosphere, Antarctica and outer space. Historically, access to most of the resources found within these domains has been difficult and they have not been scarce. However, advancement of science and technology in recent years, coupled with a growing demand for resources, is leading to increasing pressures upon them.

Look at Figure **3**. The international community acknowledges the need to conserve these resource domains, both for development and human well-being. In consequence it has adopted a number of conventions and treaties to govern the global commons.

Look again at Figure **3**. Currently the only truly universal and inclusive multilateral forum is the United Nations. Only a strong and effective UN could ensure the coordination, cooperation and coherence required in international policy-making if the benefits of the global commons are to be assured for all. However, the frameworks covering them have been described as 'complex and fractured'. Large parts of the global commons are still without regional agreement and many of the older agreements do not fully consider the environmental impacts of human activities on, for example, ecosystems. Crucially, numerous new activities, such as **bioprospecting** in the high seas, are not regulated (by UNCLOS).

Only a global governance regime, under the auspices of the UN, will be in a position to ensure that the global commons will be preserved for future generations. The post-2015 UN Development Agenda is arguably the perfect vehicle for this. This does not just mean formulating rules and regulations governing their use but providing the means and mechanisms to monitor and enforce them. In a world where many nation states and governments sign up to international agreements but then fail to enforce or even respect them, this is easier said than done.

ACTIVITIES

1 Consider how realistic it is to include science, education, information and peace into a definition of global commons. Give reasons for your answer.

2 Complete an internet search for global commons. Use a word cloud generator (such as www.wordle.net) to create several different word clouds from the results of the search. Compare and contrast your word clouds.

'Governance of the global commons is required to achieve sustainable development and thus human wellbeing. We can no longer focus solely on national priorities for economic development and environmental protection.'

(Johann Rockström, Director, Stockholm Environment Institute)

High seas	• 1982 United Nations Convention on the Law of the Sea (UNCLOS) • International Maritime Organisation and Regional Seas Conventions of the United Nations Environment Programme (UNEP)
Atmosphere	• United Nations Framework Convention on Climate Change (UNFCCC) and a multitude of international environmental treaties
Antarctica	• Antarctic Treaty System (ATS) • Protocol on Environmental Protection
Outer space	• Treaty on Principles governing the Activities of States in the Exploration and Use of Outer Space

▲ **Figure 3** *Conventions and treaties governing the global commons*

STRETCH YOURSELF

Johann Rockström writing in *Our Planet* (the magazine of UNEP) states that 'no nation or region can appropriate a larger share of the global commons without both transparently reporting this to all other nations and agreeing on mechanisms to ensure that the aggregate use of planetary space remains within safe boundaries ... In a world with growing populations and affluence, [this] will require distributing the planetary space among nations.' Rockström acknowledges that this is 'to say the least, a challenging but necessary task, which, when we succeed, will benefit humanity as a whole for generations'.

a What does Rockström mean by this?

b Is his argument realistic?

c Do you agree?

Antarctica: threats from fishing, whaling and mineral exploitation

In this section you will learn about the contemporary geography of Antarctica and its fishing, whaling and mineral resources

'Great God! This is an awful place …' proclaimed Captain Scott on reaching the South Pole on his ill-fated expedition of 1912 (Figure **1**). Antarctica certainly invites superlatives – it is a unique world of extremes. It hasn't rained in the Dry Valleys for over two million years, it is technically a cold desert in terms of its low average precipitation (in the form of ice needles and snow) but contains 70 per cent of the world's fresh water. It is also the coldest and windiest place on Earth. Ninety-nine per cent of the continent is covered in ice sheet which overspreads most of the shore and flows out to sea as slow-moving ice shelves and glaciers (Figure **2**).

Antarctica's remoteness and scale is best appreciated handling a globe. Almost 60 times bigger than the UK, if Antarctica were a country, it would rank behind only Russia in overall area. It has no native (indigenous) population, is the least populated continent on Earth – is rich in wildlife and minerals (including coal and oil). It may be repeated to the point of cliché, but this global common is 'the world's last great wilderness'.

What lies beneath?

Beneath 14 million km^2 of Antarctic ice (up to 4 km thick in places) lies a continent of about 8 million km^2 formed of two 'blocs' separated by a deep channel. East Antarctica shares similar ancient rocks to South Africa and Australia. With the similarities explained by tectonics some are exposed at the surface in the Transantarctic Mountains. Beneath the ice of West Antarctic is an **archipelago** of steep mountainous islands – much of which is exposed in the Antarctic Peninisula – which stretch towards the tip of South America.

⚫ **Figure 1** *Captain R.F. Scott and his companions on reaching the South Pole on 17 January 1912; all died of exhaustion, starvation and extreme cold on the return journey*

⚫ **Figure 2** *Antarctica from space*

South Pole

East Antarctic Plateau

Transantarctic Mountains

Look again at Figure **2**. The sea around Antarctica is frozen or covered with pack-ice for much of the year, but it still has a moderating influence on coastal temperatures compared with the continental interior. Where Antarctica's rocks are exposed, vegetation is limited by cold, aridity, wind and lack of soil. Lichens and mosses predominate, the latter particularly in the peninsula area where summers are warmer and hardy flowers can grow. Fauna is limited to minute insects and mites, which feed on the lichens and mosses. In contrast, the surrounding seas support a rich variety of marine birds (notably penguins and petrels) and seals, which feed at sea, breed on the land or sea ice, but rarely venture inland (Figure **3**).

Did you know?

The coldest temperature ever recorded on Earth was −94.7 °C, in August 2010, on the East Antarctic Plateau.

⊙ **Figure 3** *Adelie penguins on an Antarctic iceberg*

Fishing, whaling and mineral resources

Antarctica's 'discovery' in the early eighteenth century led to the development of a number of economic activities, such as fishing, whaling and sealing, but these were entirely exploitative. By 1800, the fur seals of South Georgia were wiped out and three years later they were also virtually eradicated from the South Shetland Islands! Whaling had arguably a more devastating effect.

The nineteenth century saw Norwegian, British and American exploitation of blue and right whales for oil and baleen (whalebone). Exploitation in the twentieth century widened the practice for meat and bonemeal, meat extract and frozen whale meat. By 1985, stocks were so dangerously low that most commercial whaling ceased. Fishing is following similar trends, having replaced whaling as the primary contemporary economic focus. Russian and Japanese exploitation of the Southern Ocean for rock cod and krill (central to the whole Antarctic food web) have raised serious concerns about overfishing, and so conservation is essential.

Known mineral deposits include coal, oil, manganese, titanium and even gold and silver. However, sizeable deposits that are easy to reach are rare and even then not yet economically viable to mine. Any mineral exploitation would have to overcome the seriously hostile environment – not least the major problems of inaccessibility, the extreme climate and the deep covering of moving ice sheet and glaciers.

Finally, Antarctica is home to a transient population of around 4000 scientists and their support staff, working in over 50 widely-scattered coastal and interior research stations. The nations funding this climatic, meteorological, biological, geophysical and oceanographic research, and the agreements managing the scope of their activities, are examined in 1.17.

Did you know?
Antarctica's ice sheet has an estimated volume of 24 million km^3 and in places reaches depths of over 4 km.

The Antarctic Convergence

Where does Antarctica begin and end? With a rich marine ecosystem surrounding the world's most southern place, the definition of where this wilderness begins and ends is pertinent to its protection. The Antarctic Treaty (see 1.17) covers the area south of 60 °S latitude. However, there is also a natural boundary in the Southern Ocean: a dividing line that loops all of the way around the Antarctic continent, called the Antarctic Convergence (Figure 4). You might not notice the difference if you cross this by boat, on the way to Antarctica, but here sea temperatures fall as much as 4 °C in summer. This watery division, up to 48 km wide in parts, separates cold north-flowing waters from the warmer waters of the subantarctic and it is where a significant mixing and upwelling of water creates a highly productive marine environment for both marine plantlife and animals, such as krill.

▼ **Figure 4** *Map of Antarctica and the Atlantic Convergence*

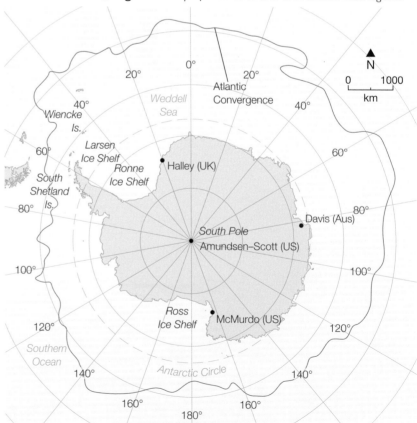

Amundsen–Scott (South Pole) 90°S Altitude 2835 m			
Month	Mean temperature (°C)	Precipitation (mm)	Mean wind speed (mph)
Jan	−26	0	1.1
Feb	−38	0	0.9
Mar	−52	0	1.4
Apr	−56	0	4.6
May	−56	2	5.4
Jun	−57	0	4.6
Jul	−58	0	5.0
Aug	−58	0	5.0
Sep	−58	2	5.7
Oct	−50	0	1.7
Nov	−37	0	1.3
Dec	−26	0	1.1

McMurdo 77°51'S, 166°40'E Altitude 24 m			
Month	Mean temperature (°C)	Precipitation (mm)	Mean wind speed (mph)
Jan	−2	15	13
Feb	−8	25	14
Mar	−17	12	17
Apr	−20	15	16
May	−22	20	16
Jun	−22	22	16
Jul	−25	15	16
Aug	−26	12	16
Sep	−23	10	16
Oct	−18	17	14
Nov	−9	10	12
Dec	−3	12	12

Davis 68°35'S, 77°58'E Altitude 18 m			
Month	Mean temperature (°C)	Precipitation (mm)	Mean wind speed (mph)
Jan	0.8	1.7	14.0
Feb	−2.4	3.6	13.8
Mar	−8.0	9.3	12.5
Apr	−13.0	9.9	11.2
May	−15.5	10.2	10.7
Jun	−15.5	8.5	12.0
Jul	−17.3	7.2	12.2
Aug	−17.3	6.7	12.0
Sep	−16.4	4.6	11.0
Oct	−12.0	4.7	12.5
Nov	−4.8	2.5	16.1
Dec	0.1	1.9	15.4

🔺 **Figure 5** *Antarctic research station data*

🔺 **Figure 6** *The Australian Davis Antarctic research station*

ACTIVITIES

S

1 Using an atlas, locate the three research stations in Figure **5** using the latitude/longitude references.

2 **a** Draw climate graphs for each station (months of the year along the *x*-axis, weather variables along the *y*-axis). For each month, plot mean monthly temperatures as points connected by a line, and mean monthly precipitation as bars. (Hint: you need two *y*-axis scales for plotting both temperature and precipitation, and each graph should have identical scales.)

  b Add additional *y*-axis scales to each graph and superimpose mean monthly wind speeds as another line graph.

3 For each station, calculate the mean annual temperature, the total annual precipitation and the mean annual wind speed. Also identify the maximum and minimum values for each variable and calculate the annual ranges.

4 **a** Compare and contrast the climate at the three named stations.

  b Analyse each of the different weather variables with reference to latitude, altitude and proximity to the Southern Ocean.

  c Explain which research station's statistics are most reliable.

Antarctica: threats from scientific research and climate change

In this section you will learn about scientific research and climate change in Antarctica

Fact or fiction?

We all love a disaster movie and Roland Emmerich's 2004 climate change blockbuster *The Day After Tomorrow* doesn't disappoint! The opening scenes show a palaeoclimatologist drilling ice cores through the surface of Larsen B – a massive floating ice shelf (see Figure **4**, 1.14). The shelf cracks, breaks away and, in saving his precious ice cores, our hero only just survives. This scene is loosely based on a true story – the research methodology is realistic and an area larger than the county of Somerset broke up in February 2002. However, the compression of time-scales, land-based hurricanes and climate-driven tsunamis dominating the remainder of the film are just plain daft! In 2015, NASA scientists predicted the complete disintegration of Larsen B by 2020 allowing three glaciers constrained behind it to accelerate, calve and melt into the Weddell Sea. Whether this will happen remains to be seen.

Think about

The British Antarctic Survey

Several countries have research organisations based in Antarctica including the UK – the British Antarctic Survey. Despite the intense cold, scientists have ideal conditions in which to study atmospheric, terrestrial and oceanic systems. Vast knowledge about climate change has come from research using deep cores extracted from ice that is up to 400 000 years old. Working for the British Antarctic Survey presents many challenges. Researchers need to be extremely fit and healthy. They need to eat energy-generating food (providing some 3500 calories a day) and wear specialist clothing (layering of lightweight clothes is considered best practice) in order to cope with the extreme cold and penetrating winds.

Scientific research

Scientific research on Antarctica accounts for almost all of the semi-permanent population across the continent. Both seasonal (Figure **1**) and year-round research stations are run by 30 nations including the UK. Indeed, this global common represents a multinational laboratory of incalculable value, particularly for conducting research into climate change. Year-round freezing temperatures mean that snow that fell hundreds of thousands of years ago is buried by subsequent layers – crucially trapping wind-blown volcanic and desert dust, tiny bubbles of ancient air and even radioactive substances. The upper layers are less compacted and may clearly show seasons within single years. However, the deeper the penetration, the greater the compaction by the weight of ice above, and so annual layers are compressed into very thin layers. Ice core drilling enables an abundance of information about past climatic conditions to be revealed and interpreted.

⬥ *Figure 1* The British Antarctic Survey's Port Lockroy research station, Wiencke Island, was established in 1944 and is now manned during the austral summer (November to March)

Climate change

The science of climate change resulting from the enhanced greenhouse effect is explained in section 5.18. It has been predominantly Antarctic research that has provided the clearest link between levels of greenhouse gases in the atmosphere and global surface temperatures (Figure **2**).

Antarctic ice core analysis is meticulous and sophisticated. For example, bubbles in the ice can be analysed for their carbon dioxide and methane content. Furthermore, we have long understood that volcanic emissions absorb and scatter the sun's energy (insolation), which temporarily cools the atmosphere. Eruptions also leave ash layers in the snow acting as 'diagnostic markers' to identify specific years of volcanic events. Even early atom bomb tests have left their (radioactive) mark. Consequently we now understand that:

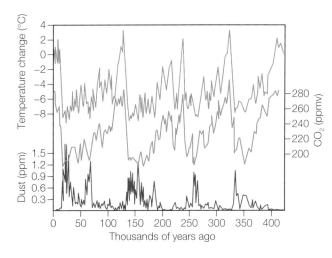

⊙ **Figure 2** *The results of an ice core drilled 3.5 km down into the Antarctic ice revealing data from four glacial cycles*

◆ climates follow natural cycles and have changed throughout geological time

◆ climates will continue to change in the future (with or without humans)

◆ human activities have contributed to contemporary climate change (global warming).

The impact of contemporary climate change is most apparent in extreme latitudes – the Arctic and Antarctica. Here global warming has driven significant changes in the physical and living environment, such as melting ice and deteriorating permafrost. But the trends are not straightforward and are even counterintuitive. Look back at Figures **2** and **4**, 1.14. East Antarctica's ice sheet is thickening – warmer seas increase evaporation to condense into cloud droplets and fall as extra snow in the frozen interior. Although this only slows sea-level rise by a tenth of a millimetre per year. However, West Antarctica's ice sheet is smaller and more vulnerable. Unlike the east, where the ice is frozen onto rock (*cold-based*), much of the glacial ice in the west is *warm-based*

(see *AQA Geography A Level & AS Physical Geography*, 4.5) and more likely to slide into the sea, which would raise global sea levels by 5 m.

It is the Antarctic Peninsula that is most sensitive to climate change. In the past 60 years, temperatures there have risen by 0.5 °C every decade – up to five times faster than the rest of the world and ice shelves have been breaking up. Although the melting of sea ice has no effect on sea level (water from a melting ice cube simply fills the displaced space), the melting of ice shelf means that land-based ice is no longer restrained. In other words, if these ice sheets slide into the sea, vast quantities of water will be added that was not there before. Research continues into the likelihood and scope of these scenarios.

ACTIVITIES

1 Study Figure **2**.
  a Describe the trends shown and any correlations between them.
  b Suggest reasons for the correlations identified.
  c Discuss the issues of reliability in only using measures of central tendency (mean, median and mode) to describe changes in the variables shown.

2 Using the website of the British Antarctic Survey (www.bas.ac.uk), explain the global significance of scientific research in Antarctica.

STRETCH YOURSELF

'The impacts of global warming are such that I have no hesitation in describing it as "a weapon of mass destruction".' (Sir John Houghton, former Chief Executive of the UK's Meteorological Office). Discuss this statement.

In this section you will learn about tourism in Antarctica

Since the first heroic expeditions to Antarctica over a century ago, which saw Roald Amundsen's Norwegian team conquer the South Pole only weeks before Scott, the continent has held a fascination for the adventurous. Polar exploration continues to this day, testing the bravest and constantly setting new records. The youngest, for example, was the Polish teenager Jasiek (Jan) Mela who reached first the North Pole and then the South Pole in 2004 – the latter the day after his sixteenth birthday (Figure **1**). Jasiek was also a double amputee!

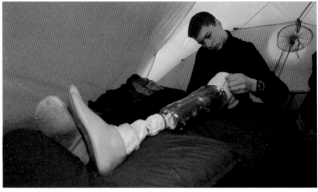

△ **Figure 1** *Jasiek Mela's remarkable achievement; North Pole aged 15 and South Pole just turned 16*

An incomparable holiday experience

Advances in transport, technology and clothing have also allowed this extraordinary wilderness to be accessible to increasing numbers of tourists (Figure **2**). Antarctica's physical isolation, spectacular unspoilt landscapes and remarkable wildlife captivate people and motivate those with sufficient wealth to embark on an incomparable holiday experience. Such are the high financial costs that it comes as no surprise to see visitors from HDEs dominating (Figure **3**). Although this is not mass tourism, total numbers have seen significant growth – from fewer than 2000 visitors a year in the 1980s, peaking at more than 46 000 in 2007–8, before falling to fewer than 27 000 in 2011–12. Numbers are now once again rising steadily.

The vast majority of tourists visit by cruise liners; many do not set foot on the continent. They are drawn by the chance to see the calving of vast icebergs, blue, killer, sperm and humpback whales, and of course guaranteed emperor and yellow-eyed penguin colonies. Now there is an increasing diversification of activities with a growth in 'adventure tourism'. Due to the availability of flights to the Union Blue-Ice runway and also the Patriot Hills runway near the Ronne ice shelf, and by using smaller ships, there are opportunities for the most adventurous to land. This opens up exciting recreational options such as helicopter flights, paragliding, photo safaris, cross-country skiing, mountaineering, diving, kayaking, camping and visits to research stations and heritage sites such as Scott's Base.

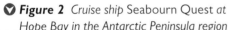

▽ **Figure 2** *Cruise ship* Seabourn Quest *at Hope Bay in the Antarctic Peninsula region*

Impacts of tourism

Increasing tourist numbers since the turn of the century inevitably led to concerns about the impact on Antarctic ecosystems. Self-regulation by the International Association of Antarctic Tour Operators (IAATO) was followed in 2009 by signatories of the Antarctic Treaty System (see 1.17) agreeing to prevent more than one ship at a time landing at any one site and also limiting the number of tourists that go ashore. But concerns persist about the accidental introduction of invasive species (such as the Mediterranean mussel) and pollution, not least the danger of collisions in some of the most dangerous waters on Earth. In November 2007, for example, the Canadian cruise ship *MS Explorer* struck ice near the South Shetland Islands and sank (Figure **4**). Remarkably, there were no fatalities, but 154 passengers and crew had to endure subfreezing temperatures in lifeboats before rescue by a Norwegian liner. Many argue that future incidents are inevitable and could end in disaster. Health and safety of tourists is among the main considerations – their waste is another. Regulations even require toilet waste to be barrelled and transported home!

Yet despite tight regulation some continue to voice concerns. For example, the fragility of Antarctic ecosystems means that any disturbance, such as footprints on fragile moss, will last for decades. Cultural heritage sites, such as old sealing, whaling and exploration stations, are under pressure. Furthermore, wildlife tends to be clustered in only a few ice-free locations, and breeds during the peak summer tourist season of mid-November to March.

Others adopt a more optimistic stance, praising the sound environmental record of Antarctic tourism. Tour operators can only realistically touch a tiny portion of Antarctica's 14 million km² of land and ice shelf and only 10 out of 200 landing sites show signs of any wear and tear. Antarctic tourism is financially exclusive, with visitors tending to be well educated, responsible, environmentally aware and guided in small groups to avoid sensitive areas. Tourist litter is negligible, unlike the waste from scientific research stations!

USA (12 308)
Australia (4087)
United Kingdom (3428)
China (3042)
Germany (2796)
Canada (1937)
France (1092)
Switzerland (1064)
Japan (831)
Others (6117)

Figure 3 *Number and proportion of tourists to Antarctica, by nationality (2014–15)*

Figure 4 *The MS Explorer sinks near the South Shetland Islands after striking submerged ice*

ACTIVITIES

1 Describe and suggest reasons for the trends in Antarctic visitor numbers over the past four decades.

2 Outline the arguments **for** and **against** the promotion of tourism in Antarctica.

S

3 a Represent the data in Figure **3** in a different way. Would a line graph be suitable for showing this data? Explain your answer.

b How might this data be represented on a map?

In this section you will learn about:
- the governance of Antarctica
- the role of NGOs in monitoring threats and enhancing protection
- the analysis and assessment of global governance

Who governs Antarctica?

As a global common, Antarctica has no government to manage its affairs or to protect its interests. It is not a nation state, although over the years seven countries have made claims to segments of it (Figure **1**) – the exact legality of the claims has never been comfortably clarified. (You only need to consider the Falklands (Malvenas) war of 1982, and the overlap between Britain's and Argentina's claims to understand why!) Any nation's claim to sovereignty or 'ownership' is disputed by all others, but the areas assigned are recognised as research zones where the individual countries have established scientific bases. Interestingly, the USA has never made a claim, but like Russia 'reserves the right to' and maintains research facilities which, under international law, is theoretically a basis to do so.

'Land of Science. Land of Peace'

This lack of sovereignty is why the Antarctic Treaty – otherwise known as the Antarctic Treaty System (ATS) – is so important. In December 1959, 12 nations, including the USA and, as it was then, the USSR (Russia), signed this international agreement not to recognise, dispute, establish or allow future claims of territorial sovereignty over Antarctica. Furthermore, and crucially, the ATS would:

- guarantee free access and research rights to all countries
- prohibit military activity such as nuclear bomb tests
- ban the dumping of nuclear waste.

The ATS also provides that any member of the UN can accede to it. Furthermore, established UN bodies, such as UNEP have 'observer' status in Antarctic Treaty Consultative Meetings to provide expertise. To date, 52 countries have signed up to the ATS and membership continues to grow.

Growing pressures

By the early 1990s, pressures were building from certain groups wanting to exploit the underground mineral resources of Antarctica. Although a convention had been agreed between signatories of the ATS to 'guarantee a fair and clean exploitation', many NGOs were acutely aware of the risk of accidents and their potentially drastic consequences for such a fragile and unique environment. With the help of influential and charismatic conservationists including Captain Jacques-Yves Cousteau (Figure **2**) and public support globally, a proposal was made to replace the unenforceable convention with a protocol aimed at truly protecting the environment.

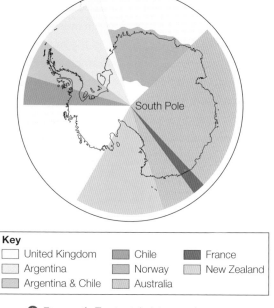

Key

☐	United Kingdom	▨	Chile	▨	France
☐	Argentina	▨	Norway	▨	New Zealand
▨	Argentina & Chile	▨	Australia		

🔺 **Figure 1** Territorial claims to Antarctica

🔻 **Figure 2** This French oceanographer and conservationist, Captain Jacques-Yves Cousteau, was a legendary undersea explorer, photographer, film maker and also the inventor of the aqualung

In 1991, 45 nations signed up to the Protocol on Environmental Protection (Figure **3**).
They agreed to:

◆ ban all mineral resource activity including exploration of the continental shelf

◆ promote comprehensive monitoring and assessment in order to minimise
human impacts on Antarctica's fragile ecosystems.

The protocol, to run until 2048, proclaimed Antarctica as a 'natural reserve,
devoted to peace and science'.

⊙ Figure 3 *Specific clauses of the
Protocol on Environmental Protection*

Clause	Purpose
Environmental impact assessment	• Any planned activities to be assessed in advance with agreement required should significant risks be apparent
Flora and fauna	• Preventing the removal of, or interference in, native flora and fauna • Prohibiting the introduction of non-native species • Identifying and designating Specially Protected Species
Waste management	• Identifying types of waste that have to be removed • Establishing rules for the storage and disposal of waste • Planning for the removal of wastes of past activities
Marine pollution	• Restricting the (marine) use of heavy fuel oil • Prohibiting the discharge of oil, noxious liquid substances and garbage in the Antarctic Treaty area* • Defining rules for the discharge of sewage, ship retention and emergency preparedness and response
Protected areas	• Identification and protection of areas of outstanding environmental, scientific, historic and/or aesthetic importance as Antarctic Specially Protected Areas (ASPA) or Antarctic Specially Managed Areas (ASMA) • Establishing an official list of Historic Sites and Monuments (HSMs)
Liability	• Establishing rules and procedures to cover potential environmental emergencies related to scientific research, tourism and so on • Ensuring operators prepare contingency plans for any such emergencies

*The Antarctic Treaty System area includes over 20 million km^2 of the Southern Ocean, extending from the Antarctic
coast to 60°S latitude. The International Maritime Organisation (IMO) designated these waters a 'Special Area' in 1990

Non-governmental organisations (NGOs) in Antarctica

Such is the scale and global significance of 'all things Antarctic' that governments alone cannot hope to monitor, understand or control every aspect. The work of NGOs is very important, not least in concentrating and providing expertise, championing causes, contributing independent perspectives, rallying public support (e.g. through the harnessing of social media) and provoking action. Few could argue, for example, that the world conservation movement has not been a positive force for good. The movement includes high-profile NGOs such as the World Wild Fund for Nature (WWF), Friends of the Earth and Greenpeace, along with scientists, naturalists, resource specialists and informed, interested citizens. They all study the Earth and intervene in environmental issues whenever and wherever they can. Indeed, the concepts of sustainability and environmental stewardship have developed directly from this movement.

On occasions the combined efforts of NGOs concentrate energies and funds to even greater effect – and this is well illustrated in Antarctica and the Southern Ocean.

For example, the work of the Antarctic Ocean Alliance (AOA), a project of the influential Antarctic and Southern Ocean Coalition (ASOC), champions the case for designating **marine protected areas (MPAs)** and **marine reserves** in East Antarctica, the Ross Sea, the Weddell Sea and the Antarctic Peninsula. They acknowledge that while many other marine ecosystems in other parts of the world have been devastated by development, pollution, mining, oil drilling and overfishing, many of Antarctica's ocean habitats remain intact with all of their predator species still thriving. Antarctica's wildlife is, however, under increasing pressure from commercial fishing and climate change – the latter already having particular impact in the Antarctic Peninsula region (one of the fastest warming areas on Earth). The AOA argues passionately that the Southern Ocean and Antarctica's relatively untouched environment provides a critical laboratory for scientists researching climate change and that the creation of large marine reserves in the Ross Sea and elsewhere would create important global climate reference areas.

The International Whaling Commission (IWC)

The International Whaling Commission (IWC) was set up in 1946 to monitor and conserve global whale stocks and oversee the whaling industry. It is a voluntary organisation with no authority to enforce its decisions and agreements. The Commission for the Conservation of Antarctic Marine Living Resources (CCAMLR) was established by international convention in 1982 with the specific objective of conserving Antarctic marine life. This was in response to overexploitation of several marine resources in the Southern Ocean throughout the 1960s and 1970s, such as rock cod, and increasing commercial interest in Antarctic krill resources – a keystone component of the Antarctic ecosystem (Figure **4**). Krill have been extensively harvested by humans to provide fishmeal, animal feed and fish oil, and are further threatened by climate change that affects the phytoplankton they feed on (see *AQA Geography A Level & AS Physical Geography*, 1.13). The 1982 IWC Whaling Moratorium decided that all whaling activity should cease from 1985–86. It remains in place today although Norway and Iceland continue commercial whaling under objection to the moratorium.

⬇ **Figure 4** *Krill are tiny crustaceans that provide food for penguins, whales, squid and other marine life*

Did you know?

It is estimated that the 379 million tonne biomass of all Antarctic krill *(Euphausia superba)* exceeds that of all humans!

Think about

The role of environmental NGOs

Large environmental NGOs need to persuade their supporters that they are living up to their donations and producing results. 'Radical' groups like Greenpeace apply pressure for policy change via positive action ('Join the movement. Become an Oceans Defender'), direct messages ('Krill-gotten gains to fund Antarctic research') and popular awareness campaigns ('World Penguin Day'). Scientific-focused groups such as WWF tackle similar global environmental issues but believe more in developing constructive relationships and working with companies. To this end, credibility and therefore image are important. Environmental campaigns need to influence others but also seek to avoid unnecessary confrontation by making best use of available scientific information, education work and involvement of local communities.

The CCAMLR debates and agrees conservation measures that determine the use of marine-living resources in the Antarctic. Notably, they practise an ecosystem-based management approach which does not exclude harvesting so long as it is carried out in a sustainable manner and takes account of the effects of, for example, fishing on other components of the ecosystem. Specific achievements of CCAMLR include:

- challenging illegal, unreported and unregulated fishing
- establishing the world's first high seas Marine Protected Area (MPA) – the South Orkney Islands Southern Shelf Marine Protected Area which covers 94 000 km^2 of the Southern Atlantic Ocean
- reducing seabird mortality (arising from fishing operations) from thousands of birds annually to almost zero within those fisheries that are regulated
- establishing the CCAMLR Ecosystem Monitoring Programme (CEMP) to detect and record significant changes in critical components of the marine ecosystem. This serves as a basis for the conservation of Antarctic marine-living resources and distinguishes between changes due to the harvesting of commercial species and changes due to environmental variability (both physical and biological)
- managing vulnerable marine ecosystems (VMEs) such as seamounts, hydrothermal vents and cold water corals by regulating bottom fishing on the high seas.

Analysis and assessment of global governance in Antarctica

The original Antarctic Treaty System (ATS) had an important clause inserted allowing any dissatisfied country to call for a review conference after 30 years – none did. Even today, over 50 years later, the Treaty has prevailed with no major problems. Each party has enjoyed peaceful cooperation and freedom of scientific research – and that research has contributed significantly to the protection of the global environment and our knowledge of the Earth. As the ATS has matured, it has become recognised as one of the most successful sets of international agreements, setting an example of peaceful cooperation for the rest of the world (Figure **5**).

Figure 5 *'Land of Science. Land of Peace'; flags of the founding members of the Antarctic Treaty surround a symbolic 'barber pole' topped with a shiny metallic sphere, which marks the South Pole*

However, the decision-making that takes place within the ATS can be problematic. Decision-making by consensus does not mean that everyone must agree, but that no one can voice disagreement. So one country, if it feels strongly about an issue, can stop a resolution from going forward (this is true for many similar international bodies). Furthermore, without legal penalties for violating agreements, most parties are essentially on their honour to abide by their obligations – whether this is under the ATS, the 1991 Protocol on Environmental Protection or CCAMLR. Much is agreed and achieved, albeit not always quickly because the consensus-based decision-making process can often be slow to produce results. Mutual trust clearly works in Antarctica and the Southern Ocean, and coupled with the expertise, lobbying and prompting of NGOs – particularly when grouped such as within the AOA and ASOC – participating governments to all Antarctic agreements are moving forward in the protection of this remarkable global common.

ACTIVITIES

1 Prepare a case for sustainable development in Antarctica. Your discussion should consider:
   - environmental considerations relevant to the world's last great wilderness
   - future global mineral and energy resource requirements in the context of contemporary and future climatic change.
   - existing and further development of Antarctica's tourist potential.

2 Comment on the view that Antarctic governance represents an inspirational example of international cooperation working for the greater good.

3 The *Rainbow Warrior* is a purpose-built campaigning ship for Greenpeace. What does the name of the ship suggest about the actions of Greenpeace in Antarctica?

STRETCH YOURSELF

'Maintaining Antarctica as a global common is a luxury tomorrow's world cannot afford.' Discuss.

S In this section you will learn about constructing a critique of globalisation using imagery and headlines

Study each of the images, read the text and then answer the activities that follow.

What makes a good newspaper headline?

Newspaper headlines are far more than a quick title to sum up a piece of writing. Used skilfully, headlines lure the reader into the article or story. They suggest or infer much in just a few words and should bring a positive answer to the question 'would this make me want to read on?' Being able to write effective headlines is a useful skill to summarise a topic – less is sometimes more.

The following are suggestions for writing a catchy/memorable headline.

◆ Use numbers – sometimes peculiar numbers work best (but it needs to be accurate).

◆ Use interesting adjectives – such as strange, painstaking.

◆ Use simple action verbs – such as use rather than utilise.

◆ Use 'trigger words' – for example, what, why, how, when?

◆ Use key word(s) – geography is full of processes, features and specialist terms – but avoid too much jargon!

◆ Make a promise – challenge the reader to read the article (e.g. 'Why you should be afraid of the TNCs that feed us').

◆ Play on words – for example, 'Aid floods in following river disaster'.

▲ **Figure 1** *Dimensions of globalisation*

The power of imagery

Think of a great memory – perhaps a holiday memory from last summer or a memory from early childhood (it might even be of your favourite geography lesson)! Now try to describe that memory in words … it is probably not easy – the colours, emotions, scale and so on might necessitate many lines of text. On the other hand, a powerful image has the potential to say a thousand words. This is one reason why images, which are so easily copied from the internet, should never be pasted into an answer without due diligence and thought. Any images used should work alongside the text, including an appropriate caption and figure number. Look through this chapter again – to what extent would the pictures work without any caption at all?

▼ **Figure 2** *Global systems*

Figure 3 *Global trade*

Figure 4 *Transnational corporations*

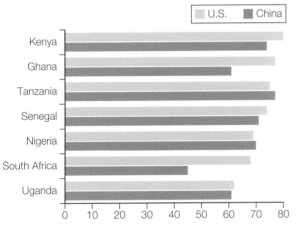

Figure 5 *Attitudes towards Chinese and US investment in sub-Saharan Africa; surveys of seven African countries and the percentage of their population that has a favourable view of USA and China*

ACTIVITIES

1 Choose at least three images. For each one, write two headlines that give different views of the image. For example, headlines from the *Daily Mail*, the *Sun*, *New Scientist*, *National Geographic* would all concentrate on different aspects.

2 Match the most appropriate image to each group listed below. Justify your choice.
   a United Nations
   b Environmental NGO
   c Elite tourist company

3 Look again at the pictures and artwork in this chapter. Identify three of these images (by their page and figure number) that you feel 'say a thousand words'. Give reasons for your choices.

4 A storyboard is a sequence of drawings or pictures, typically with some directions and dialogue, representing the shots planned for a film or television production.

Imagine that you are filming a documentary entitled 'Globalisation – the benefits and costs'. Using images sourced from the internet (including graphs and maps), produce a storyboard (with six or more images) for the documentary. Remember to include appropriate captions.

Figure 6 *Trade in food*

Now practise…

The following are sample practice questions for the Global systems and global governance chapter.

They have been written to reflect the assessment objectives of Component 2: Human geography Section A of your A Level.

These questions will test your ability to:

◆ demonstrate knowledge and understanding of places, environments, concepts, processes, interactions and change, at a variety of scales [AO1]

◆ apply knowledge and understanding in different contexts to interpret, analyse and evaluate geographical information and issues [AO2]

◆ use a variety of quantitative and qualitative and fieldwork skills [AO3].

---

1 Outline the spatial organisation of a named transnational corporation (TNC). (4)

2 'Rather than being a leveller, the forces of globalisation have served to reinforce unequal power relations between nation states.' To what extent do you agree with this view? (9)

3 Study Figures 1 and 2. Using these resources, describe and explain recent trends in the number of tourists visiting Antarctica. (6)

▼ **Figure 1** *Antarctic tourist trends, 2011–2017 (IAATO)*

	2016–2017 season (estimate)	2015–2016 season	2014–2015 season	2013–2014 season	2012–2013 season	2011–2012 season
Seaborne tourism with landings	31 493	27 607	25 341	25 526	23 305	20 271
Seaborne tourism, no landings	8680	8109	9459	9670	9070	4872
Air and cruise combination, with landings	3154	2353	1471	1848	1587	860
Air and land tourism, Antarctic interior	558	409	431	361	354	516
Total	43 885	38 478	36 702	37 405	34 316	26 519

While there were sizeable increases in the 1990s and first eight years of the twenty-first century, tourism to Antarctica has been flat over the past few years. The worldwide downturn in the economy has largely been responsible for this and this is affecting decisions both at the consumer level as well as the tour operator level. For example, some of the expedition vessels that were in the vanguard of Antarctic tourism 15–20 years ago are now being phased out of the industry, and financing new builds or even refurbishment of existing ships to Antarctic standards has not kept pace.

During the 2015–16 Antarctic tourism season, the total number of visitors travelling with IAATO member companies was 38 478 – an increase of just under five per cent compared to the previous season. IAATO numbers have not reached the peak of the 2007–8 season (46 265), although the trend has been a slow increase over recent years.

(Adapted from iaato.org, 2016)

▲ **Figure 2** *Is the increase in Antarctic tourism a concern?*

4 What do you understand by the term 'global norm'? (4)

5 'Global systems impact on all our lives but for small-scale producers that impact can be crushing.' To what extent do you agree with this view? (6)

6 Assess the degree to which the development of new technologies, for example communication and information systems, have facilitated the route to a global economy? (9)

7 Explain how marketing patterns used by transnational corporations are key to their role as agents of globalisation. (4)

8 Outline some of the major trading relationships between *either* China *or* India and smaller, less developed economies, such as those in sub-Saharan Africa, southern Asia or Latin America. (6)

9 Assess the role of *either* the UN *or* the WHO. To what extent has it been successful in its role in global governance? (9)

10 What do you understand by the term 'global commons'? (4)

11 Study Figure **3**. Using this resource, outline the advantages and disadvantages of onshore, offshore and nearshore approaches to recruitment and the supply of labour in the global economy. (6)

🔻 **Figure 3** *Onshore, offshore or nearshore: what's your best option?*

If you are looking to add experienced developers to your team, you've probably already noticed it's not that the US doesn't have any highly skilled developers. We just don't have enough.

So, how to plug the labour gap? If the lowest hourly rate is your priority, then it cannot be argued that 'offshore' outsourcing offers the best deals for labour cost (often as low as US$20 per hour); however, keep in mind that the price may end up rising again if it comes to miscommunication, high travel costs, and production delays. Offshoring usually refers to working with teams in far-away countries such as India, China, or Ukraine.

The advantage of 'onshore' outsourcing is that you're working with a highly skilled American team in your own country, but this option comes with a high price tag (hourly rates of far more than US$100 per hour are common).

'Nearshore' offers a mix of onshore and offshore benefits. It decreases your cost while still providing you with some of the perks of onshore outsourcing such as regular communication during business hours. Outsourcing to Mexico is a unique way of outsourcing because it includes more benefits than any other nearshore or offshore company can offer.

(Adapted from iTexico blog. iTexico is a global software company based in Austin, Texas and Silicon Valley in the USA, and Guadalajara, Mexico)

12 Explain the pattern and processes involved in the world trade of one food commodity or one manufacturing product. (6)

13 Analyse the roles of international governmental organisations and NGOs in the governance of Antarctica and assess their effectiveness. (20)

New York mourners commemorate the anniversary of 9/11 at Ground Zero – how is our perception of this place changing?

Your exam

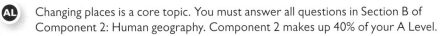 Changing places is a core topic. You must answer all questions in Section B of Component 2: Human geography. Component 2 makes up 40% of your A Level.

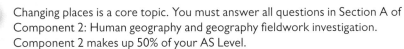 Changing places is a core topic. You must answer all questions in Section A of Component 2: Human geography and geography fieldwork investigation. Component 2 makes up 50% of your AS Level.

Your key skills in this chapter

In order to become a good geographer you need to develop key geographical skills. In this chapter, whenever you see the skills icon you will practise a range of quantitative and relevant qualitative skills, within the theme of 'changing places'. Examples of these skills are:

- Using quantitative data, including geospatial data, to investigate and present place characteristics 2.8, 2.13

- Analysing critically the impacts of different media on place meanings and perceptions 2.1, 2.3, 2.6, 2.7, 2.10

- Using qualitative approaches to investigate and present place characteristics, from face-to-face interviews to analysis of media texts 2.7, 2.11, 2.13

- Using different types of data to develop critical perspectives on data categories and approaches 2.4, 2.9, 2.10, 2.11

Fieldwork opportunities

The themes introduced within the Changing places topic are applicable wherever you live; after all, every settlement, whether large or small, rural or urban is a changing place. So don't discount fieldwork on your doorstep. Lessons learnt undertaking fieldwork locally could inform your distant place study work and may give you a new perspective on where you live. There are many excellent opportunities for local fieldwork investigations, as well as desk-based enquiries, like analysing different media images of your place and using 'big data', freely available online.

1 Is your town a clone town, border town or home town?

National and international chain stores predominate in the retail experience and urban planners are accused of using a 'cookie-cutter' approach to planning new public spaces. Is your place a 'clone town'? Why not use the method of conducting a land-use survey designed by the New Economics Foundation. It is quick and allows a calculation of the extent to which your local high street is distinctive or 'could be anywhere', allowing you to compare it to other UK towns where the same fieldwork exercise has been completed.

2 How has your place changed in living memory?

In-depth interviews with people who have lived in an area for decades will give you an insight into how that place has changed, as well as how people make sense of these changes and attach meanings to places. This makes them much more than just a dot on a map. In the process, you might be shown old photographs of places you recognise. Why not photograph them today, from the same viewpoint, and observe the changes that have occurred.

3 Comparing your place with another place in the UK

Using relevant 'big data' sources (e.g. the latest census or Index of Multiple Deprivation), compare and contrast your place (a small town, village, suburb etc.) with somewhere else of a similar size. Consider what sort of indicators you could use to compare quality of life as well as age make-up and level of ethnic diversity. Find out how have both places have changed since the last census.

In this section you will learn about:
- how humans perceive, engage with and form attachments to places
- how place-meanings are bound up with different identities, perspectives and experiences

What do we mean by place?

The concept of **place** is one of the key aspects of geography. Is there more to a place than simply a dot on a map? Is *your place* a house, a street or a city? What other places do you feel attached to? Maybe a location you've been to on holiday or visited with your school has a special significance for you?

Mount Snowdon: a special place

Mount Snowdon (1085 m) is the highest peak in England and Wales and more than 360 000 people make the three-hour climb to the summit every year (Figure **1**). But a significant number travel up the mountain by train on the Snowdon Mountain Railway.

Reaching the summit is a memorable event. Spending time there, looking out across the spectacular landscape of Snowdonia, it is easy to see why many people feel this is a special place. Cut into the stonework of the café and visitor centre near the summit are the words of Welsh poet Gwyn Thomas:

'Copa'r Wyddfa: yr ydych chwi, yma, Yn nes at y nefoedd'

Translation: 'The summit of Snowdon: here you are nearer to heaven.'

'It is essential that you check the Snowdonia weather forecast before you leave home.

- While you can still climb safely in mist and rain, you may not see much further than your nose from the top, in which case you may decide a lower-level walk is more fun.

- Some snow on the summit is ok, but winter conditions – strong wind, heavy rain and especially ice are far more serious.

- If things change, do not be afraid to turn back. Do not carry on regardless.'

🔺 **Figure 2** *An extract from 'A beginner's guide to safely climbing the highest mountain in England and Wales'. (www.walkingclub.org.uk)*

Think about

If some places are special, others suffer from the problem of **placelessness.** This is the idea that a particular landscape, for example, an airport terminal, 'could be anywhere' because it lacks uniqueness. Human geographers propose that this occurs when global forces have a greater influence on shaping a place than local factors. British high streets are increasingly criticised for their uniformity where chain stores predominate. Is your place a *clone town*?

🔻 **Figure 1** *Walkers descend from the summit to Snowdon's café*

'This is a fabulous land of faeries and giants and kings. For centuries Welsh princes held council here. It is a land rich in alpine flowers and rare ferns left behind the retreating ice age and it is dotted with ruins that chronicle the history of long lost communities…'

🔺 **Figure 3** *An extract from 'The journey to 1085 metres' (www.snowdonrailway.org.uk)*

What is the tourist gaze?

The **tourist gaze** is organised by business entrepreneurs and governments, and consumed by the public. This is as true of cultural sites as it is of adventure tourism to the ends of the Earth (*AQA Geography A Level & AS Physical Geography*, 4.17).

Even 'death sites' like Ground Zero (Figure **4**) and Auschwitz-Birkenau are marketed and managed by tourism professionals. They choose what visitors are allowed to access – they mediate our experience of the place.

However, it is also worth noting that such tourist sites will have a different meaning and importance for each individual visitor. Each visitor's senses will be attuned differently, and they see the location through the 'spectacles' of their prior experiences. Religious beliefs, moral code, family history, ethnicity, and education all influence how a tourist perceives (makes sense of) a death site.

Sometimes, people's differing perceptions of a place can lead to conflict. Have you ever been somewhere where your experience of it was at odds with that of other people?

9/11 Memorial, New York

All places are changing, all the time, but few have gone through such dramatic changes in the last 15 years as 'Ground Zero', the site of the former World Trade Center in New York.

'When I got there I found there's a cold, stark, deep water feature that just goes down, down, down. It was not a place of comfort for me… And the gift shop, "Visit mass grave, buy a T-shirt". I ask you? New York just means total horror.'

(The grieving mother of a victim of the 9/11 attacks)

'Breathtaking. Words cannot express how beautiful this place is. I would challenge anyone to walk around the museum without shedding a tear. I had the most amazing day, thank you.'

(A day-tripper who took the tour)

▲ **Figure 4** *The North Pool of the 9/11 Memorial*

◄ **Figure 5** *Different experiences of the 9/11 Memorial*

ACTIVITIES

1 a Do you think the sense of achievement on reaching the summit of Snowdon is different if you have walked up, rather than using the train?
   b Make a list of groups of people for whom, without the railway, climbing Snowdon might not be an option.
   c Does mode of travel matter? Does it affect your response to the destination?
2 Read Figures **2** and **3**, which both describe aspects of the same place, Snowdon. How do they differ?
3 Read the contrasting accounts of the 9/11 Memorial in Figure **5**.
   a Explain why the different place-meanings that people attach to it could lead to conflict.
   b How might tour operators or authorities at the World Trade Center site respond to the differing needs of individuals within the cosmopolitan audience they serve?

STRETCH YOURSELF

Identify three different places that are special to you. Describe their geographical location and why they are important. You might use photos to illustrate your work.

In this section you will learn about:
- the concept of place and its importance in human life
- the mutual links between place-meaning and identity
- how and why people may feel like insiders in some places and like outsiders in others

The concept of place

The geographical concept of place has three different aspects:

- **Location** – where a place is on a map, its latitude and longitude coordinates, for example 40°47'N 73°58'W (Central Park, New York City)

- **Locale** – each place is made up of a series of locales or settings where everyday life activities take place, such as an office, a park, a home or a church. These settings affect social interactions and help to forge values, attitudes and behaviours – we behave in a particular way in these places, according to social rules we understand.

  '[Locale means] not just the mere address but the where of social life and environmental transformations.' (Agnew, 1987)

- **Sense of place (place-meaning)** – the subjective (personal) and emotional attachment to place, its meaning.
  (A summary of the definition proposed by John Agnew, 1987)

It is sense of place (place-meaning) that we will examine in sections 2.2 to 2.5.

The importance of place in human life and experience

Attachment, home and identity

'There's no place like home, there's no place like home…'
(Dorothy in *The Wizard of Oz*)

As a newborn baby, the first environment we find ourselves in is the arms of a parent. And it is likely to be that parent with which we will form an important attachment (emotional bond), between the age of six and nine months. At this age a baby begins to notice the absence of their primary carer, as anyone knows who has been asked to 'hold the baby for a minute'! (Figure **1**)

The Humanist geographer, Yi-Fu Tuan, has described the way that our understanding of the environment and our attachment to it expands with age. Indeed, he suggested that our geographical horizons expand in parallel with our physical ability to explore the world: a baby cannot travel but puts anything and everything in its mouth, and then learns to crawl, toddle and eventually walk.

Stop the press! Concept update:

Geographers agree that a locale need *not* be tied to a particular physical location, so a vehicle (e.g. the school bus) or an internet chat room may be a locale that structures interactions between people.

⬥ *Figure 1* '*To the young child the parent is his primary "place". The caring adult is for him a source of nurture and haven of stability*'. (Yi-Fu Tuan, 1977)

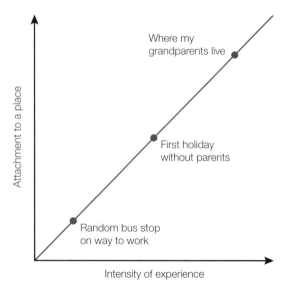

⬥ *Figure 2* *The relationship between intensity of experience and depth of attachment to a place*

The depth of feeling (attachment) we have for a place is influenced by the depth of our knowledge and understanding of it – this attachment increases with age, as we learn more about our home.

However, research also shows that our attachment to a place is influenced by the quality or intensity of experience we have there. So, the more enjoyable the experience or the greater degree to which we feel safe in a place (where all of our needs are met – think again about the needs of a child), the more attached we feel to it (Figure **2**). It becomes *home*.

Identity and place

Our sense of (a) place, the meaning we give to a location, can be so strong that it features as a central part of our **identity**.

We can see that the people quoted in Figure **3** strongly identify with different places. Notice that they differ in terms of the scale of place that they identify with – from regional (Yorkshire) to continental (Europe).

Now, consider the possibility that all of these statements could have been made by the *same person*. Perhaps this person thinks of their identity in layers, or as having a number of facets that derive from distinct aspects of their family history, upbringing and experience.

> I am European in outlook

> I'm a Yorkshireman, through and through

> I'm British

> I am Indian

🔺 **Figure 3** *People identify with different places*

Sense of place (homeland) → Identity
e.g. Britain ← e.g. British

🔺 **Figure 4** *The sense of place (place-meaning) we give to the country of our birth or homeland influences our identity*

Changing places, changing identities

Changes in the nature of places – be they social, economic or environmental – affect people and their identity. Consider football fans whose confidence takes a hit after watching their national team experience a crushing defeat in the semi-final of the World Cup (Figure **5**).

If a major employer or industry fails, those made redundant will miss the social interactions associated with the locale of the factory or office, and must re-evaluate their role in society (Figure **6**).

🔽 **Figure 5** *Brazilian fans watch their team being defeated 7–1 by Germany in the 2014 World Cup*

▶ **Figure 6** *In 2015, Sahaviriya Steel Industries, announced the closure of Redcar Steelworks in the north-east of England with the loss of 1700 jobs. In 2016, Tata Steel announced its intention to sell its entire UK business. This would result in the loss of a further 1200 jobs in the North East and 4000 at Port Talbot in South Wales.*

Insiders and outsiders have different perspectives on a place

'[In Jordan] I remember sitting in the garden with my mother at sunset. A cool breeze was starting to blow. My mother would make tea. We would sit under the jasmine tree and talk. That physical, psychological place is home to me. When I got to Durham, I started looking for a jasmine tree.'

(A Jordanian's reflections on feeling homesick in Britain)

▶ **Figure 7**

Social and spatial exclusion: what does it look like?

All places are shaped by people and understood by them in different ways. It follows that anybody whose behaviour varies from what is seen as 'normal' may feel uncomfortable. The dominant groups, who have the economic, social and cultural power in a location or within a society, may make such 'wrongdoers' feel **out of place**.

Human geographers are interested in finding out about groups in society that are excluded not only spatially but also socially, politically or economically, and the reasons for this separation.

Excluded groups may include ethnic minorities; immigrants; local nationals (born and brought up in a place, but who feel separated from it).

▼ **Figure 8** *Characteristics of people who feel insiders and outsiders in 'Country **X**'*

	Insider	Outsider
Place of birth	Born in **X** or their parents were born there	Not born in **X**, they are an immigrant and/or their parents and grandparents were immigrants
Status (citizenship etc.)	Permanent resident. Holds a passport of Country **X** Can work, vote, claim benefits like free housing and healthcare	Temporary visitor Holds a foreign passport and/or limited visa to stay in **X** May not be able to work, vote, claim benefits May be travelling for business/in search of work, pleasure, safety (an asylum-seeker)
Language capability	Fluent in the local language	Not fluent. Does not understand local idioms (variations or slang)
Social interactions: behaviour and understanding	Understands unspoken rules of the society of **X** Conforms to local norms	Frequently makes *faux pas* or misunderstands social interactions
State of mind	Safe, secure, happy – feels at home or 'in place' in country **X**	Homesick, alienated, in exile – feels 'out of place'

▼ **Figure 9** *Anti-homeless spikes in sheltered area outside flats in Southwark, London. Public spaces are increasingly being designed to prevent their use by rough sleepers.*

ACTIVITIES

1 Define what geographers mean by the term *place*. (Hint: there is more than one aspect to it.)

2 How does our sense of place (our attachment to places and the meanings we give to them) change as we grow up?

3 What is the relationship between our identity and our place of birth or homeland?

4 Study Figure **8**.

   a Have you ever felt like an outsider? Where was this?

   b How did you feel excluded?

Are some Britons excluded from rural England?

The number of black and Asian people who are members of organisations like the Youth Hostel Association, the Ramblers or the National Trust is very small in Britain. Geographers have tried to explain this, highlighting the way that the countryside of the *south* of England in particular has been held up as a significant national icon, to inspire patriotism.

Historically, most immigrant populations arriving in Britain have moved into urban areas – London, Birmingham, Manchester, Bradford and Leeds. As a result, these areas are home to large populations of black and Asian Britons today, who feel little connection to images of a southern rural idyll.

Rural England and national identity

From the end of the nineteenth century, the decline of the industrial north led to the growth in importance of London. The capital became the new *economic* hub of the country and the British Empire. But, at the time, the realities of polluted London life were seen as undesirable and socially chaotic. Instead, the people involved in national propaganda turned to the countryside as a symbol of an *ideal* Britain.

'The soft hills, small villages around a green, winding lanes and church steeples of the English southern counties came to represent England and all the qualities the culturally dominant classes desired… the squire and the villager each knew their place and were content with it.'
(Gillian Rose, 1995)

Towards the end of the nineteenth century, John Constable's paintings grew in popularity and that appeal endures to this day. In 2005, *The Hay Wain* (Figure **10**) It was voted the second most popular painting in a British gallery in a poll organised by BBC Radio 4. Constable's depictions of the Suffolk countryside are frequently referenced by campaigning conservation groups and also politicians (see 2.9).

Underrepresentation in National Parks

In 2001, a UK National Parks visitor survey showed that less than one per cent of visitors to National Parks were of a black or ethnic minority background (the percentage of the UK population is around 10 per cent). It may be argued that, while the media is dominated by historic images of a rural Britain stuck in a time that predates twentieth-century immigration, few black and Asian Britons are inspired to spend their leisure time in the countryside.

Mosaic is a national project that builds links between black and ethnic minority communities and organisations such as the National Parks authorities and the Youth Hostel Association (Figure **11**), with its network of youth hostels around the country.

Think about

If some black and Asian Britons feel excluded in Britain's countryside, might it also be true that some white families feel excluded or 'pushed out' of inner urban areas? For example, in the London borough of Tower Hamlets, the percentage of the population who are of Bangladeshi (Asian or Asian British) origin is larger than the percentage of white British (data from Census 2011).

⊘ **Figure 10** *Constable's* The Hay Wain *hangs in the National Gallery in London*

⊘ **Figure 11** *The YHA actively encourages young people from black and ethnic minority groups to volunteer in the countryside*

◁ **5** Read the text on this page.
   a Why do ethnic minorities living in Britain feel excluded from rural places?
   b What do you think organisations involved with rural areas could do to boost black and ethnic minority visitor numbers?

6 The disabled are another group who are often socially and spatially excluded. Is your school fully accessible to disabled students? How could you find this out?

In this section you will learn that:
- ◆ we humans divide the world up into different categories of place
- ◆ our understanding of distant places is socially constructed and affects how we relate to people who live there
- ◆ our understanding of (and the meaning we attach to) experienced places and media places is different

Far places and near places

Congratulations!

Today is your day.

You're off to Great Places!

You're off and away!

(*'Oh, the Places You'll Go!'* by Dr Seuss)

Exploration, difference and distance

If home is a place we know well and feel secure in, a prop for our identity, it can also be a tie. Travel and exploration is something we crave even if it can be a little scary.

'Place is security, space is freedom.' (Yi-Fu Tuan, 1977)

Anthropologists, who travel to the far-flung corners of the Earth, investigate the customs and cultures of human communities. They have found that everyone, wherever they live, recognises the division between 'us' and 'them' (Figure **1**). 'We are from here' and 'they are from there' is universal.

National identity, difference and xenophobia

Students of politics argue that some feeling of belonging to a place is necessary for a society's solidarity to grow. This sense of place is established or reinforced not only by looking inward to the group, but also by looking outward. People actively compare themselves with others who live in **distant places**, specifically those who they feel are different, alien or exotic.

'They do things differently there'

Try to make a list of terms or phrases in English that include the word 'French', for example, French windows. For more ideas, see Figure **2**. Do all of these things really originate from France?

▲ *Figure 1* *In Thailand, Western tourists are seen as different to the local population. Farang is the Thai word for white people or Westerners. It is not generally used as a term of disrespect and derives from the Thai word for the French, farangset.*

English terms or phrases that reference the French	French terms or phrases that reference the English
... if you'll excuse my French (please excuse me for swearing)	*[foodstuff] à l'anglaise* (something cooked in the English manner, simply without a sauce)
French cricket (a simplified version of cricket in which the batsman's legs are the stumps)	*Filer a l'anglaise* (to go AWOL/leave without permission or without saying goodbye)
French plait (variation on a hair plait, known in France as a *tresse africaine*)	*Un coup de Trafalgar* (underhand trick or a nasty surprise)

▲ *Figure 2* *Despite being neighbours in Europe, both the English and French alike see themselves as distant and different from each other*

Racism, conflict and colonial power

The phenomenon of perceived distance between 'us' and 'them' and between places that are **near** and **far**, prompts a wide range of different human behaviour– from the use of mildly mocking terms, like 'whinging Poms' (the Australian name for the English), at one end of the spectrum to racially motivated hate crime at the other. On the international stage, racist ideologies have been used to justify atrocities committed in wars and by colonial powers, including the British.

A different approach to the 'other'

In contrast, the inspiration for the international Fairtrade movement has been to reduce inequalities between 'us' and 'them', approaching all growers and producers, wherever they are located, with greater respect. Our co-existence with the '**other**' throws up challenging questions (Figure **3**) about how places and people should relate to each other today.

'If History is about time, Geography is about **space…** Space [unlike time] is the dimension of the simultaneous… this means that space is the dimension that presents us with the existence of the other. Space presents us with the question of "How are we going to live together"?' (Doreen Massey, 2013)

Experienced places and media places

Topophilia: '[the] human love of place … diffuse as a concept, vivid and concrete as personal experience.' (Yi-Fu Tuan, 1974)

How do we acquire a sense of place?

Today people travel a lot. We have access to faster modes of transport and more leisure time than earlier generations.

You may feel a deeper emotional attachment to a place that you have visited in person and felt you understood than somewhere you have heard about on the news. We cannot go everywhere, although as geographers we might like to! We depend on media representations of some places to help us make sense of the world, but do we really *know* these places? If we go on a virtual field trip, using the World Wide Web, is the sense of place (place-meaning) we gain less valid than if we had got our boots muddy?

'You had to be there': The role of direct experience

Experiencing a place – actually visiting it or living there – stimulates all of our senses. We taste the food and smell the drains! We hear the hum of the insects or the drone of the motorway. We sweat in the heat or wish we had packed more clothes. These environmental stimuli are rich. As a result, we acquire a deeper understanding of a place and, perhaps, perceive its true nature.

▲ **Figure 3** Other *people live in faraway places. But just how different are they?*

Think about

Our understanding of what is *near* and what is *far* depends on how we travel and also how distance is measured (time or miles/km). If we use a fast method of travel, or if we use the internet to maintain contact with people in distant places, perhaps this division of the world begins to break down (Figure **4**). With the forces of globalisation, some geographers propose that space is reducing in importance and that 'the near is often an expanding domain' (Levy, 2014). What do you think?

▼ **Figure 4** *The internet makes the world a smaller place*

Genius loci: the true spirit of a place or not?

In their profession, town planners aim to evoke a sense of place. Ancient civilisations believed that places such as Mount Olympus were inhabited and protected by spirits or gods. The term **genius loci** (literally 'spirit of a place') is often used in planning to describe the key characteristics of a place with which new developments must concur (see 2.5).

However, the idea that every place has a *true nature* is a matter of some debate. Human geographers like Doreen Massey, Peter Jackson and others have written about the way in which all place-meanings are socially constructed. Furthermore, they assert that the most widely held meanings benefit, and are reproduced by, the most powerful groups in society. Different people notice different things about the same place (see 2.1) and react differently to it. A single place may create *topophilia* (a strong attachment to it) in some people and *topophobia* (a dread or adverse reaction to it) in others. For example, the landscape of a National Park (see 2.2).

So, for the researcher, perhaps having a direct experience of a place is not as important as you might have first thought?

Sense of place in fiction

Novelists find themselves with a similar dilemma to geographers. They can spend time and money visiting a place in order to correctly set a scene in their book (Figures **5** & **6**). But with a few clicks of a computer mouse, they can reach the same location and take a virtual walk up and down the same streets. Will the reader really notice the difference?

Media representations inform our everyday life

'We live in an age in which photography rains down on us like sewage from above. Endless Instasnaps on your phone, everywhere.' (Artist Grayson Perry, BBC Reith Lectures, 2013)

In the so-called 'information age', we are bombarded with images and other forms of representation of the world. We benefit from this wealth of data but have to sift through it to try to make sense of it all (Figure **7**).

The representations of places that feature in the media often give contrasting images to those presented by official cartography, such as Ordnance Survey maps, or statistics, for example, census data. This is because their purpose and target audiences differ, as discussed in 2.6 and 2.7.

⬆ *Figure 5* *'My inclination is to go if I can, because research is as much about reassuring the author as persuading the reader.' (David Nicholls, author)*

⬇ *Figure 6* *A statue of Juliet stands below that famous balcony in Verona, Italy*

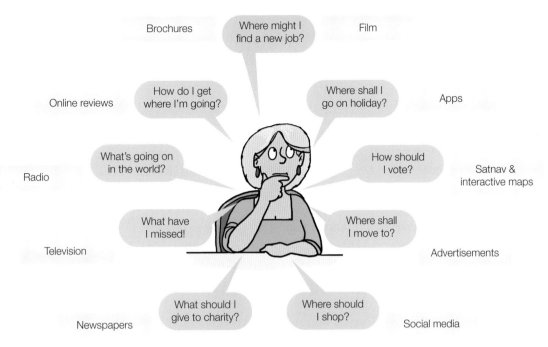

🔺 **Figure 7** *Media representations of places are part of our everyday lives*

Passage 1: Bologna

'Almost too soon we were in Bologna, one of those cities where the airport is disconcertingly close to the centre, so that you might comfortably walk there with your shopping. But I had learnt my lesson in Florence and took a taxi. My guidebook sang the city's praises, but the taxi skirted the old town on the northern ring road and what I saw was squat, modern and pleasant, with a fragment of an ancient wall in the centre of a roundabout then the dull warehouses of the airport.'

Passage 2: Verona

'Marching along a marble-paved shopping street, I followed the crowd through an alleyway into a packed, cacophonous courtyard beneath a stone balcony – Juliet's balcony, supposedly. It looked as if it had been glued to the wall, and sure enough my guidebook informed me with a sniff that it was only built in 1935, though given Juliet was a fictional character, this seemed to be missing the point. 'Romeo, Romeo, wherefore art thou, Romeo!' shouted wags from around the globe.'

(Extracts from *Us*, David Nicholls, 2015)

ACTIVITIES

1   What is meant by the term the 'other' in the section entitled 'Far places and near places'?

2   Can you think of examples of groups or organisations working to break down social, cultural or economic barriers that separate places and people into 'us' and 'them'?

3   Find a partner who has visited a place you have also been to.
   a   Compare your experiences of this place in terms of your different senses: what you saw, heard, smelt, touched and tasted.
   b   How do your memories differ?

4   Read the two passages above from the novel *Us*.
   a   The author only actually visited one of these places. Which do you think it is? Explain your answer.
   b   Why is the place-meaning we attach to a place we have visited in person different to somewhere we have only 'visited' online?

5   Make a list of different media images of places that you have been exposed to today.

STRETCH YOURSELF

Find out about a theory called **Orientalism** proposed by Edward Said. Why and how did Europeans define themselves as different to 'the East'? The Orient has long been viewed as being exotic, decadent and corrupt. What actions has this view been used to justify in regions like North Africa and Asia?

In this section you will learn about the different factors that help to shape the character of places and the communities that occupy them; both local influences and relationships with other places.

Places are unique

As we have learnt earlier in the chapter, our understanding of the world is closely tied to our appreciation of different places (2.2 and 2.3), and based on our direct experiences or their depictions in the media, we can describe what makes places unique. The fact that every place is unique may inform decisions about how each one might be developed or redeveloped. Landscape architects, who design public spaces and landmarks for clients like councils, national parks and multinational organisations, are taught to think about the *genius loci* of a place. This means that they should consider a place's key characteristics and its context – where it is located.

Factors that contribute to the character of places

Different aspects of local, regional, national and international geography influence the character of places.

Physical geography shapes places

Contrast Aberdeen, the 'Granite City' (Figure **1**), with the village of Abbotsbury in Dorset (Figure **2**). The local stone used to build houses in a place can give the built environment a particular colour and contributes to its character.

> **Think about**
> What are the key characteristics of the place where you live? How might they influence your designs for a local park or another public space? What sort of local groups and other organisations might like to have their say in what you plan to develop?

⊙ *Figure 1 Granite-built tenement flats in Aberdeen*

⊙ *Figure 2 Cottages built of local Corallian limestone in Abbotsbury, Dorset*

Likewise, the rock that remains underground shapes a place, affecting its topography –the shape and height of the land. You might hear somebody describe a county according to its topography. A generalisation like 'Derbyshire is rugged, Dorset is quaint' could be inspired by comparing the architectural style of these places – Dorset is peppered with charming thatched homes (Figure **2**), whereas housing in the Peak District is termed 'robust and simple', even 'plain' by the Peak District National Park Authority. However, this stark comparison of counties is more likely to be inspired by the contrast between the gentle rolling countryside of Dorset and the drama of areas such as the Dark Peak region of Derbyshire, named after the dark-coloured bedrock of Millstone Grit (Figure **3**). The local geology influences not only the look but also the physical experience of travelling around a place as well as many other aspects of a place including drainage, soil fertility and land use.

● **Figure 3** *Stanage Edge stands above the village of Hathersage in Derbyshire*

Demographic and economic characteristics also shape places

Dialects, ducks and data

Ay up mi duck? is a common greeting in Derbyshire, meaning 'Hello, what's up friend?'. The phrase was given international kudos when actress Angelina Jolie addressed a fellow actor who was from Derby in such a way at an award ceremony in Hollywood.

Accents and local **dialects** vary greatly within the UK. These contribute to our understanding of residents and contribute to a sense of place. However, they may also prompt the stereotyping of local people, hiding the diversity within the population of a county, city, town or village.

Census data about a population, broken down by age, employment status, education and home ownership, may arguably tell you more about a local community than analysing the dialect (although social scientists draw on both qualitative and quantitative sources).

Figure **4** compares census data about the populations of two different places in Derbyshire. Hathersage and Eyam is a ward in the Peak District National Park in North Derbyshire. The ward of Sinfin is much more populous, being located on the outskirts of the city of Derby. Sinfin's total population (15 128) is almost four times that of Hathersage and Eyam. But in what other ways do these two communities differ?

	Hathersage and Eyam	Sinfin	England
Health (%)			
Very good health	51.0	43.2	47.2
Good health	33.2	34.8	34.2
Not good: day-to-day activities limited a lot	6.4	9.5	8.3
Employment status: aged 16–44 (%)			
Employed full-time	35.1	34.2	38.6
Unemployed	2.2	7.5	4.4
Education of people aged 16 and over (%)			
With five or more GCSEs grade A–C	13.6	15.6	15.2
With no formal qualifications	14.5	31.7	22.5
Homeownership (%)			
All households who owned their own accommodation outright	46.2	18.7	30.6
All households who owned their own accommodation with a mortgage or loan	28.6	26.2	32.8
Ethnicity (%)			
White	98.6	71.0	86.0
Mixed or multiple ethnicity	0.8	5.9	2.2
Asian or Asian British	0.5	14.9	7.5
Black or Black British	0	6.4	3.3
Other	0.1	1.8	1.0
Number of households	1663	5760	22 063 368

● **Figure 4** *Comparison of demographic make-up of Hathersage and Eyam ward with Sinfin ward (Census 2011)*

Endogenous factors

The local demographic characteristics and the physical geography of a place are both **endogenous factors** that help to shape its unique character. Figure **5** illustrates the range of components of local geography that we may consider when trying to unpick why a place is the way it is.

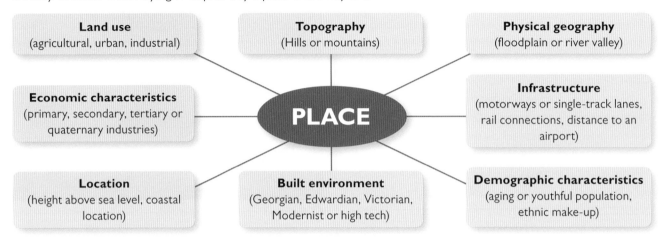

| Land use (agricultural, urban, industrial) | Topography (Hills or mountains) | Physical geography (floodplain or river valley) |

PLACE

Economic characteristics (primary, secondary, tertiary or quaternary industries)

Infrastructure (motorways or single-track lanes, rail connections, distance to an airport)

Location (height above sea level, coastal location)

Built environment (Georgian, Edwardian, Victorian, Modernist or high tech)

Demographic characteristics (aging or youthful population, ethnic make-up)

⬆ **Figure 5** *Endogenous factors that contribute to the character of a place*

Exogenous factors

However, not all influences on places are *local* in origin. Human geographers and other social scientists are also interested in **exogenous factors** that affect the character of places. These are defined as the relationships with other places. For example, a village may supply workers to a nearby town or a town may be the source of day-trippers for a tourist destination. These relationships are shown by the movement or flow of different things across space:

- people
- resources
- money and investment
- ideas.

Migration within the EU

As a member of the European Union (until the Brexit that was voted for in the 2016 referendum is achieved), the UK welcomes immigrants from the 28 member countries. Likewise British people may live and work abroad anywhere within the EU. Following the enlargement of the EU in 2004, when 10 new countries joined (the majority of whom were located in Eastern Europe), flows of people into the UK from other EU countries peaked at 1.5 million between 2004 and 2009. Two-thirds of these immigrants were Polish. Industries such as fish processing in Scotland and farm work in East Anglia benefited from this influx of labour and many still do. But EU immigrants have not taken up residence in an even pattern across the British Isles (Figure **8**), and the character and community of some places have been impacted more than others, from new shops on the high street (Figure **6**), to schools struggling to cope with large numbers of children for whom English is a second language.

Think about

Do you think different endogenous factors work *independently* to shape a place's character, as Figure **5** implies? Can you suggest any improvements to the diagram? Try to relate it to a real place – this will help you to tease out the interactions between these local, influential factors.

▶ **Figure 6** *Polish supermarkets have sprung up on Britain's high streets over the last decade*

Like other exogenous factors that shape places, the pattern of immigration into the UK itself changes. For the first time since 2004 when the EU was the enlarged, 2009 saw more nationals from the eight central and eastern European states *leave* the country than arrive. Of course, with the departure of these people a considerable amount of money earnt in the UK also departed; a flow back into the countries from which the migrants originally came.

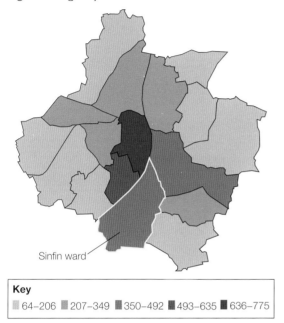

Key

64–206	207–349	350–492	493–635	636–775

▲ **Figure 7** *Number of job seekers in Derby's wards (2007)*

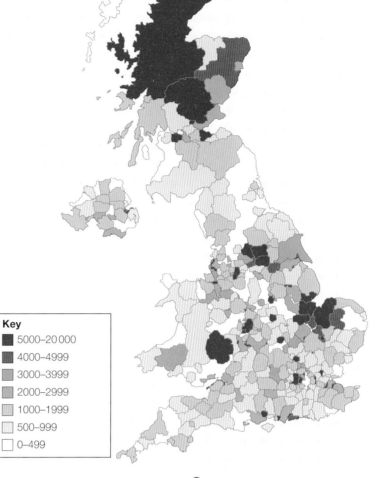

Key

■	5000–20 000
■	4000–4999
■	3000–3999
■	2000–2999
■	1000–1999
■	500–999
□	0–499

▲ **Figure 8** *Total numbers of Eastern European migrants registered for work in each UK local authority (2004–2007)*

ACTIVITIES

1 Describe two ways in which the local geology of an area can affect the character of a place.

2 Study Figure **5**. Aside from its geology, what other aspects of a place's physical geography contribute to its character? Make a list.

3 'The endogenous factors that shape a place include its demographic characteristics.' Comment on this statement.

4 Study Figure **4**.

   a Compare the demographic character of Hathersage and Eyam with that of Sinfin, using only Figure **4**. (In your description of these populations, you might also refer to differences between each ward and England as a whole.)

   b What can you infer from the home ownership data about the age make-up of these two wards in Derbyshire? Check your answer using the Office for National Statistics' Neighbourhood Statistics website.

S

5 Search the internet for the 'Office for National Statistics' quiz 'How well do you know your area?'. Test your knowledge of the demographic make-up of your community. Then, find out more about your place (e.g. your ward) at www.neighbourhood.statistics.gov.uk where you can also produce bespoke choropleth maps. Enter your postcode and the area of study (e.g. ward). Then choose from the many datasets available to create a map (see Figure **7**). You can change the colours and the number of divisions to ensure your map presents this quantitative data effectively. (Also see 2.8.)

STRETCH YOURSELF

According to the *English Index of Multiple Deprivation* (IMD see 2.8), five of the eight areas within Sinfin (Figure **7**) are amongst the 10 per cent 'most deprived' in England; both Hathersage and Eyam are in the 20 per cent 'least deprived'. But what does deprivation look like in a community? Make a list of the data sets that might be used to create the ranking in the IMD. Start with a review of the information in the UK census – it's a statistical goldmine! Also search the internet for 'Townsend Index' - another measure of deprivation.

In this section you will learn that:
- all places are changing and are socially constructed
- different forces of change have both current and historic impacts

Re-examining place

We know that places are changing – they always have been and always will be. Shanghai is a famous example of rapid change, which is associated with China's rapid economic growth. Figure **1** illustrates how the city, in particular Pudong District, was transformed in just twenty years.

Other cities across the world are also experiencing rapid expansion (see 3.2–3.5). But is change occurring in *every* place we try to describe and understand as geographers?

Place in the English language

In recent years, human geographers have criticised earlier theories of place. They argue that the concept has (historically) implied stability, even inertia – a lack of change. This is certainly true if we look at common uses of the word *place* in the English language:

- 'Know your *place*.'
- 'They looked out of *place*.'
- '… a special *place* in my heart.'

In all three of these examples, we can conclude that the social position of the person or people referred to is permanent, unchanging.

Place as a social construction

A group of geographers dubbed the 'social constructionists' (that includes David Harvey and Doreen Massey) has suggested that all places must be understood as:

- *dynamic* not static
- socially constructed.

We can easily see that places change over time but we need to examine a few examples in detail to understand what is meant by the term **socially constructed**.

▲ *Figure 1 Spot the difference? View of Shanghai's Pudong District in 1987 (top) and 2013 (bottom)*

Landscapes of power

In 2014 there were 3.7 million members of the National Trust and 17 million people paid to visit National Trust properties that year. Many of these properties were houses or manors of substantial grandeur. Other historic seats of power, such as Milton Abbey (Figure **2**), can be rented out as wedding venues. How did these lone stately homes come to have such commanding positions within the rural landscape? At what cost have these sweeping vistas been won? How should we respond to the marketing of such places that reinforces notion of permanence and longevity?

What, at first glance, might appear to be a permanent or unchanging feature of a place may, in fact, be a land use created by the economic power of the aristocracy. The creation of most, if not all, of these landscapes involved social processes that would be seen as unacceptable today.

◀ *Figure 2 The grounds of Milton Abbey were designed by 'Capability' Brown'*

Milton, Dorset

Lancelot 'Capability' Brown (1716–83) described himself as a 'place-maker'. He was an influential landscape architect with a 'naturalistic' style. His use of open grassland was a break with traditional formal garden design, such as that seen at Versailles.

In the 1770s, Brown helped Joseph Damer, the owner of Milton Abbey in Dorset, to create a place for Damer's new neighbours, the residents of the town of Milton (formerly Middleton) to move into.

Milton's long history was of no concern to Damer – he felt Milton 'spoilt the view'!

With the residents of Milton gone, the buildings were demolished and the spoil transported to the south-east to create the village of Milton Abbas. Some of the original residents of Milton were offered new homes in what we might see today as, a pretty, picture-postcard village (Figure **4**).

🔻 **Figure 3** *A map of Milton in 1769, prior to its demolition*

Milton grew up around a tenth-century church and monastery, with links to the first Kind of England, King Athestan, who also granted an annual fair.

The abbey – site of Joseph Damer's new home.

An act of Parliament was needed to move the grammar school to Blandford.

Milton, originally Middleton, was a market town, with almshouses, a brewery, traders and inns.

Damer convinced the community of Milton to leave. Many of them depended on the estate for their living or their lease.

Damer had flooded one of the properties of William Harrison, a lawyer. Harrison successfully sued with the court ruling that demolition would have to wait until Harrison died.

Some people were forced out by his direct actions, such as local road closures that made trading impossible.

🔻 **Figure 4** *Annual village fair in Milton Abbas, Dorset*

Poundbury, Dorset

Elsewhere in Dorset, another new aristocratic project is currently under construction. Poundbury, a new town on the edge of Dorchester, has been created with a more philanthropic approach than Joseph Damer had with Milton. Its patron is HRH the Prince of Wales, and the land on which it is located is part of his Duchy of Cornwall.

'For a long time I have felt strongly about the wanton destruction which has taken place in this country in the name of progress; about the sheer unadulterated ugliness and mediocrity of public and commercial buildings, and of housing estates, not to mention the dreariness and heartlessness of so much urban planning.' (A Vision of Britain, HRH the Prince of Wales, 1989)

Prince Charles is famously outspoken about British architecture. In 1984, he described the plans for a new wing of the National Gallery as a 'monstrous carbuncle on the face of a much-loved and elegant friend'. The design he loathed was dropped. Prince Charles also denounced the National Theatre, built in the 1970s on London's South Bank as 'a clever way of building a nuclear power station in the middle of London without anyone objecting' (Figure **5**).

Keen to influence current thinking in place-making and design, Prince Charles has written about his ten principles for the design of buildings and urban environments, which include:

◆ 'Developments must respect the land. They should not be intrusive; they should be designed to fit within the landscape they occupy.'

◆ 'Materials also matter.'

◆ 'The pedestrian must be at the centre of the design process. Streets must be reclaimed from the car.'
(*The Architectural Review*, 2014)

But it is in Poundbury, in Dorset, that he has gone beyond critique by seizing the opportunity to create. The foundation stone for this urban extension of 2500 dwellings to the west of Dorchester was laid by Prince Charles in 1993. The development is now more than one-half built (Figure **6**).

⬤ **Figure 5** *National Theatre or nuclear power station?*

⬤ **Figure 6** *Poundbury's eclectic street scene can be explored remotely using Google Street View*

Think about

Is change a positive or a negative thing? What comes to mind when you read the phrase 'changing places' – change for the good or for the bad? In fact, the phrase has a double meaning. Not only are all places changing all the time (and we have to cope with the impact of this on our lives) but also you and I can get involved.

Forces of change

The examples of Milton Abbas and Poundbury demonstrate that individuals within the aristocracy have, historically, wielded considerable power in the place-making process and, to some extent, Britain's royal family still does. However, individuals need not be high-born to shape the character of places. And there are other players involved in the social process of place-making, including companies and community groups. Figure **7** illustrates the wide range of **forces of change** that have an impact.

Figure 7 Forces of change in the place-making process

- Local community groups e.g. New Era Estate residents
- Transnational corporations (TNCs) e.g. Tata Steel, Tesco
- Individuals e.g activists, aristocrats, celebrities
- National government
- **CHANGING PLACES**
- Local government
- International institutions e.g. European Union
- Global institutions e.g. World Trade Organization
- National Institutions e.g. National Trust

New Era Estate, Hackney: 'Social housing not social cleansing'

In recent years, local community groups have become a force for change in housing policy in London. They've used the power of protest, social media and celebrity to further their cause. One example of this phenomenon is the campaign run by the residents of the New Era Estate, in Hoxton, East London (Figure **8**).

Community groups have been galvanised by the actions of other influential forces. The growing investment in London property by transnational corporations, like Westbrook Partners, and wealthy foreign individuals is recognised as fuelling a dramatic rise in house prices. The issue of whether or not rents remain affordable for the existing population of areas like Hoxton will help to shape their character in years to come.

New Era Estate victory: Residents prevent takeover of Hackney estate

Figure 8 Protestors from the New Era Estate outside the offices of US investment company Westbrook Partners, which threatened to increase their rents in 2014

ACTIVITIES

1   Make a copy of Figure **7**.
   a   Think of a verb to go with each force of change to show *how* it influences the place-making process. Write this alongside the arrow, e.g.

   Local populations ——*vote*——➤ Changing places

   b   Add a named example of each of these forces that, in your opinion, is the most influential in your local area. For example if you live in London you might write: Local government, *e.g. Mayor of London*
   c   Compare your ideas to those of other people living around you. Find out who or what they think are the most influential forces of change in your area. (Think about how you phrase the question! See Figure **9**, 2.9)

2   Read the text about Poundbury.
   a   Using web-based resources, take a virtual tour of Poundbury *and* investigate the Prince's aims for the town in terms of its street scene and also its community.
   b   What do critics of the development say?

3   'Past and present connections, within and beyond localities, shape places.'

   With reference to the examples outlined in this unit, explain what is meant by this statement. You should refer to current and historic forces of change at regional, national, international or global scales, and, in particular, outline the social nature of the place-making process.

STRETCH YOURSELF

What is the link between Figure **1** and Figure **6**, 2.2. How have places like Redcar and Scunthorpe in the north-east of England been affected by the decisions of multinational corporations and the UK and the Chinese governments? (See 1.5)

In this section you will learn that external agencies (governments, companies) as well as local community groups attempt to influence or create specific place-meanings in order to shape the actions of others

Rebranding and place-meaning

'He who pays the piper calls the tune'

Creative people are paid by companies, councils and even national governments to help give new place-meaning to a location. This is the process of **rebranding** and it is undertaken in order to boost footfall (shoppers, visitor numbers) or the employment prospects of an area. The people behind the catchy slogan, hashtag or map of a heritage landscape have a particular task – to change our perception and subsequently, they hope, our behaviour.

Figure **1** summarises such a rebranding campaign in Plymouth ahead of the *Mayflower400* festival in 2020, celebrating the 400th anniversary of the sailing of the *Mayflower*. Plymouth City Council aims to retain skilled workers by enhancing the city's cultural amenities. '[We're] selling the soft stuff… as much as the work' said Tudor Evans, Leader of PCC.

Inspiration (linked to existing place meanings)

Plymouth's maritime history: e.g. 2020 will be the 400th anniversary of the *Mayflower*'s departure from the city

Its maritime industries: they include some of the largest private-sector employers in Devon and cornwall, supporting HM Naval Base located in Devonport Dockyard

Its site and situation: Plymouth Hoe affords views of the Plymouth Sound, a natural harbour that connects the city to the English Channel

Its architecture: much of the city centre is built of Portland limestone

Aims and scope of the rebranding project

A joint venture between Plymouth City Council's Arts and heritage and its Economic Development departments. Key aims:
- To connect Plymouth with its past (local place-meaning)
- To promote promote and enhance its image as a cultural and economic hub within Britain and further afield (national/ international place-meaning)
- To secure its ecomomic future.

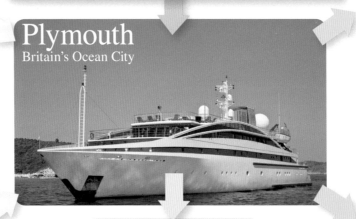

Plymouth
Britain's Ocean City

Outcomes

'*The vibrant city of Plymouth is shaped by the relationship betweens its place and its people, forging an identity characterised by resilience, adventure, vision and community.*'
The rebrand is linked to a Mayflower400 festival in 2020 and the construction of a £25 million international history centre, supported by lottery funding –
'*an international cultural and learning destination for the city*':
- to house the city's museum, art gallery, records and the south-west film and TV archives.
- be a **trailhead attraction** within the city's new cultural district.

Audiences

Marine and manufacturing industries: current and potential employers based in Devonport and the wider south-west region

Creative industries: potential employers in advertising, architecture, design, media, the arts, publishing, toys and games

Young people: who might choose to study or relocate for work to the city, which has two universities, from around the country and abroad

The local population: Plymolians contributed to the vision (#LetsTalkPlym) via the Council's online planning portal and consultations about what local people value about Plymouth and what they would like see in a new museum

Figure 1 *Plymouth: Britain's Ocean City*

Rebranding for tourism

We will look at two examples of rebranding that have been designed to influence people's perception of holiday destinations.

◆ The rebranding of Llandudno in North Wales as 'Alice Town', with associated art, walking trails, maps and apps, is an example of rebranding to promote **heritage tourism**.

◆ The 2015 rebranding initiative by VisitBritain to find new Chinese names for landmarks across the country, illustrates how simply renaming places can be used to boost tourist numbers.

Llandudno and Alice Liddell

Llandudno, 'Queen of the Welsh Resorts', was developed in the second half of the nineteenth century as a purpose-built holiday resort by the Mostyn family, who still own much of the land today. Mostyn Estates, along with the town council, strictly regulate any new developments on the promenade and require the regular repainting of historic hotel frontages. As a result, day-trippers feel the place still has the air of Victoriana about it.

The real Alice in Wonderland

Lewis Carroll was a friend of the family of Alice Pleasance Liddell, for whom he wrote *Alice's Adventures in Wonderland*. Alice's family owned a house, Penmorfa (which was demolished in 2008), in Llandudno and holidayed every year in the resort.

It is possible that Carroll's incredible tale, first published in 1865, about a girl who followed a white rabbit into its burrow, was inspired by Alice's own stories of her adventures on the Great Orme – the prominent limestone headland that dominates Llandudno.

Reinventing 'Alice Town'

The closure of the town's Rabbit Hole Museum in 2009 might have meant that Llandudno's link with *Alice in Wonderland* was lost. But in 2012, to mark the 160th anniversary of the birth of Alice Liddell, Conwy County Council commissioned four large wooden sculptures of characters from the book (Figures **3** and **4**). These larger-than-life structures were greeted with a mixed reception locally, being heralded as 'a great asset to the town' by some, while others suggested they were 'garish', 'pointless' and not in keeping with the local character.

▲ **Figure 2** Location map of Llandudno, North Wales

▲ **Figure 3** On the Alice Town Trail, Llandudno

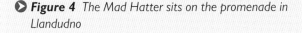
▶ **Figure 4** The Mad Hatter sits on the promenade in Llandudno

Enhancing the experience

A guided walk around the new sculptures and other highlights of Llandudno, the 'Alice Town Trail', was the idea of local entrepreneurs, who were then able to gain funding from Llandudno Town Council, Conwy County Council and Visit Wales to make it happen.

The Alice Liddell Innovative Community Enterprise Ltd (ALICE Ltd) was founded with the ethos 'to firmly establish and market the *Alice in Wonderland* connection in Llandudno worldwide, benefitting the local community'. A range of 3D 'Alice' apps brings the story to life using augmented reality and digital animation (Figure **5**). They also guide you around the town trails, following bronze-cast rabbit footprints mounted in the pavements.

⌃ **Figure 5** *The White Rabbit app combines augmented reality with a town trail*

Benefits for Llandudno

All profits from trail-linked maps and apps sold are reinvested in community projects. New businesses like The Looking Glass Ice Cream Parlour have created jobs in the town.

An annual 'Alice Day' in May involves ten local schools and begins an established Victorian extravaganza weekend that has boosted visitor numbers for many years in this part of North Wales (Figure **6**). 'Our aim was, in part, to remind local people of the history and some areas of the town they might never have visited, to help them rediscover its beauty… because it really is a wonderland' said Barry Mortlock, Director of ALICE Ltd.

⌃ **Figure 6** *The Victorian extravaganza weekend in Llandudno*

Lessons from Potter's Lake District

Conwy County Council's initiative may have been inspired by the way in which the Lake District National Park has profited from the promotion of the area's association with author and illustrator Beatrix Potter.

Today you can visit Potter's home in the village of Near Sawrey (Figure **7**) and a gallery of her paintings in Hawkshead. The award-winning family attraction 'The World of Beatrix Potter' in Bowness, on Lake Windermere, has welcomed more than three million visitors over the last 20 years, with a 15 per cent increase in visitor numbers in 2013–14.

⌃ **Figure 7** *Beatrix Potter's home, Hill Top, is popular destination for visitors to the Lake District*

Great Mandarin names for Great Britain

'For centuries, the British roamed the world, slapping English names on just about everything.

'Now the Chinese are returning the favour…' (Visit Britain, 2015)

A campaign to bridge the cultural divide

VisitBritain's 'Great Names for Great Britain' social media campaign invited Chinese people to invent Mandarin names for key UK landmarks (Figure 8), tapping into an existing trend among the Chinese of giving literal names to favourite people and places. For example, The Shard was given the name *Zhai Xing Ta* ('the tower that allows us to pluck stars from the sky') and the Cerne Abbas Giant in Dorset became *Bai Se Da Luo Ben* ('big white streaker'). VisitBritain estimates that the campaign, on the Weibo and WeChat social media platforms, reached nearly 300 million potential Chinese tourists.

A growth market for UK Plc

Chinese visitors to Britain spend approximately £500 million a year and VisitBritain aims to 'double the value of that market and ensure this growth is spread across the nations and regions'.

Chinese visitors already stay longer in Britain than in European competitor destinations and are high spenders – for every 22 additional Chinese visitors to Britain one additional UK job in tourism is created.

在英国, **Hadrian's Wall** 画出了古罗马帝国的北境线, 却与众多美景趣事一样, 缺少一个响亮的中文名。

英国等你 来命名

UK Visas & Immigration 申请英国签证, 前所未有的简单

GREAT 英国 *You're invited*

在地图上留下你的智慧, "名" 扬历史 访问visitbritain.com/greatnames #英国等你来命名#

visitbritain.com/greatnames

▶ *Figure 8* *VisitBritain's 'Great Names for Great Britain' campaign in 2015 generated new Mandarin names for 101 British attractions and boosted interest in travel to the UK. Hadrian's Wall became Yong Heng Zhi Ji (wall of eternity) and The Kelpies sculptures were given the name Kai Po Ju Ma (glorious armoured giant horses).*

Communities challenge place-meanings

The images we associate with a place, including an entire country or continent, are also those that can influence the actions of investors and donors – from the individual right up to multinational institutions such as international charities and UN agencies.

#TheAfricaTheMediaNeverShowsYou

What place-meaning do you associate with Africa? In 2015 individual activists, along with community organisations and a handful of government agencies, all came together from across Africa in an attempt to change the images with which it is frequently associated. They took part in a spontaneous social media event that questioned the dominant image of Africa in the media which, they felt, was at odds with their own experience of it.

The depiction of Africa as a 'dark continent', a crisis-ridden land of hopeless poverty that is in constant and desperate need of aid (Figure **9**), has been widely criticised but still persists in the Western media. You need only look at the sort of stories in the news in any given week – drought, disease, famine and war.

Place-meanings matter

The Nigerian novelist Chimamanda Ngozi Adichie (Figure **10**) has spoken about how she was affected by the prevalence of such negative perceptions of Africa, when she studied in the USA:

'My [American] roommate had a single story of Africa, a single story of catastrophe. In this single story there was no possibility of Africans being similar to her in any way, no possibility of feelings more complex than pity, no possibility of a connection of human equals.' (*The danger of a single story*, TED 2009)

A crowd-sourced campaign

More recently, inspired by a different place-meaning of Africa and grounded in their respective lived experiences of different locations within it, many Africans shared images on social media linked to *#TheAfricaTheMediaNeverShowsYou* (Figure **11**). The hashtag quickly started to 'trend' and images posted were subsequently republished by numerous mainstream media outlets, including the UK's *Guardian* and *Daily Mail* newspapers and Canada's Global News TV network.

▲ **Figure 9** *Live Aid raised an estimated £150 million as a direct result of the concerts*

▶ **Figure 10** *'The single story [of Africa] creates stereotypes, and the problem with stereotypes is not that they are untrue, but that they are incomplete.' (Chimimanda Ngozi Adichie, author.)*

⬤ **Figure 11** *Images of Africa that show progress and diversity went viral in 2015*

ACTIVITIES

1   Read the 'Llandudno and Alice Liddell' example.
   a How might republicising the link between Llandudno and Alice Liddell benefit the town?
   b How have local entrepreneurs sought to develop this link?
   c Why do you think such developments have been greeted with a mixed response from the local community?

2   Read the 'Great Mandarin names for Great Britain' example.
Imagine you work for the national tourism agency, VisitBritain. Write a 100-word summary of the social media campaign for the government's Department of Culture, Media and Sport, to justify its £1.6 million budget.

3   Read the '#TheAfricaTheMediaNeverShowsYou' example.
   a Why did the participants take part in this media event?
   b Do you think the campaign was a success? Explain your answer.

4   Here are some examples of recent attempts at rebranding UK places:
   • People Make Glasgow (launched ahead of the 2014 Commonwealth Games)
   • Britain's Ocean City (Plymouth)
   • Europe's Youngest Capital (Cardiff)
   a Suggest reasons why *one* of these brands was chosen for its location and the audience at which it was aimed.
   b Is there a place close to you that has been given a new name or image? Find out why.

STRETCH YOURSELF

1   Analyse media representations of Africa. Make a note of 20 news headlines in any given week that include the word 'Africa' or the name of an African country. Use software like Wordle or Tagxedo to find which words are used most. You could also categorise the headlines or different paragraphs within the news stories into positive or negative. Compare your findings to 20 stories about Asia or South America published over the same period of time.

2   Examine the tone and content of VisitBritain's 'Great Mandarin names for Great Britain' campaign. Is the self-mocking tone a product of a modern critique of Britain's imperial past and its actions in far-flung places, or a recognition of the influence of a new economic (and potentially cultural) global superpower? Or a bit of both?

In this section you will learn that:
- places may be represented in a variety of different forms, in diverse media
- finding out about the purpose and wider context in which a particular representation of a place was produced can help us evaluate its reliability.

Critically evaluating representations of places

As discussed in section 2.3, we rely on many different images of places to inform and construct the meaning we attach to them. For example, poetry about places can inspire patriotism. Take a look at Figure **1**. These meanings affect our decision-making – whether we visit a place, invest in a place or even care about a place. And as we saw in 2.6, place-meanings created by people matter, particularly to the economy. Before examining the case studies that follow, let's first consider how **reliable** the representations of places are. Is there any way we can evaluate this?

We're going to advance

We're going to a dance

USA

'O beautiful for spacious skies,
For amber waves of grain,
For purple mountain majesties
Above the fruited plain!
America! America!
God shed his grace on thee
And crown thy good with brotherhood
From sea to shining sea!'

Extract from *America the Beautiful*
(Katharine Lee Bates, 1895)

England

'And did those feet in ancient time
Walk upon England's mountains green
And was the holy Lamb of God,
On England's pleasant pastures seen!'

Extract from *Jerusalem*
(William Blake, 1808)

Figure 2 The telephone game

Reliability

Secondary sources supply information via another person's experience, their eye or lens. Each secondary source is an act of interpretation. Sometimes the message gets lost in translation (Figure **2**). Even photography, which you might assume is the most straightforward form of representation, has its problems (Figures **3** and **4**).

Figure 1 Patriotism and poetry

Figure 3 The photographer frames their subject. But who or what is just off camera?

Be it the work of a film director, artist, composer or novelist, we should always remember that these sources offer us a *subjective* perspective – a personally curated view of a place. All are selective and therefore open to accusations of inaccuracy. What have they left out?

We, in turn, interpret the data we are given and add our own subjectivity. For example, we might like or dislike a piece of art depicting a place. Our feelings about that place or the piece of art may affect our views on how reliable the art is as a source. As was noted in 2.1, our views are shaped by our own experiences, education and background.

When investigating place-meaning, it is always a good policy to use more than one source or **text** (Figure **5**).

Provenance and textual analysis

Unfortunately, you cannot avoid being human or living in a world of representations drawn by mere mortals. All researchers face the same problems of viewing places through the eyes of other people and, furthermore, being themselves subjective. How can we stay alert to these limitations?

⬣ **Figure 4** *A photograph can be digitally altered to change our perception of a place*

◆ Consider whether the source gives a positive or negative impression of a place and why that might be. Look for symbols or stereotypes and metaphors in the text. Think about the author/artist's choice of vocabulary, colour or camera angle.

◆ Find out about the context in which the source or text was produced and about its creator – this is its **provenance.** When and by whom was it produced? What was its purpose? Does it support or contest the views of dominant groups or powerful ideologies of the time?

◆ How does it compare to other available texts about the place? Was it produced earlier or later than them, inspired by them or as a reaction to them?

◆ Look for subtexts or hidden texts – what is the source that is being studied *silent* about? This means, what was the author aware of as being relevant, but has chosen to leave out of his or her work, such as men, women, the economy, or the environment?

◆ How does the text relate to wider relevant geographies or processes in society, such as industrialisation, deindustrialisation, globalisation or the emancipation of women?

Social scientists suggest that as you uncover more about the context and purpose of a text you should reread the text and reappraise it. Also, try to reflect on your views of it and how they are shaped by your own experience, upbringing, and the time in which you live.

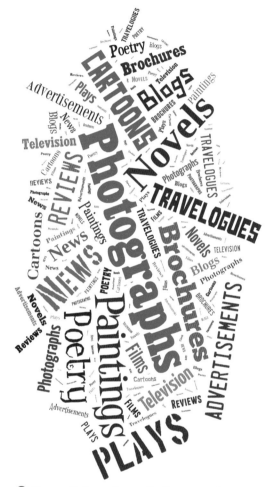
⬣ **Figure 5** *Texts for qualitative research*

ACTIVITIES

1 Write a definition of the following, in your own words:
   • text • reliability • provenance

2 You will be investigating your home or study area in some detail as part of the *Changing places* topic. Start this work by making a list of texts you could analyse in a review of media representations of your home area.

3 What are the advantages/disadvantages of different media forms:
   a for the researcher evaluating data sources or texts?
   b for the publicist or promoter tasked with changing people's perceptions of a place?

Ⓢ 4 Choose one of the two texts in Figure **1** and find out about its provenance. You will need to use online resources and read the poem in its entirety. Using the method described above, investigate the image of the country it presents.

S In this section you will learn about:
- the value of big data, geospatial data and geographical information systems (GIS)
- examples of quantitative sources of data that you could use to inform a place study

Exciting times

For human geographers these are *exciting times* because, in the new information age (see 1.1 and 1.3), data about people is often geolocated (**geospatial data**), giving an insight into the way we live and how geographic communities differ.

◀ **Figure 1** *Barack Obama's 2012 presidential campaign team used big data analysis about voter preferences to good effect*

What is 'big data'?

Big data is big in terms of its scale (the number of responses, the variety of data sets and the size of populations) and it requires huge amounts of computational power (servers, algorithms). It has huge potential. Lots of social and economic data sets fall into this category. There are different definitions of big data but they each have several things in common:

◆ Volume: the data is not a sample, it is a record of whole datasets/population of users.

◆ Velocity: often real-time information, for example, purchase transactions.

◆ Digital footprint: may be a cost-free by-product of digital interaction, for example, Tweets, Facebook posts at any given moment.

Because a lot of big data has a spatial element (everything happens somewhere and most of the time these activities are geo-tagged), analysts claim to be able to use it to make predictions that relate to the population of an area. In turn, their predictions may inform more cost-effective allocation of resources or, in the case of election campaigns, help a candidate to win by targeting key voters (Figure **1**). However, this is still a new area of computer science and statistical analysis and, as with any quantitative source, predictions made using big data are only as good as the quality of the data and people's understanding and interpretation of it.

Concerns have been raised by some groups of people about the idea that our every move or browse online can be monitored and is traceable using big data. However, the vast scale of data produced on a daily basis should allay the fears of most about 'Big Brother' watching you.

Think about

Allow this app to view your location?

With Global Positioning System (GPS) technology in the phones we use to interact with the World Wide Web, organisations (governments, companies, political parties) can map the data they collect, creating a boom-time for the discipline of digital cartography and data visualisation. What would you like to map? How would you present it in a new app and for what purpose?

Analysing the UK Census: a 'big' snapshot

To make sense of data collected in a census, statisticians organise the responses geographically – they aggregate the data. This gives anonymity to those who completed it and allows wider conclusions to be drawn about populations and places at different spatial scales.

For example, in terms of the sort of scale you will need for your place studies, data from the England and Wales Census can be analysed by output area, neighbourhood or ward (Figure **2**).

Type of small area – census 2011	Average population	Avge no. of households	Total no. in E&W
Output area (OAs)	309	130	181 408
Neighbourhood or lower layer super output area (LSOAs)	1614	672	34 753
Ward (electoral districts)	6543	2726	6543

⬆ **Figure 2** *Different types of small areas used by the ONS for detailed analysis*

So, pick the scale you'd like to begin with and visit www. neighbourhood.statistics.gov. uk to explore and map the data (as in activity **5**, 2.4). Also try the interactive map interface DataShine Census (Figure **3**). DataShine Scotland has data from the Scottish Census.

Both sites allow you to create choropleth maps with additional overlays of spatial information, of the kind presented on an OS map – settlements, rivers, roads, railways. The overlays help users to make sense of places presented. Note that the more detailed DataShine Census maps present data by census output area, instead of aggregating the data by ward as is the case in the ONS mapping tool (activity **5**, 2.4).

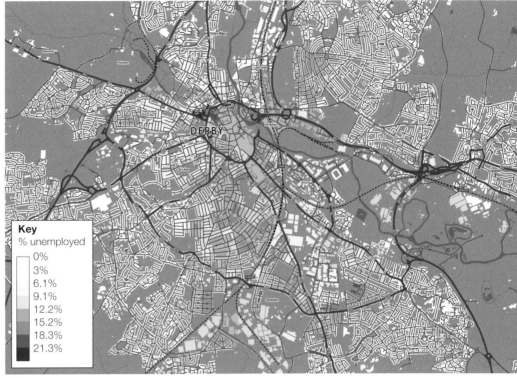

Key
% unemployed

0%
3%
6.1%
9.1%
12.2%
15.2%
18.3%
21.3%

🔻 **Figure 3** Map of Derby showing the percentage of unemployed people within the economically active age group across the city. DataShine Census is an interactive mapping tool you can use to present data from the 2011 Census.

The English Index of Multiple Deprivation (IMD) 2015

The concept of **deprivation** was introduced in the 'Stretch Yourself' in section 2.4. The *English Index of Multiple Deprivation (IMD)* is published by the Department for Communities and Local Government on a regular basis and informs national and local government decision-making and associated patterns of investment. It ranks over 32 000 neighbourhoods (LSOAs) across the whole country according to a combination of seven domains of deprivation (see Figure **4**). Each of these domains is based on a 'basket' of indicators. As far as is possible, each indicator is based on

the most recent data available although, in practice, most indicators in the 2015 data relate to the tax year 2012/13, for example, the number of recipients of income support benefits in a given area that year.

The **deciles** are produced by ranking the 32 844 lower-layer super output areas (neighbourhoods) and dividing them into 10 equal-sized groups. Decile 1 represents the most deprived 10 per cent of areas nationally and decile 10, the least deprived 10 per cent.

| Income 22.5% | Employment 22.5% | Education 13.5% | Health 13.5% | Crime 9.3% | Barriers to housing and services 9.3% | Living environment 9.3% |

🔻 **Figure 4** The English IMD is a system of ranking neighbourhoods based on seven indicators of deprivation that are combined according to a weighted formula

A few notes of caution should be observed when interpreting the map and ranks of individual neighbourhoods:

◆ The ranks (and deciles) are relative: they show that one area is more deprived than another but not by how much.

◆ When interpreting maps, the eye is drawn to large swathes of colour. This can be misleading as a geographically large local authority district may have a smaller population than a geographically smaller district.

◆ The neighbourhood-level indices provide a description of areas but not of individuals within those areas.

◆ The indices identify aspects of deprivation, not affluence. The rich aren't mapped.

The IMD is of interest to geographers, both in terms of the spatial pattern of (poor) quality of life it depicts, and also in terms of how it defines deprivation, incorporating a spatial element in the domain of 'barriers to housing and services' (see panel).

The complex pattern of deprivation across England

Most deprived neighbourhoods and least deprived neighbourhoods were spread throughout England (see Figure **5** and the interactive *IMD Explorer* online).

As was the case in previous indices, there are concentrations of deprivation in large urban conurbations – historically industrial, manufacturing or mining areas, coastal towns and large parts of East London. Figure **6** presents the local authorities most affected by relative deprivation.

Local authority district	Number	Per cent
Blackpool	19	20.2
Knowsley	13	13.3
Kingston upon Hull	20	12.0
Middlesbrough	10	11.6
Liverpool	26	8.7
Great Yarmouth	5	8.2
Barrow-in-Furness	4	8.2
Burnley	4	6.7
North East Lincolnshire	7	6.6
Manchester	18	6.4

Geographic distance: a barrier to services?

The IMD 2015 considers physical distance to be a contributory factor in areas of deprivation. The following is incorporated into the ranking for every neighbourhood:

◆ Road distance to a post office

◆ Road distance to a primary school

◆ Road distance to a general store or supermarket

◆ Road distance to a GP surgery

Figure 5 *Map of multiple deprivation across England (IMD 2015)*

Figure 6 *Local authority districts with the highest proportion of neighbourhoods (LSOAs) in the most deprived one per cent of areas nationally (based on the IMD)*

Deprivation over time and space

There are 98 neighbourhoods that have been ranked among the most deprived one per cent in each IMD update (in 2004, 2007, 2010 and 2015). These are places where the community has experienced the poorest quality of life in the UK for over a decade.

The areas most affected by persistent relative poverty are in Merseyside and Greater Manchester. In contrast, no neighbourhoods in London were ranked among the most deprived one per cent in each and every update.

▶ **Figure 7** *East Village in E20, a new postcode for an area of Newham that has benefited from Olympic investment*

▼ **Figure 8** *Age of housing mapped across the Queen Elizabeth Olympic Park and surrounding areas in East London*

Rather, of the local authorities where relative deprivation has fallen significantly between 2010 and 2015, the top four are in London (Hackney, Tower Hamlets, Greenwich and Newham). These are four of the six London boroughs that hosted the 2012 Olympics. Their track record is, perhaps, testament to the power of rebranding and regeneration in the place-making process (Figures **7** and **8**).

ACTIVITIES

1. Write your own definition of 'big data'.
2. Use the www.datashine.org.uk and www.neighbourhood.statistics.gov.uk websites to investigate the results of the 2011 Census for your place.

   Compare these two GIS software sites in terms of the clarity of maps created; ease of construction; ease of interpretation.
3. Read the text about the IMD and look at the IMD Explorer site http://dclgapps.communities.gov.uk/imd/idmap.html

   a. Who might be interested in the Index of Multiple Deprivation? Why?

   b. Look at Figure **4**. Comment on the different domains (data sets) used to create this index of neighbourhoods.

   c. What are the limitations of the IMD?

STRETCH YOURSELF

Using online resources locate the CDRC map of 'dwelling age' created by combining data from the Government's Valuations Office Agency and the ONS (also see Figure **8**). Use the map to find out about the different ages of housing in your local area. How might you use this information to direct further research for your place study or make sense of other data you have already collected?

In this exemplar of a local place study you will learn about:
- the developing character of Great Missenden, in Buckinghamshire
- the range of data sources that you could use to investigate your locality

Place study

Dear reader, I live here

The author of this chapter is female, white, English, educated, middle-class and lives in the Chilterns. As a researcher I am, therefore, 'situated'; I have a particular view on the world, specifically Great Missenden (Figure **1**).

My **place study** is a (subjective) representation of this place, although it incorporates multiple sources of data, which I have tried to evaluate critically. Remember who the author is when reading these pages. You too will need to reflect on, and challenge, your own subjective perspective when investigating your place or local study area.

Connections: past, present and future

Great Missenden is a large village located in Buckinghamshire in the south-east of England, with a population of 10 138 (Census 2011). It is located in the Chiltern Hills, a chalk escarpment, at the head of the Misbourne Valley.

At the northern end of the village is Mobwell Pond, the source of the Misbourne, a chalk-fed river. The upper part of this river flows intermittently after winter rains and dries up during the summer – a physical feature known as a *winterbourne* (Figure **2**). Some of the rarest species living in chalk streams ,such as brown trout and the water vole, are specially adapted to living in winterbournes and 'although altered by many centuries of human activity such as milling, watercress growing, ornamental landscaping and abstraction for public water supply, the Misbourne is still a valuable habitat.' (Chiltern Chalk Streams Project)

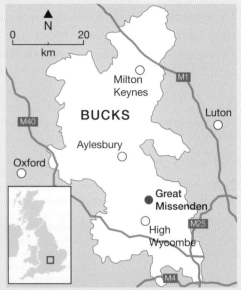

▲ **Figure 1** Location map of Great Missenden, Buckinghamshire

An Area of Outstanding Natural Beauty

According to the Chilterns Conservation Board, the 1965 designation of the Chilterns as an Area of Outstanding Natural Beauty (AONB) was given for a range of endogenous factors:

'the natural beauty of its landscape and its natural and cultural heritage. In particular … to protect its special qualities which include the steep chalk escarpment with areas of flower-rich downland, woodlands, commons, tranquil valleys, the network of ancient routes, villages with their brick and flint houses, chalk streams and rich historic environment of hillforts and chalk figures.' (The Chilterns Conservation Board)

The AONB designation put the Chilterns 'on the map' – it had become a landscape of national importance. Today, there are 46 AONBs across Britain.

▲ **Figure 2** Villages of brick and flint construction contributed to the Chilterns' designation as an AONB

What is the Chilterns AONB? An Area of Outstanding Natural Beauty, a landscape of national importance.

Resident population
80 000 (and 10 million live within an hour's drive)

Jobs supported by visits
2770

Ten Mile Menu @TenMileMenu · Oct 9
Another amazing Friday morning in the #Chilterns. Our favourite time of the year. #Autumn

Visits per year 55 million leisure visits (Tourism South East, 2007)

Designated by? UK government – in law all public bodies 'shall have regard to the purpose of conserving and enhancing the natural beauty of the area of AONB.' (Countryside and Rights of Way Act, 2000)

Recreational value 'One of the most popular areas in Europe for walking, cycling and horse-riding.' (Chilterns Conservation Board)

Area
833 km²

Economic value of visits per year
£80 million (2013)

🔺 **Figure 3** *What is the Chilterns AONB?*

Flora and fauna

An aspect of the local physical geography worthy of particular note is its woodlands. Two-thirds is ancient woodland (defined as areas that have seen woodland cover for over 400 years) – the Chilterns have the highest proportion of ancient woodland of any comparable area in the country.

In 2010, when the Chilterns Conservation Board undertook an assessment of the ecological value of local woodland. Their report affirmed the importance of the total area of woodland in the AONB and its *enduring spatial pattern*. They argued that the impact of the 'fragmentation of habitats and colonies of flora and fauna' that would result from the construction of a new railway line (HS2) should be evaluated as part of any environmental impact assessment. Woodlands physically connect.

This is also true of hedgerows in the Chilterns that function as corridors for wildlife, aiding the dispersal of species, as well as being important habitats for bird nesting and feeding. There are around 4000 km (2500 miles) of hedgerow in the AONB.

The cultural value of woodland in Great Missenden is addressed in a range of sources in section 2.10.

Connected by road and rail

Great Missenden is connected to central London by rail and serves as a home for thousands of daily commuters who work in the capital. Prior to arrival of the railway in 1892, the village was first and foremost a farming community.

'...a self-supporting, more or less closed world' (*One Hundred Years of Great Missenden*, Valerie Eaton Griffith)

Even earlier, it had functioned as a coaching stop between London and the Midlands, with 12 public houses and inns on its High Street.

Figure 5 (on page 98) shows how the population of the parish of Great Missenden has grown and also how the main type of employment has changed from agriculture in 1811 to 'professional' in 2011.

S The British Film Institute is the home of the national archive of film. 'Britain on Film' (www.bfi.org.uk/britain-on-film) is a free-to-access, geo-located, online archive of 2500 historic films (including home movies, documentaries and news footage) that provide a fascinating insight into earlier times, from the Victorian era up to the 1980s. Why not look up your place?

🔺 **Figure 4** *Great Missenden was primarily a farming community prior to the arrival of the Metropolitan railway (image from* In Milton's Country *(1924) via BFI's Britain on Film)*

Place study

The census and diversity

The community is not particularly ethnically diverse – 96.3 per cent of the population of Great Missenden is white, whereas in England this figure is 79.8 per cent of the total population. Residents aged 65 years or older (elderly dependants) outnumber those aged 0–15 years (young dependants) by 2.5 per cent. Across England the situation is reversed young dependants outnumber the elderly by 2.5 per cent.

Great Missenden ward is among the 10 per cent least-deprived neighbourhoods in the country. A walk around the local area or an online virtual tour illustrates that there is a wide range in the size and value of housing and associated levels of personal wealth. However, interview data shows that a significant percentage of young people on reaching employment age can no longer afford to stay in the village (see 2.11).

Connected by tourism and recreation

Today, the village is at the heart of the Chilterns Area of Outstanding Natural Beauty (Figures **3** and **7**) and a honeypot site for tourists. Walkers, cyclists and day-trippers are drawn to the natural environment of chalk hills and valleys, open farmland and beech woodland.

The popular Roald Dahl Museum and Story Centre (Figure **6**) is sited close to where the children's author lived for 36 years. There are numerous shops and services on the High Street geared to the needs of visitors: pubs, restaurants and cafés (7), hairdressers and beauty salons (7), gifts shops (5) and estate agents (3).

Faster connections and the impact of HS2

Great Missenden is undergoing economic change as a result of national and international external forces. High-speed rail has been expanding across the European Union since the 1980s and in 2007, the first high-speed services were launched in China. But in the UK, high-speed rail was limited to a single stretch of track, the Channel Tunnel Rail Link, between London and the tunnel itself.

In 2009, the proposal of a second UK high-speed line (HS2) was announced by the British government. Ministers advised that this service on a brand-new line would address capacity issues on the West Coast Main Line and reduce travel times to Birmingham, Manchester, Leeds and Edinburgh. It was suggested by politicians that the new HS2 line would help to reduce the North–South economic divide, allowing business people to reach key cities in the north of England faster, and therefore giving a boost to northern businesses.

◆ **Figure 5** *Census data for Great Missenden Parish*

Table 1: Percentage of occupations of all people in employment (Census 2011)

	Great Missenden %	Chiltern %	England %
Managers, directors and senior officials	21.7	17.4	10.9
Professional occupations	24.0	23.1	17.5
Associate professional and technical occupations	17.2	15.8	12.8
Administrative and secretarial occupations	12.0	11.1	11.5
Skilled trade occupations	7.7	8.9	11.4
Caring, leisure and other service occupations	6.7	8.2	9.3

Table 2: Occupations of all families (Census 1811)

	Number of families
Agriculture	164
Trade, manufactures or handicraft	113
Occupations other than the above	18
Total number of families	295

Table 3: Population of Great Missenden (Census 1811–2011)

1811	1576
1851	2097
1911	2555
1951	4464
2011	10138

▲ **Figure 6** *The Roald Dahl Museum on Great Missenden's High Street*

In 2010 the government stated that the preferred route would include a section running through the Chilterns AONB and the parish of Great Missenden. The new line would be less than a mile from the centre of the village.

Great Missenden Parish Council, the Chiltern Conservation Board and Chiltern District Council all declared their opposition to the development on the grounds that it would have a detrimental impact on the natural environment of the Chilterns and bisect the AONB (Figure **7** – the AONB is shown in yellow).

In 2013 HS2 Ltd's Environmental Statement spelt out its reasons for selecting this route from among the four possible routes between London and Birmingham:

◆ the first route via Heathrow was '...discounted due to its cost and longer journey times'.

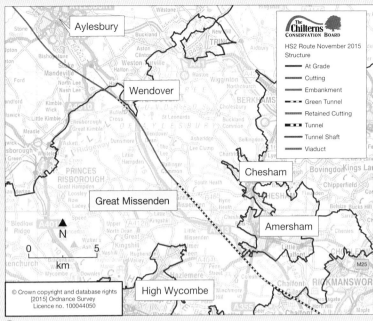

▲ **Figure 7** The route of HS2 runs through the Chilterns AONB

◆ 'Each of the three remaining options would have different environmental effects, but overall the difference between the three options in environmental terms was found to be marginal. The preferred route was selected because it would offer shorter journey times, with associated economic benefits and for less cost.'

So, the government selected the shortest and straightest route, which, it recommended, would be perfect for speeds of up to 400 kph (284 mph).

Even before the construction start date of 2017, Great Missenden witnessed significant impacts on the local economy, in common with many other places in England along the route. Since the government's announcement, businesses and home-owners have been directly affected in terms of falling property prices, compulsory purchases and a decline in business.

There are visible signs of this **planning blight** and local protests against the HS2 project and its effects (Figure **8**).

ACTIVITIES

1 **a** Using only the data in Figure **5**, describe how the population of Great Missenden has changed over time.
   b What further information would you like to provide a fuller picture of demographic change in this place?

2 Figure **9** presents responses to a question about change in Great Missenden that was posted on a local Facebook group.

   How have Great Missenden's connections to other places shaped its changing economy? You should consider connections at the regional, national and international scales in your answer.

3 Great Missenden is a casualty of the social inequalities between 'the north' and 'the south' of England. To what extent do you agree with this statement?

▼ **Figure 8** Annie Bailey's Restaurant and Bar (closed in 2013) and its amended signage, located on the proposed route of HS2

▲ **Figure 9** Crowdsourced answers to the question 'Who or what is changing Great Missenden today?' gathered via a local Facebook group; words that appear larger were mentioned by a greater number of respondents.

In this place study you will learn about:

- qualitative and quantitative data sources that can be used to inform a place study
- the importance of critically evaluating the representations of a place presented in different media forms, and their impact on place-meanings and perceptions

Place study

The contested character of Great Missenden

The place-meaning of Great Missenden and its surrounding landscape has become contested as part of the debate in the national media about HS2. This place has been represented in different media by government ministers, local artists, writers and musicians. Here is a selection of sources or texts for you to consider.

Source A

Extract from 'Roald Dahl's Countryside Trail – a family activity walk' leaflet produced by the Roald Dahl Museum and Story Centre.

Atkins Wood

This is a classic Chiltern beech wood. Beech trees are common in South Bucks because they were planted in the nineteenth century for the chairmaking industry in High Wycombe. The towering trees and dark forest floors typical of local woods provide the setting for the pheasant poaching scenes in *Danny the Champion of the World*:

'I flattened my body against the ground and pressed one side of my face into the brown leaves. The soil below the leaves had a queer pungent smell, like beer'.

Source B:

High-speed rail will be a 'pleasant surprise' for many

'... but some oppose the line on purely aesthetic grounds, arguing that the new line will be a scar on the landscape of rural Buckinghamshire, despoiling an area of outstanding natural beauty. 'Have you looked at the route? ...Between Great Missenden and the HS2 route are the A413, the Chiltern Railway and a line of pylons. So this is not some Constable country.'

(Philip Hammond, the then Secretary of State for Transport, *The Telegraph*, 11 December 2010)

Source C

HS2 and the environment

'I have to concur one hundred percent ... on grounds of geography. Constable country is in the Dedham Vale AONB on the Suffolk–Essex border, whereas the Chilterns AONB lies just north-west of London.

I think what he really meant was that the Chilterns was somehow an 'inferior' AONB to Dedham Vale and could be trashed with impunity.

I personally believe that the rolling, wooded countryside of the Chilterns is very beautiful, but Mr Hammond obviously prefers something wetter.'

(Blog by Peter Delow, 19 April 2011)

Source D

Song entitled 'Oak Tree Lament (Stop HS2)' by Dirty Mavis (2011)

2 hundred years I stand... on the edge of the Misbourne Plain
Survived the harshest winters and summers without rain
My roots are deep... and you should weep
For this morning strangers came,
With axe and saw, and cut me to the floor
So hang your head in shame
Men say that time is money, half an hour could be saved
And for this bold lie I am to die and thrown down in my grave
My roots are deep and you should weep
At the cause of so much pain
So scrap this plan leave me stand on the edge of the Misbourne Plain.
(*Chorus*)
On the English Hills runs a line that kills
Leaves my country scarred and raw
So if you love my leaves and shade then please.... prepare yourself for war
For when there's no more oil.... and you're in the soil
And the breeze whispers to you ....
Did you fight to save this England did you stop the HS2?

Source E:

HS2 opponents are NIMBYs who only care about house prices

The Mayor of London says people who oppose HS2 railway are 'pretending' to have an environmental objection.

(Boris Johnson, the then Mayor of London, *The Telegraph*, 28 April 2014)

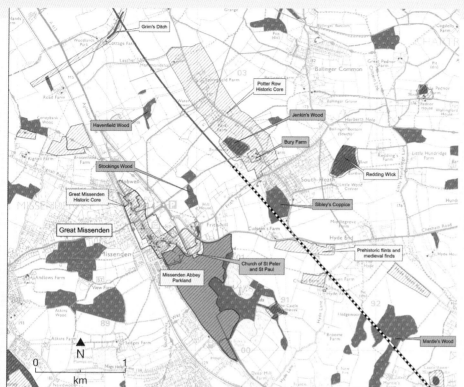

Source F

Environmental map of Great Missenden and the surrounding area by Chilterns AONB Board

Historic Environment

- ▨ Archaeological Notification Areas
- Listed Buildings
- Scheduled Monument
- Parks and Gardens
- Conservation Areas
- Ancient Woodland

Historic Routeways

- —— Pre 18th Century
- —— 18th or 19th Century

HS2 Route November 2015
Structure

At Grade	Retained Cutting
Cutting	Tunnel
Embankment	Tunnel Shaft
Green Tunnel	Viaduct

Source G

'Train perimeter on my patio'. Artwork that incorporates lasers to illustrate the route of the rails (HS2//Intervention by Mark King)

Source H

Guerrilla knitting adorns trees in Sibley's Coppice, Great Missenden

ACTIVITIES

1 Most of the sources and figures in 2.9 and 2.10 are **qualitative data**.

  a Which, if any, provide **quantitative data**?

  b As researchers, should we approach quantitative and qualitative sources differently?

2 **Myths** are socially constructed versions of reality. They may be influential and widely held ideas, with a long history. Often, they are passed off as common sense by those whose interests they serve.

  Read the statements **i** and **ii** (below) giving a view of both sides of the HS2 debate. Then review sources **A–H**.

  i Privileged home owners in the Chilterns (selfishly) prioritise their way of life above the wider good of the country.

  ii The natural environment of the Chilterns is important and should be conserved.

  a State which data source(s) support each myth.

  b In which of the sources, if any, has the place-meaning of Great Missenden been managed or manipulated?

  c Can you compose another myth about Great Missenden based on the data presented here?

3 Select a source for each side of the HS2 debate. Use the textual analysis techniques described in 2.7 to evaluate each text.

  a What is its provenance?   **b** How reliable is it?

STRETCH YOURSELF

Your place may not be rural or the site of a hotly debated development. But the way it, and places like it, are represented in the media is of interest. For example, if you live in a suburb, consider what the dominant representations are of suburbia. How do they affect the place-meaning of 'the suburbs' in the UK today?

S In this section you will learn:
- how to carry out an interview and what to do with the data
- more about economic and demographic change since 1939

Place study

'Tell me about...'

There are a range of oral sources that you could use to inform a place study that, when transcribed, become rich texts to analyse and reflect on. These include interviews (Figures **3**, **5** and **6**), reminiscences and songs (Source **D**, 2.10). This section discusses the advantages and disadvantages of collecting qualitative data as part of a place study by interviewing, and gives some hints about how to go about, what can be not only a very informative process but also a privilege.

'There's nothing quite like finding yourself meeting – and connecting with – someone from a very different background and set of experiences … Interviewing is a recipe for some highly memorable fieldwork experiences!' (Richard Philips and Jennifer Johns, 2013)

Why interview?

One-to-one interviews have a number of advantages when undertaking a place study. Interviews can be used to collect primary data that gives an in-depth understanding of people's lives and their lived experience of a place, both in the past and present. In contrast, with their limited number of tick boxes, questionnaires push respondents into answers, which may not fit their actual experience. They give little opportunity for explanation, or space for you to record any extra information, if offered.

An interview allows respondents to raise issues that the interviewer may not have anticipated. In the example of the place study of Great Missenden, the author undertook a series of interviews with people who had lived and/or worked in Great Missenden for more than fifty years. The issue of immigration came up, in particular the movement of Polish people into the area following the Second World War. This was raised as a significant place-shaping event, despite not featuring in any other texts gathered during the author's research. Similarly, the author's understanding of the changing pattern of employment in the area was reshaped by comments about light industries that played an important role in the local economy after the Second World War (Figures **5** and **6**). There is little physical evidence remaining today.

🔺 **Figure 1** *Some interviewees produce photographs, like this one of Great Missenden High Street in 1953, on the occasion of the coronation of Queen Elizabeth II*

Disadvantages of interviewing

Interviewing can be a time-consuming business. As a technique for data collection it may be further criticised on two counts. Firstly, in terms of its lack of objectivity compared with other approaches used by human geographers, and secondly with regard to sample design.

Objectivity and subjectivity

As previously noted in 2.7 and 2.9, we are all human and our understanding of the world around us is partial and situated. So, perhaps a discussion of whether questionnaires generate more objective data than interviews is really missing the point! The data produced by these two techniques may be as similar as chalk and cheese. What is true is that bad question design is not limited to either technique. For example, the following question might be used in either an interview or a questionnaire but the way the question is phrased is important. Should we ask:

| What impact will the construction of HS2 have on the area? | or | What impact will the disruption caused by HS2 have? |

Is it possible to draw wider conclusions from interview data? An interview is a chance to get a deeper insight into one person's view of a place and how their experience of living, shopping, playing and working there may have changed over time. The data it generates is rich but very personal, so the conclusions you can draw from their observations are, perhaps, more limited.

We should also consider that your understanding, as the interviewer, of what is said will be subjective. You should reflect on how your own upbringing, social and economic status may colour your analysis of their words – be aware of how you are situated.

Remember that your primary role is to listen. You will need to build a rapport with the interviewee – putting them at their ease will help the conversation flow. However, declaring your own opinions on a subject upon which the interviewee doesn't then feel able to speak freely, will obviously hamper your research.

A representative sample?

'The aim in recruiting informants for interview is not to choose a representative sample, rather to select an illustrative one'. (Gill Valentine, 2005)

A small sample of three, five or even ten interviews *won't* be representative of the population of a place the size of Great Missenden, home to more than 10 000 people. Instead, the sample you select should be informed by the theories you want to explore.

The author chose to investigate economic change in Great Missenden, specifically employment, and so selected a range of people (male and female) to interview who had different careers – some were 'white collar' professionals, some 'blue-collar' skilled tradesmen. Both business-owners and employees featured in the sample.

Part of the picture

Remember that interviewing need not be used in isolation – it can form part of a multi-method approach to a research question. Valentine (2013) uses the analogy of 'triangulation', the use of different bearings to give a correction position, in her discussion of using oral sources alongside quantitative data to draw wider conclusions about phenomena in human geography.

Getting started

Look at Figure **2**. Once you have an idea of who you'd like to interview and what you'd like to ask, you should begin the process of finding people to talk to. A local councillor, faith leader, chair of the Women's Institute may be good starting points when seeking out interviewees. In social science these people are known as **gatekeepers** because, once they understand and endorse your research project, they may be able to recommend other interviewees and could encourage others to take part.

Do	Don't
… dress up, a bit. Wear clothing that meets your interviewee's expectations of the meeting. You want them to take you and your research seriously.	… presume everyone you approach will want to be interviewed. People are busy and may not feel confident or competent to answer your questions.
… reassure them that you won't be using real names in your report. You're more likely to get their honest opinion.	… forget to be polite, avoid leading questions and avoid being judgemental. Your role is to listen and collect data for analysis.
… find somewhere quiet, where your interviewee is at ease. Have a few visual aids – photographs (Figure **1**), a map – that might to provide a focal point for the conversation.	… conduct the interview somewhere uncomfortable, noisy or where you may be interrupted. A group interview or focus group may be harder to steer and keep to the point.
… prepare and practise your interview technique. Try out your questions in advance on family and friends.	… just ask the first thing that comes to mind or allow the conversation to move away from the topic in question.
… be flexible in your approach: topics may come up that you want to investigate further; interviewees may suggest other contacts for you to meet ('snowballing') – go with it!	… stick rigidly to your list of questions; interviewees will be more interested in talking about some topics than others.
… use eye contact and body language to show you are interested in what your interviewee has to say.	… treat your interviewee as an object; build a rapport by sharing your own relevant experiences. Write a thank you letter with a brief summary of your findings.
… record the interview if they are happy to be recorded – you won't be able to write fast enough. But do take some notes as a backup.	… turn up unprepared. Take spare pens/pencils and check equipment for charge and memory capacity beforehand.

▲ *Figure 2* *Dos and Don'ts of interviewing*

Analysing and presenting your material

Once you've conducted the interviews, listen to them again. You will need to transcribe a few sections of these conversations that you think are key to giving a deeper understanding of the place under study and its changing character, in terms of either:

◆ demographic and cultural characteristics; or

◆ economic change and social inequalities.

Given the wealth of the material you are likely to have at this stage, you will need to be *selective*, similar to a news editor who only has space for one or two leading stories on the front page. While romantic, perhaps beguiling, the quotes in Figure 3 might not form the basis on which to build an analysis of changing employment patterns:

The power of a quote

Focus on key questions that produced interesting responses that give a new insight into what it is, or was, like to live in a place.

Quotes can be used to give a little local colour, illustrate a point or, conversely, refute something that, when presented in statistical form might otherwise be taken as a universal truth. Consider Figure **4**, an exchange between the author and two respondents (in their eighties) about the decline of the number of businesses in the High Street:

Dividing text up into themes and subthemes (or codes)

It can be useful to give each new theme within a conversation a title or **code** – make a list as you listen. Timings of the coded sections can also be added to your notes to make it easier to find relevant sections you want to return to. Having done this for a number of interviews, you might then choose one major theme, common to all, to unpick in further detail – it is these sections of your conversations that you will need to transcribe in full. You might select (as the author has) changing employment as a key theme, and then split the text into different subthemes (with sub-codes) that relate to particular employers. On second reading you might then subdivide further – fact versus opinion and/or personal anecdotes versus media-based knowledge (Figure **5**).

A: There were three butchers on the High Street then… and Bob Colgrove. He made lovely brawn. You don't even know what brawn is, do you?

▲ *Figure 3* Off at a tangent

B: I started working (at the greengrocers) when I was 10. We were only allowed to use the van once a week because of petrol (during the war) so I had to cycle (miles) all over… to Lee Common, Little Missenden. It was a different time.

Q: How do you think the construction of the bypass affected Great Missenden?
A: It made a big difference to us kids; we couldn't go sledging because it cut the hills in half, especially opposite the Nag's Head (pub) – that was a good one!

Q: Do you think it adversely affected businesses, particularly shops on the High Street?
A: But I don't know what Missenden would be like if it hadn't been built… You used to have everything come through, the old timber bogies come through with the big horses on their way to Amersham…

B: No, I don't think it was the bypass that reduced trade; it was just how things go. You had Tesco and all those others coming in. Life moves on…

▲ *Figure 4* The effect of transport on business transcript

▼ *Figure 5* Themes and subthemes (codes and subcodes) relating to changing employment in Great Missenden

Employment in the parish (E in)
— Agriculture (A)
— Manufacturing industry (M)
— Electro-plating and enamelling – Gerhardi's (G)
— Fact (fa)
— Experience (exp)
— Hearsay (hea)
— Media knowledge (med)
— Opinion (op)
— Building trade – Wrights Yard (W)
— Service industry (S)
— Shops and the High Street (H)
— Grocers and home delivery service (D)
— The Abbey book shop (A)

Employment out of the parish (E out)
— Local towns (T)
— London & commuters (L)

Once you have categorised your data, collate all of the comments under the relevant titles (themes and subthemes) so that you have all of the data in an organised form.

◆ How does this qualitative data relate to other data sources?

◆ Does it support or contradict your findings from other sources?

◆ How reliable are your oral sources? Did everyone say the same thing about a particular change and its effects? If not, can you suggest why not?

Great Missenden: changing patterns of employment

C: When I moved back to Great Missenden, (I found that) most of my favourite pubs were people's houses.

Great Missenden was an agricultural hub, coach stop and 'market town' up until the 1920s and 1930s, and is now a tourist honeypot and dormitory settlement. For the author, the qualitative research has proved invaluable in filling the gap between these two points in history.

So, what type of economic activity characterised this place between these time periods? Great Missenden had a significant service sector, focused in and around the High Street, including its original twelve pubs. It also had a thriving manufacturing sector including Gerhardi's – an electro-plating and stove enamelling factory, owned by a family who had fled Russia in the early twentieth century (Figure **6**). The Gerhardis moved their business from the south coast of England to Great Missenden in 1940 and supplied the aircraft industry during the war. There is now housing where the factory once stood, behind the High Street.

Wright's Yard, a major business in the local construction industry, employed almost 250 people before the Second World War and continued up to the 1980s. Today, the only business with the name 'Wright' on the High Street is a funeral director's. This local manufacturing sector has been lost to the economy which has in turn affected the job prospects for some young people in Great Missenden (Figure **7**).

D: More than one hundred young men and women did their war work there (in the Gerhardi factory). Plating needs large baths of acids… and that's not without danger. Accidents could and did happen. They were harder times than today.

A: They employed quite a lot of people (at Gerhardi's). It's hard to say exactly how many … we never took any notice. It was all girls that worked there.

B: In those new houses (where the Gerhardi factory once stood), of course, you aren't allowed to grow vegetables in the soil there now, because of all the Cadmium poisoning.

A: We (Wright's Yard) were the biggest in Missenden. We had the biggest lot of apprentices as well… and they came from all over the place.

▲ **Figure 6** Employment transcript 1

Q: Where do you think people work that live in the village today?

E: Do you know, I don't know. We've really got nowhere. Wright's were very big employers and of course Gerhardi's. I can't think of anywhere … maybe the restaurants?

A: In Missenden? You've got to run to London for work.

Q: And with the arrival of all of these commuters has the community changed?

A: We're a dying breed. All the locals are going; they're either dying off or the people who've grown up here, that are younger, well, they're having to move out…

B: My two sons moved away. They did their apprenticeships at Wright's and now one lives in Aylesbury and the other's in South Africa.

▲ **Figure 7** Employment transcript 2

ACTIVITIES

1. Read 'Objectivity and Subjectivity' on page 102.
   a. How might questions that feature in a questionnaire differ from those written for an interview?
   b. Which of the two questions suggested about HS2 might be deemed to be a leading question?
2. Look at Figures **6** and **7**.
   a. What themes can you pick out? Break down these comments into content linked to different codes (see Figure **5**).
   b. Although the unemployment rate in Great Missenden is well below the national average, what conclusions can you draw about changes in employment in Great Missenden since the mid-twentieth century?
   c. Suggest a further data source to support your conclusions to question **2b**.

In this exemplar of a distant place study you will find out about:
- the factors that have helped to create Detroit's social and economic geography
- the wide variety of available sources that may be used to investigate a distant place

Place study

'From the advent of the automotive assembly line to the Motown sound, modern techno and rock music, Detroit continues to shape both American and global culture. The city has seen many of its historic buildings renovated, and is bustling with new developments and attractions that complement its world-class museums and theatres. The city offers a myriad of things to see and do. Detroit is an exciting travel destination filled with technological advances and historic charm.'

(Extract from a visitor guide to Detroit)

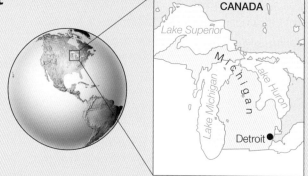

⬢ **Figure 1** *The Renaissance Center (left), Detroit, head office of General Motors (2014)*

Detroit's location and early development

The city's colonial history

Detroit is situated on the Detroit River, which links Lake Huron and Lake Erie (Figure **2**), two of North America's Great Lakes that connect to the Atlantic Ocean via the Saint Lawrence River. French colonialists founded Fort Pontchartrain du Détroit in 1701, which subsequently became Detroit, finding its physical geography to be of great advantage. The different countries that have controlled Detroit across its history are represented on the city's flag (Figure **3**).

- The bottom right and top left quarters both have 13 features (stripes or stars) to represent the first colonies of the USA.

- The five fleur-de-lis represent the French Standard.

- Britain, another colonial master of the city in the late eighteenth century, is shown by the three lions.

A transport hub

In the nineteenth century, shipping and shipbuilding brought wealth to the city. The so-called Gilded Age mansions that were built to the east and west of Detroit's downtown area demonstrated the economic rewards available in this emerging transport hub of the north-east.

⬢ **Figure 2** *Location of Detroit, USA*

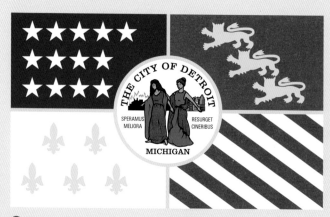

⬢ **Figure 3** *Flag of the city of Detroit*

Economic success and expansion

The Steel Belt boom

Detroit saw significant population growth in the nineteenth century, but it was not until the twentieth century that its expansion (both in terms of population and land area) finally took off. Home-grown companies manufactured new standardised consumer products for customers across the USA and, eventually, for export around the world. For example, Ford and General Motors used innovative production processes such as the assembly line that were replicated worldwide (Figures **4** and **5**).

Southern Michigan was part of a larger region within the USA known as the Steel Belt, where established waterways and canals, roads and railroads helped to connect iron ore mines with coal resources from the Appalachian Mountains. With the economic decline of the late twentieth century, this area became known as the **Rust Belt** (Figure **9**).

New workers needed

During the twentieth century, millions of African Americans travelled from the rural southern states to the urban north-east and Midwest to apply for the new jobs created in the automotive industry – this was known as the 'Great Migration'. It was driven, in part, by the racism that African Americans experienced in the southern states at this time. The Ku Klux Klan, white supremacist group, had a great deal of support in the region but the authorities also officially favoured whites. African American workers felt pushed out by the lack of economic opportunities apart from labouring on plantations.

Growth of the city and the suburbs

Detroit welcomed many African Americans into the city alongside southern and eastern Europeans. Figure **6** shows the growth in population that took place during the twentieth century and how the city's African American and white populations changed.

In the post-war period the total population of Detroit's wider **metropolitan** area grew as life became increasingly dependent upon the car, and urban areas sprawled in the original *Motor City*.

△ **Figure 4 Fordism** *at work: the Cadillac assembly line in General Motor's Detroit car plant (1954)*

△ **Figure 5** *Worker on the assembly line at Chrysler's Jefferson North Assembly Plant, Detroit (2011)*

▽ **Figure 6** *Graph of the city of Detroit's population from 1900 to 2010 (US Census)*

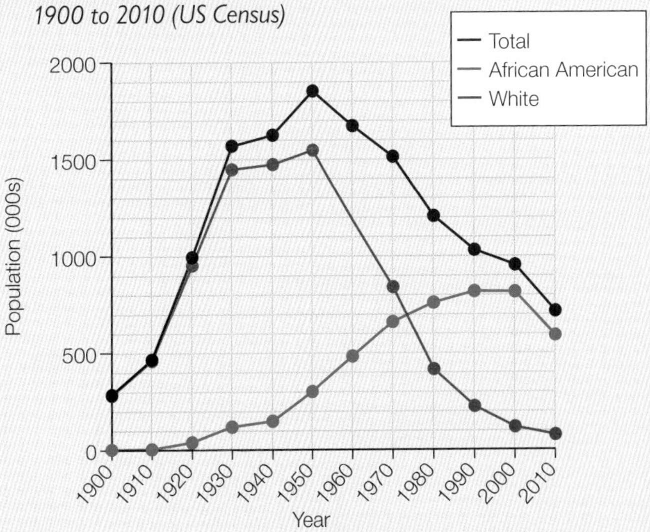

Changing world, changing fortunes
Competition and fuel insecurity

From the 1970s onwards, the economic tide was turning as international oil crises prompted drivers to buy vehicles with greater fuel economy. Competitors from Asia such as Honda, Datsun (Nissan) and Toyota produced more desirable models, causing Detroit's big employers to suffer a decline in sales and profits. Ford, General Motors and other employers linked to the automotive industry responded by cutting jobs and shutting down less efficient plants. With rising numbers of unemployed, Detroit saw a fall in taxes raised by the city authorities who, in turn, had less to invest in public services.

Forces of change or more of the same?

Despite this change in the economic climate, Detroit still backed the auto industry. Controversially, the mayor sanctioned the compulsory purchase of land to enable new hi-tech car plants to be built in the city – the General Motors Hamtramck Assembly Plant was built in an area formerly known as Poletown, so-named because of the high number of Polish people in the local community. The proposals saw not only protests against the development, but also some support from the locals.

The first model to roll off the production line at Hamtramck in 1985 was perhaps optimistically named the Cadillac *Eldorado* (after a mythical city of gold). But the city that had enjoyed great growth and wealth earlier in the century struggled to reinvigorate itself. The Ford-financed Renaissance Center (Figure **1**), completed in 1977, was an exercise in place-making: urban rebranding. It was planned as an iconic development that would attract business back into the rundown downtown district. However, it has been widely criticised for creating office and retail space that was physically disconnected and protected from the rest of the city centre (its design protected visitors from the rising tide of crime). The Renaissance Center is now the headquarters General Motors.

America's Rust Belt region

By 2008 Toyota had become not only the leading global producers in the auto industry but also the leader in global sales, overtaking General Motors. In contrast, Detroit (twinned with the city of Toyota, Japan) had become the confirmed capital of the Rust Belt (Figure **9**).

Figure 7 *A four-storey-high Uniroyal Tire advertises this auto-industry product made in Detroit. Built in 1964 and originally a Ferris wheel, the giant tyre is still an iconic Detroit landmark.*

Figure 8 *'If you come here at night it glows.' The Uniroyal tire plant closed in the 1980s, forming part of the* **drosscape** *of Detroit. In 2012, the first step in revitalising Detroit's waterfront began, including the excavation of this site to a depth of nine metres.*

Figure 9 *The USA's successful Steel Belt became its Rust Belt in the late twentieth century*

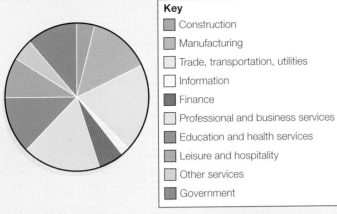

Key
- Construction
- Manufacturing
- Trade, transportation, utilities
- Information
- Finance
- Professional and business services
- Education and health services
- Leisure and hospitality
- Other services
- Government

▲ **Figure 10** Employment in Detroit by sector (2013)

Rank	Company/organisation	No. of Employees
1	Detroit Medical Center	11 497
2	City of Detriot	9591
3	Quicken Loans	9192
4	Henry Ford Health System	8807
5	Detroit Public Schools	6586
6	US Government	6308
7	Wayne State University	6023
8	Chrysler	5426
9	Blue Cross Blue Shield	5415
10	General Motors	4327

▲ **Figure 11** Top 10 employers in the cities of Detroit, Hamtramck and Highland Park (2013)

Detroit to revive: one neighbourhood at a time [2003]

Razing the city to save the city
There have been 3000 derelict homes razed to the ground already this year... [2010]

Sixty Detroit postal workers attacked by stray dogs say... [2013]

Motor City population declines 25% [2011]

Homes still selling for $1 dollar in Detroit [2012]

How Detroit became the world capital of abandoned buildings [2012]

Detroit in US largest-ever city bankruptcy [2013]

Detroit Residents plant 15 000 trees in a day [2014]

Half of Detroit property owners don't pay tax [2013]

▲ **Figure 12** Detroit news headlines

ACTIVITIES

1 Compare the opening quote about Detroit with Figure **1**. Who do you think created each image of this place and for what purpose?

(S)

2 Study Figure **6**.
   a Describe the growth of the population of Detroit from 1900 onwards.
   b Compare the growth of the number of African American people resident in the city to the growth of the total population.
   c Using the information from this section describe the contribution of two endogenous and two exogenous factors (see 2.4) to Detroit's economic and demographic boom in the first half of the twentieth century.

3 Look at Figure **10**. What were the top three employment sectors in Detroit in 2013? List them in order.

4 Study Figure **11**. In 2013, how many of Detroit's top city employers were part of the car industry? Name them.

5 Read the news headlines in Figure **12**.
   a Sort the headlines into positive and negative stories about the city.
   b Choose one headline and try to write the first paragraph of a news story using what you know about Detroit.

6 It is hoped that tourism may give Detroit an economic boost in years to come. Using the information in this section explain why tourists may or may not wish to visit Detroit.

STRETCH YOURSELF

Detroit is twinned with seven cities:

- Chongqing, China
- Dubai, UAE
- Kitwe, Zambia
- Minsk, Belarus
- Nassau, Bahamas
- Toyota, Japan
- Turin, Italy

Choose one city and find out more about it. Can you explain why city authorities made this international link and how it might have benefited Detroit in the past or will in future?

In the second part of this distant place study you will find:
- further quantitative and qualitative sources, specifically with regard to Detroit's changing demographic and cultural characteristics
- an introduction to Alter Road, a locality comparable in size to Great Missenden

Place study

Racial disintegration in Detroit

A brief history

Figure **2** lists ten key events in Detroit's history since 1900 that have shaped the city's economy as well as its social and cultural geography. A number of these events relate not only to racial integration, but also disintegration – the growing and entrenched *segregation* of white and African American residents. Among these events is the foundation of Motown, which is widely recognised as having played an important role in promoting the racial integration of music in the 1960s and 1970s (Figure **1**).

Founded in 1959, the Motown record label became another successful brand of the city of Detroit. Motown fused the predominantly 'black' soul music with the predominantly 'white' pop sound. The company signed bands such as the Supremes, the Jackson 5 and the Spinners (known as the Detroit Spinners in the UK).

However, race riots in the 1940s and 1960s demonstrated that racial harmony was not always possible in the pursuit of the 'American Dream' in Detroit. In the 1970s, city authorities were accused of supporting the racial segregation of schools and housing, reinforcing **ghettos** and the racial divide for which the city is well-known today.

▲ *Figure 1* *60s Detroit soul band, the Spinners*

1903	Henry Ford founds Ford Motor Company in Detroit.
1944	Ford, retooled for the Second World War, shows the world how to produce a B-24 bomber in an hour (or 650 a month) – a key part of Roosevelt's 'Arsenal of Democracy'.
1959	Berry Gordy Jr founds the Motown Record Company in Detroit.
1967	Race riots result in action by the National Guard to quell the disturbance. More than 7200 people are arrested and 2000 buildings destroyed over five days.
1973	Detroit elects its first African American mayor, Coleman A. Young.
1977	A major redevelopment of downtown Detroit known as the Renaissance Center opens. This seven-skyscraper office and retail complex (a 'city within a city') includes the GM headquarters.
1984	Mayor Coleman declares Halloween a 'vision of hell' as youths set 800 fires across the city.
1991	Crime rates peak across the city at 2700 violent crimes per 100 000 people.
2009	General Motors files for bankruptcy but the company survives after a government-backed reorganisation.
2013	Governor of Michigan declares a financial emergency in the city.

▲ *Figure 2* *Ten key events in Detroit's history*

▲ *Figure 3* *8 Mile was a critical and box office success; 'a rap movie masterpiece'* (Vibe *magazine*)

Mapping a segregated city

Figure **5** is a map of Detroit produced using 2010 US Census data, which shows the pattern of the predominantly African American population of the city, surrounded by the white population of Detroit's suburbs. Each dot represents 25 residents of the city: red shows the distribution of white residents, blue for African American residents, green for Asian, orange for Hispanic, yellow is for all other ethnic groups.

The distinct dividing line between red and blue areas on the north side of the city is in fact a highway – 8 Mile Road (Figure **4**). *8 Mile* is also the name of a 2002 film featuring the music artist Eminem (Figure **3**). The film is a semi-autobiographical tale: Eminem stars as an aspiring white rapper seeking to make it big in Detroit's African American music scene. Eminem's character, Jimmy 'B-Rabbit' Smith, lives north of 8 Mile Road.

⚫ **Figure 4** *Map of the districts of the city of Detroit and its surrounding suburbs*

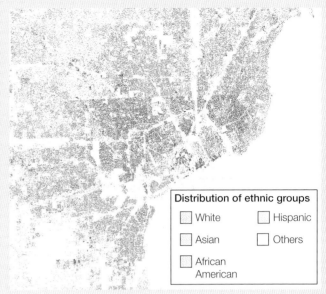

⚫ **Figure 5** *Race and ethnicity, Detroit (2010)*

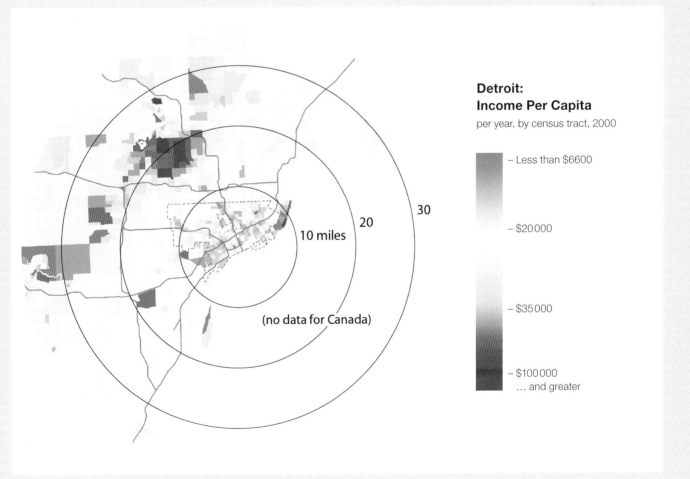

⚫ **Figure 6** *Distribution of wealth in the city of Detroit and its wider metropolitan area*

Detroit: a home for racial segregation?

Social processes create spatial patterns

White flight ...

In the second half of the twentieth century, Detroit's troubled times prompted middle-class white families to move out of the city to the suburbs of metropolitan Detroit. This migration was a choice made possible by their higher incomes, as they could afford to commute back into the centre. Eventually, many jobs followed this relocation to the suburbs: a new socio-economic, spatial pattern was created by this process of 'white flight'.

... and racial integration?

In recent decades, a similar movement of African American middle-income families from the city to the suburbs has taken place. Though fewer in number, they also sought better schools, less crime and a higher quality of life. Social commentators have debated whether or not this movement might reduce the level of segregation around, if not within, the city of Detroit.

Segregation at a local scale

However, in contrast to the integration optimists, some researchers have sounded a note of caution. They theorise that racial mixing in these outlying areas may only be temporary, with African Americans moving into older neighbourhoods that are already falling out of favour with the white middle classes:

'Much to their chagrin, many new black suburbanites found that integration was just a phase between when the first blacks moved in and the last whites took their children out of the public schools.' ('A Dream Deferred', *New York Times*, 2011).

Moreover, in affluent white districts that border the city, such as Grosse Pointe Park, there is evidence of the ways in which residents seek to maintain the divide between the city of Detroit and their homes, by using physical barriers (Figure **7**).

Detroit's other Berlin Wall: Alter Road

Not yet mapped on GPS navigation systems, unofficial barriers block the road bemusing and frustrating both drivers and pedestrians who attempt to cross Alter Road into the suburb of Grosse Pointe Park from East Side, Detroit (see Figure **4**). These barriers range from wooden fences and concrete walls (Figure **7**) to heaped Christmas trees and overgrown scrub. Like 8 Mile Road, Alter Road has become a visible demarcation line of the social divisions within Detroit.

In what appeared to be an official endorsement of the growing barrier between neighbours, the authorities of Grosse Pointe Park undertook a range of measures in 2014 to restrict movement of traffic along Kercheval Avenue, a major commercial thoroughfare that crosses Alter Road (see Figure **4**). Construction cones, reinforced with snow banked up against them, appeared earlier in the year. Although this blockade was quickly removed, a roundabout and farmers' market impeded the flow of traffic again within a matter of months.

This barrier of wooden sheds and concrete curb stones became a national story:

'(Alter Road) represents a huge dividing line separating the haves and have nots...' (*Los Angeles Times*, 2014)

Figure 7 *A wall has been erected to prevent travel on foot or by vehicle from East Side, Detroit into Grosse Pointe Park.*

News articles at the time noted the disparity between both the ethnic make-up and average wealth of the residents of these two areas:

♦ Majority ethnic group: 82 per cent of the residents of the City of Detroit are African American; 85 per cent of the residents of Grosse Pointe Park are white.

♦ Median household income: City of Detroit $26 955; Grosse Pointe Park $101 094.

A I'm aware that the homeless people I walk by on the way to work are predominantly African American, unlike the tech entrepreneurs in Madison Heights (a suburb of Detroit). I don't assume that everyone starts on a level playing field. I studied physics and view the world as a complex collection of systems; if everyone started on an equal footing then economic success should be randomly distributed, and, obviously, it is not. (Entrepreneur recently moved to Detroit)

B I remember last year, a performer from Midtown donned black face to put on a performance as Michael Jackson. The video they posted got a lot of likes, to my disgust. When I said something about it, I was told I was being overly sensitive and that there was nothing wrong with blacking up. Of course, these were the same group of people who thought it was funny to utter racial slurs as jokes at a comedy open mic night … (Detroit resident)

C I'm a Detroiter, and I believe the only way to overcome the race issue is to stop talking about it. Why should we continue to perpetuate this idea that we're different because our skin color/ethnicity/culture is different from our peers? Are whites not capable of this? Are blacks not capable of this? Are 'Latinos' not capable of this? It's time to stop telling people they're different or less important because they're black, or that they're somehow more important because they're white. Everyone is human, we all bleed red. Something like a 'racial map' only intensifies the division we already see in this great city. (Detroit resident)

D The (predominantly white) people of Grosse Pointe are likely to say that their relative wealth is the money that built Detroit. They mean it's old money, because it's where the descendants of the great auto-industry – particularly the Fords – still live. When, in fact, it was the labour of working-class Detroit that built Detroit. It's the lion's way of telling history. (Social worker, Detroit)

▲ **Figure 8** *Blogging on race and renewal in Detroit*

ACTIVITIES

1 Study Figures **4**, **5** and **6** along with associated text.

   a Describe the spatial pattern presented in Figure **5**. Use information about the layout of Detroit in Figure **4** in your description.

   b To what extent does the pattern you have described in your answer to activity 1a correlate with the spatial pattern of different income groups presented in Figure **6**?

   c What are the strengths and possible weaknesses of Figures **5** and **6** as sources of data for a place study of Detroit?

2 Find the lyrics of 'Lose Yourself', a rap co-written by Eminem for the film *8 Mile*. What can you infer from the words about the quality of life for some people growing up in the city of Detroit or just beyond its northern border?

3 Why is Eminem's music seen by many people as an important force of change with regard to racial integration in Detroit, and the USA more generally?

4 Read the Alter Road example.

   a Outline the reasons why, in 2014, a farmers' market was built blocking Alter Road at a major intersection.

   b Explain why these sheds made news headlines, nationally and internationally.

5 Using online map resources (such as Google Earth) explore the areas just either side of Alter Road (a trip along Kercheval Avenue is instructive). Refer to Figure **4**.

   a How does the district of East Side, Detroit differ from the suburb of Grosse Pointe Park in terms of the quality of the built environment?

   b How do these observations inform your understanding of Alter Road as a place?

6 Read the different blog comments about race and renewal in Detroit in Figure **8**.

   a Make a list of issues raised that relate to race.

   b How and why do the opinions of these bloggers on racial issues differ?

   c Why might a 'racial map' of Detroit intensify the divisions already seen within this place, as blogger **C** notes?

STRETCH YOURSELF

How might the classification of people as *insiders* or *outsiders* be used to describe or understand the situation of African American or white residents of the city of Detroit and those of its wider suburbs today?

Now practise…

The following are sample practice questions for the Changing places chapter.

They have been written to reflect the assessment objectives of Component 2: Human geography Section B of your A Level, and Component 2: Human geography and fieldwork investigation Section A of your AS Level.

These questions will test your ability to:

◆ demonstrate knowledge and understanding of places, environments, concepts, processes, interactions and change, at a variety of scales [AO1]

◆ apply knowledge and understanding in different contexts to interpret, analyse and evaluate geographical information and issues [AO2]

◆ use a variety of quantitative and qualitative and fieldwork skills [AO3].

1   Study Figure **1**. Critically assess the link between identity and sense of place proposed in this article. (6)

You could say theirs is the *Generation of Three E's*. There is *Erasmus*, the European Union program that organises and subsidises student exchanges among universities across its 28 countries and elsewhere. There is *EasyJet*, the budget airline that lets them hop between European cities as simply and cheaply as it can be to trek across town. And there is the *Euro*, the currency used in most of the member countries.

Young adults are now grappling with what Britain's vote to exit the European Union means for their profoundly European way of life. For them, it is perfectly normal to grow up in one country, study in another, work in a third, share a flat with people who have different passports and partner up without regard to nationality.

'It means that we are not going to be sisters and brothers of a big project,' said Antoine, 24, a Frenchman whose résumé and network of friends provide a crash course in European geography.

'At best, we are going to be allies' — friends, but no longer family. 'Britain feels less like home.'

(Adapted from an article in the *New York Times*, July 2016)

🔺 *Figure 1* *'Brexit' Bats Aside Younger Generation's European Identity*

2   'Today, media places are as real as experienced places for many.' Evaluate this statement. (6)

3   With reference to a named place or places, outline attempts by local or community groups to reshape place-meaning. (4)

4 Explain why some people might feel like outsiders in a locality you have studied. (4)

5 'In the age of globalisation, can we really say that every place is unique?' Examine evidence for and against this argument. (20)

6 Study Figures **2** and **3**. Suggest reasons for any differences that you observe between the images of Glasgow that each figure presents. (6)

Overwhelmingly, the number one response from contributors [to the public consultation] was that it's the people of Glasgow that make the city great, which, ultimately, led to the development of 'People Make Glasgow' as the city's new brand.

One of the real strengths of this brand is its flexibility. It can be adapted for different audiences and sectors to highlight Glasgow's key strengths and reflect the city's increasingly diverse economy.

'People Make Glasgow' will provide a platform to show the world Glasgow's strengths and act to attract investment, growth and opportunity to our city. (Adapted from an article by the official Glasgow City Marketing Bureau, published in The Drum, November 2013)

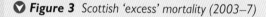
Figure 2 *It's the people that make it: how Glasgow positioned itself as a dynamic destination*

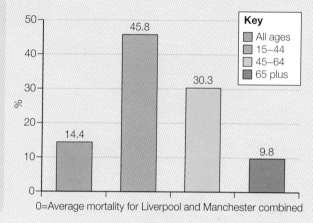
Figure 3 *Scottish 'excess' mortality (2003–7)*

0=Average mortality for Liverpool and Manchester combined

7 Explain why some places are perceived as being 'near' while others are perceived as being 'far'. (4)

8 Summarise the factors, relationships and connections that have shaped the character of a place that you have studied. (4)

9 'People shape places and places shape people.' Examine such processes with reference to change in your local place. (6)

10 Explain the importance of attachment to place in human development. (4)

11 'What is the best way to find out about people's lived experience of a place? Answer: Participant observation, it's living it: walking in their shoes, working their hours and drinking their coffee.'
Evaluate the techniques that you used to undertake one or more of your place studies. (6)

12 'The past might be a foreign country but it continues to shape the one you live in.'
With reference to a locality you have studied to what extent do you agree with this statement? (6)

13 For a place you have studied that has successfully been rebranded, evaluate the strategies used in the place-making process. (20)

3 Contemporary urban environments

Slum housing in the Tondo district of Manila, the capital of the Philippines. The Philippine economy is expected to continue to grow rapidly. How much will the effects of this growth trickle down to places like Tondo?

Your exam

 'Contemporary urban environments' is an optional topic. You must answer one question in Section C of Component 2: Human geography, from a choice of three: Contemporary urban environments or Population and the environment or Resource security. Component 2 makes up 40% of your A Level.

 'Contemporary urban environments' is an optional topic. You must answer one question in Section B of Component 1, Physical geography and people and the environment, from a choice of two: Contemporary urban environments or Hazards. Component 1 makes up 50% of your AS Level.

Your key skills in this chapter

 In order to become a good geographer you need to develop key geographical skills. In this chapter, whenever you see the skills icon you will practise a range of quantitative and relevant qualitative skills, within the theme of 'contemporary urban environments'. Examples of these skills are:

◆ Using atlases and other map sources 3.1, 3.5

◆ Interpreting digital imagery and remotely sensed images 3.7, 3.14

◆ Presenting data and interpreting graphs 3.6, 3.12, 3.14

◆ Analysing quantitative and geospatial data, including applying statistical skills 3.2, 3.3, 3.7, 3.15

Fieldwork opportunities

It's easy to forget the way physical processes continue to operate in our towns and cities. Urban locations might appear to be entirely shaped by people. However, the topics of urban climate (heat islands, air quality) and urban drainage (catchments, flooding and conservation) allow us to refocus on the inescapable physical parameters of the places where we live and the need to monitor our impact on them. Of course, the way we humans relate to each other in urban areas is also a valid topic for investigation. Does urban form itself have a role to play in shaping those interactions?

1 Monitoring the urban microclimate

Either as a desk-based exercise or in the field, monitor the climate of a small area within your town or city. Find out how the climatic conditions contrast with a nearby rural location. Are the differences in temperature between rural and urban greater in the day than at night? Which location experiences more precipitation? Can you suggest reasons for this variation? You will need to establish the exact site of the weather stations (if using online data rather than your own) and visit them, if only virtually, to consider local factors that might affect the weather. At a very local scale, could you monitor wind speeds experienced on different sides of, or at different distances from, a tall building such as a shopping centre, office block or apartment block?

2 Using interviews or questionnaires to investigate a local planning issue

Is your local urban area facing an issue such as what to do with its rubbish or how to sort out its ailing transport network? Find out about local opinion on landfill versus incineration or what people think their town needs to do in order to sort out the morning rush-hour. How do these opinions compare with current urban planning policy in your local area? For a questionnaire, think about sample design before you start to ensure you question a representative sample of people who live and/or work there.

3 An urban transect of land use, with a difference

Travel north to south or east to west across an urban area logging changes in land use in order to assess whether zones of similar businesses or housing exist and, if so, whether they are laid out across the town or city in a pattern predicted by models and urban theory. While you are there, why not take a photograph every tenth step or a shot of the most interesting aspects of the street scene. You could create a film of what it's like to walk that way and give others a sense of place. Using online resources, take a look at the Urban Earth project by 'guerrilla geographers' for inspiration.

In this section you will learn about urbanisation, its importance in human affairs and global patterns of urbanisation since 1945

The rise of the city!

In 1945, less than one-third of the world's population lived in cities or urban areas. By 2008, more people lived in urban areas than in rural areas and this proportion is expected to rise to around two-thirds by 2030. In the UK, for example, urban life is now a reality for nearly 85 per cent of the population. This unprecedented move from a rural to an urban society has enormous potential to bring positive change but it also carries risk.

Urbanisation and urban growth

Do you live in an urban area? Looking out of your bedroom window you might think that this is a straightforward question. A town or city is an urban area – but today's city boundaries are far from clear. You might live in the extended suburbs far from the shopping or commercial centre (CBD) and still officially live within an urban or metropolitan area. The metropolitan county of Greater London constitutes nearly all of Middlesex, parts of Kent, Essex, Hertfordshire and Surrey.

Note the difference between the following terms:

◆ *Urban growth* – the increase in the total population of a town or city.

◆ *Urbanisation* – the increase in the proportion of the population living in urban centres.

◆ *Urban expansion* – the increase in size or geographical footprint of a city.

A city may experience urban growth and expansion, but if this growth is matched by population increases in rural areas, then urbanisation is not occurring.

The importance of urban centres in human affairs

Cities are not only important as centres of population but they also influence and shape our lives at every level. They are important for:

◆ the organisation of economic production, for example, concentration of financial services

◆ the exchange of ideas and creative thinking, for example, universities

◆ social and cultural centres, for example, theatres and national stadia

◆ centres of political power and decision-making, for example, seat of government.

Look at Figure **1**. This shows a number of important features of an economy and the percentage of each feature that is concentrated in some cities.

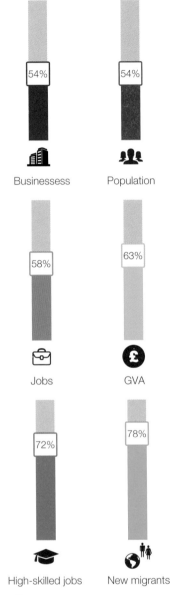

Businessess 54%
Population 54%
Jobs 58%
GVA 63%
High-skilled jobs 72%
New migrants 78%

🔺 **Figure 1** *Some economic features of the 64 largest cities in the UK*

Did you know?
The city status of Maza in North Dakota was removed by US authorities in 2002. It had a population of five!

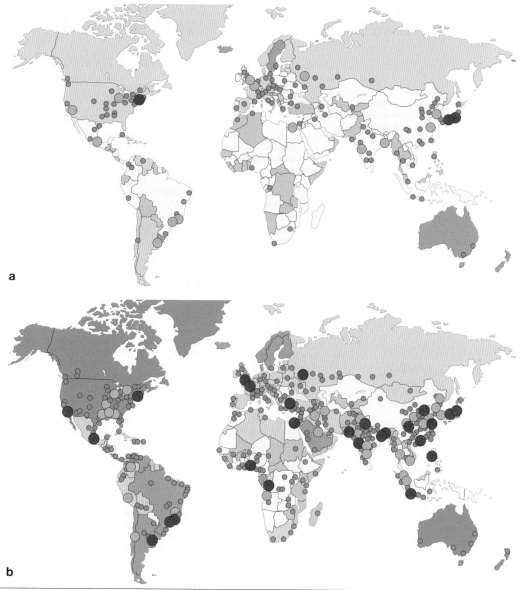

a

b

Percentage of urban area population

■ 81–100	▨ 21–40
▨ 61–80	□ 0–20
41–60	

City population (millions)

● ≥10
1≥ ● >10
1≥ ● >5

▲ **Figure 2** *Urban growth and urbanisation* **a**) *1970* **b**) *2014. The pace of urbanisation has changed the face of our planet over the last 50 years.*

ACTIVITIES

1 a With reference to named places, explain the difference between urbanisation and urban growth.

b Why are HDEs likely to continue to see urban growth but reducing rates of urbanisation?

2 Study Figure **1**. Why do you think medium-sized cities are sometimes more attractive for migrants than either the largest cities or smaller towns?

Ⓢ

3 Study Figure **2**.

a Compare and comment on the percentage urban population 1970–2014.

b Describe the location of the world's largest cities in 1970. How has this pattern changed by 2014?

STRETCH YOURSELF

Look at the United Nations website (esa.un.org/unpd/wup/Maps) and look at the 'Percentage urban and urban agglomerations by size class' map. Describe and suggest reasons for how the pattern has changed on the 1990 and 2030 maps.

In this section you will learn about urbanisation, suburbanisation, counterurbanisation and urban resurgence

One step forward, one step back...

If you blink you might miss it. There are around 1.3 net additions to the worldwide urban population every second – but where do they all live? Like an ecosystem, urban metropolises have expanded and adapted to meet the changing needs of their populations. All cities are different but they do broadly follow four stages of growth and change.

Look at Figure **1**. It shows the four main forms of urbanisation (sometimes also referred to as the cycle of urbanisation). In most cities in HDEs, all four processes are taking place at the same time, although it is likely that for a period of time one process will dominate (Figure **2**). In LDEs, urbanisation continues to be the main urban process but many Asian cities are already beginning to show the effects of suburbanisation and even counterurbanisation.

The nature of urbanisation

Urbanisation is the process of change by which places become more urban. This simple definition hides a multi-strand process that is dynamic across space and time and is conditioned, in part, by the level of development of a country or region. For example, demographic changes (the increasing percentage of urban population) may be linked to structural changes (such as the development of industrial capitalism) and changes in social attitudes and behaviours (such as use of social media). Studies of the experiences of HDE urban areas would suggest that three processes are related to urbanisation – suburbanisation, counterurbanisation and urban resurgence. While it is helpful (and usual) to consider all four processes as a sequence (see Figure **1**), it is also important to appreciate that they may also occur concurrently, 'out of order' and are not necessarily a feature of all cities in all places. For further research on the generalised experiences of urbanisation in countries around the world, 'the urbanisation curve' is a useful starting point.

🔽 **Figure 1** *The four processes or cycle of urbanisation for a typical western European city*

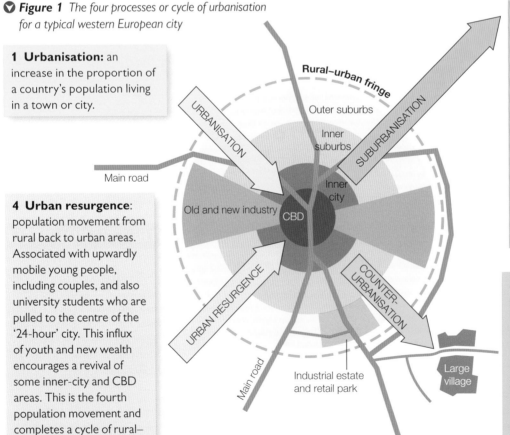

1 Urbanisation: an increase in the proportion of a country's population living in a town or city.

4 Urban resurgence: population movement from rural back to urban areas. Associated with upwardly mobile young people, including couples, and also university students who are pulled to the centre of the '24-hour' city. This influx of youth and new wealth encourages a revival of some inner-city and CBD areas. This is the fourth population movement and completes a cycle of rural–urban movements.

2 Suburbanisation: the **decentralisation** of people, employment and services towards the edges of an urban area. This outward growth of lower density urban development, or urban sprawl, is closely linked to the development of transport networks, particularly roads and in London the extension of the underground network.

3 Counterurbanisation: population movement from large urban areas to smaller urban settlements and rural areas. People move as a combined result of the *push* problems of the city (e.g. crime, congestion, land degradation) and the *pull* of rural life (e.g. bigger living space, 'safer' environment).

In the 1980s, and 90s the most dominant form of urbanisation in UK cities was counterurbanisation. This followed the suburbanisation of the 1960s and 70s when increasing car ownership allowed the population to be more mobile. More recently, many UK cities have experienced reurbanisation or an urban resurgence. Some, such as Sunderland, have struggled to do so, and there is very little evidence of resurgence.

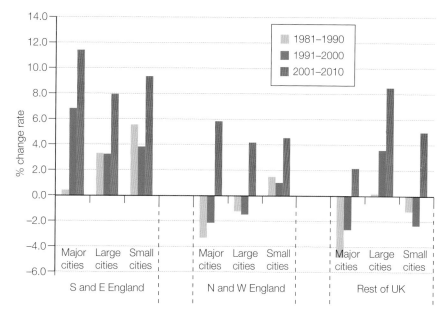

▶ **Figure 2** *Population change rate, by region, for the UK's three city sizes 1981–2010.*
Major cities: *London, Birmingham, Leeds, Liverpool, Manchester, Newcastle, Sheffield, Belfast, Glasgow*
Large city: *population 275 000+*
Small city: *population 75 000–125 000*

Urban resurgence: Ubisoft, Montreal

Ubisoft is a large computer-animation firm that decided to locate in the Mile-End neighbourhood centre of Montreal, Canada. As a high tech industry, it has all the characteristics of a footloose industry but chose to locate in an old textile factory in the downtown district of Canada's second largest city. Ubisoft argue that their employees work at all hours and it was important that they located in the centre of a 24-hour city where employees had good access to shops and services.

Many of their 2700 employees are recruited from the nearby McGill University and École de Technologie Supérieure. Ubisoft argue that they do not want a long commute – many walk or cycle to work from city centre neighbourhoods.

Ubisoft, unlike the manufacturing industries of the last century, needs relatively little floor space – the power engines of this industry are desk-sized laptops or PCs – and can therefore afford the higher downtown land prices.

ACTIVITIES

1   Around 13 per cent of the land area in England is designated as Green Belt. The purpose of Green Belt policy is to prevent urban sprawl by keeping land 'permanently open'.
   a Where on Figure **1** would Green Belt be marked? Explain your answer.
   b Which of the urban processes would Green Belt encourage? Conversely, which would it act as a barrier against? Justify your answer.

Ⓢ
2   Study Figure **2**.
   a Compare the urban population change rate for south and east England with north and west England.
   b Suggest reasons why large cities in the rest of the UK are tending to attract a far greater percentage of population than major cities.
   c Explain why the urban population growth rate for the south and east of England is significantly higher than the rest of the country.

STRETCH YOURSELF

Champion and Townsend (1990) discussed a so-called 'north–south drift' and 'urban–rural shift'. In other words, the rural south was the most dynamic quarter and the urban north the least. Discuss the validity of this viewpoint today. (The UK Government's Foresight Future of Cities Project is a good starting point for research.)

In this section you will learn about the emergence of megacities and world cities

If you travel through Shinjuku Station in Tokyo, prepare to be squeezed! It is the world's busiest station with an average of 3.64 million people passing through daily. These people are 'helped' onto the train by *Oshiya*, people who are employed to literally push passengers into the crowded trains!

Megacities

Some cities keep on growing. In 1955, Tokyo-Yokohama became the second city in history to record a population greater than 10 million. Today, at over 5000 square miles, the Greater Tokyo area is the second largest on the planet in terms of size of urban land mass. In terms of population, it is officially the world's largest city with a staggering 38 million people. Such city regions or agglomerations with populations of over 10 million people are referred to as *megacities* and are at the top of the urban hierarchy (Figure **1**).

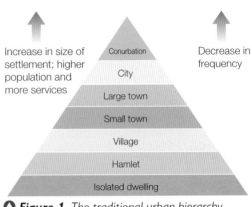

▲ **Figure 1** *The traditional urban hierarchy*

Where are all the megacities?

Look again at Figure **1** in 3.1. In 1970, only three megacities existed. By 2014, this number had increased to 28.

Now look at Figure **2**. In comparison to the relative slowing down of urban population growth in HDEs, urbanisation in EMEs and LDEs has been growing at an increasing rate. The United Nations forecasts a 1.1 billion increase in urban population by 2030. Approximately 95 per cent of this growth is expected to be in LDEs and EMEs by further expansion of megacities, although the growth of smaller cities down the urban hierarchy is also likely to be significant. For example, China has 19 megacities but the highest growth rates in the last decade have been in the medium-sized cities (1–5 million people).

▼ **Figure 2** *Distribution of world population by settlement size and development grouping*

Development grouping	Type of settlement and number of inhabitants of urban settlement	Total population (in millions)			Percentage distribution			Growth rate (percentage)	
		1975	2000	2015	1975	2000	2015	1975–2000	2000–15
HDEs	10 million or more	36	67	69	3.4	5.7	5.7	2.5	0.2
	5 million to 10 million	62	45	51	5.9	3.8	4.2	−1.3	0.8
	1 million to 5 million	145	219	250	13.9	18.5	20.6	1.6	0.9
	500 000 to 1 million	69	91	96	6.6	7.6	7.9	1.1	0.4
	Fewer than 500 000	422	481	503	40.2	40.5	41.4	0.5	0.3
	Rural areas	315	285	246	30.0	24.0	20.3	−0.4	−1.0
	Total population	1048	1188	1214	100.0	100.0	100.0	0.5	0.1
EMEs	10 million or more	33	195	306	1.1	4.0	5.1	7.1	3.0
	5 million to 10 million	64	110	197	2.1	2.3	3.3	2.1	3.9
	1 million to 5 million	182	485	756	6.0	10.0	12.7	3.9	3.0
	500 000 to 1 million	106	209	277	3.5	4.3	4.7	2.7	1.9
	Fewer than 500 000	425	943	1314	14.0	19.4	22.1	3.2	2.2
	Rural areas	2217	2925	3091	73.2	60.1	52.0	1.1	0.4
	Total population	3026	4867	5940	100.0	100.0	100.0	1.9	1.3
LDEs	10 million or more	0	12	21	0.0	1.9	2.3	–	3.6
	5 million to 10 million	0	5	32	0.0	0.8	3.5	–	12.2
	1 million to 5 million	6	44	84	1.6	6.8	9.3	8.2	4.3
	500 000 to 1 million	6	15	17	1.7	2.3	1.9	3.8	1.0
	Fewer than 500 000	39	91	162	11.1	14.1	18.0	3.4	3.9
	Rural areas	298	477	586	85.6	74.0	64.9	1.9	1.4
	Total population	348	645	902	100.0	100.0	100.0	2.5	2.2

World cities

Look at Figure **4**. While the growth of megacities is increasingly associated with LDEs and EMEs, the majority of **world cities** are found in HDEs. These cities are not necessarily the largest in terms of population (not all have populations of greater than 10 million) but are disproportionately important in the global economy. It is through these cities, with their capital, knowledge, expertise and political stability that many aspects of the global economy are channelled.

The Globalisation and World Cities Research Network (GaWC) have looked at the economic interconnectedness of global cities – how cities are linked economically to other cities. The most interconnected cities, such as London and New York, are given the highest ranks and are called *Alpha* cities (see Figure **4**). *Beta* cities occupy a second tier of rank and *Gamma* cities are in the third grouping. This economic grouping of cities allows relatively small, populated cities to feature, providing that they have sufficient services so as not to be overly dependent on larger world cities (e.g. traditional centres of manufacturing such as Auckland and Sao Paulo).

⬥ **Figure 3** *At just 2 m², Tokyo Capsule rooms (or Pods) are the world's smallest hotel rooms but they are cheap. In the 1980s, the 3.4 km² of Imperial Palace grounds in of Tokyo was said to be worth more than all the real estate in California!*

⬥ **Figure 4** *Distribution of GaWC-ranked world cities and the links between them; the deeper the red shading the greater the interconnectivity*

ACTIVITIES

1   Why do cities keep on growing?

2   What is the difference between a megacity and world city? Is it inevitable that a megacity will develop into a world city?

3   Study Figure **4**.
   a   Using named examples, describe how world cities are interconnected.
   b   Why are the majority of the world cities still located in HDEs?

STRETCH YOURSELF

Refer to *Demographia World Urban Areas*, an annual online report for urban areas with a population of more than 500 000, to write a report on world urbanisation. Make particular reference to megacities and world cities. Include in your answer:

• distribution of megacities and urban areas by size

• changes in population size, density and geographical area

• possible reasons for distributions described above.

In this section you will learn about the role of world cities in global and regional economies

World cities and economic growth

In many ways, the countryside might be a more pleasant place in which to live, but it is in cities where most economic activity takes place. World cities drive regional, national and global economies, support prosperity and create jobs. Interconnecting world cities act as funnels for economic growth (Figure **1**). This growth then flows to other regions or cities that, in turn, act as centres of further economic growth and so on.

However, not all world cities are growing – for example, New York's population has actually been declining in recent years as the so-called middle classes have moved to the suburbs and beyond. Technology has also played a role as a counter-force to urbanisation. Other cities, particularly in LDEs where urbanisation is occurring at its most rapid, are unable to stimulate economic growth. This is partly a result of underinvestment in **infrastructure**, but also reflects a lack of skilled workers and entrepreneurs.

What makes world cities so productive?

As a general rule, city productivity increases with city size. This is because more able, creative and educated workers are pulled into cities and also because the productivity of an individual increases with city size. This is all part of what economists call **agglomeration economies** or the idea that the presence of many people, services, industry and so on in a small area has productivity gains. For example, in densely populated areas/cities there are reduced costs for moving goods, labour can be pooled between different firms and ideas flow more quickly. However, world cities particularly succeed because of good governance structures. Where fragmentation of administration exists, such as in some megacities, barriers to productivity emerge, such as lack of coordination and management of transport networks, land use and infrastructure. (The movement of activity, usually industry, away from agglomerations is known as *deglomeration*).

❤ *Figure 1* World cities are 'hubs' or funnels of economic activity

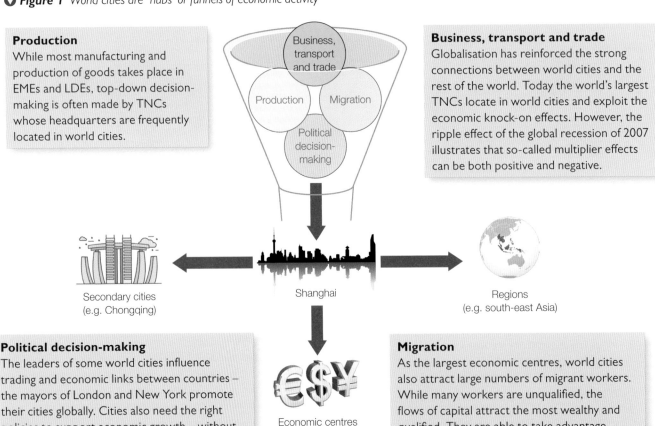

Production
While most manufacturing and production of goods takes place in EMEs and LDEs, top-down decision-making is often made by TNCs whose headquarters are frequently located in world cities.

Business, transport and trade
Globalisation has reinforced the strong connections between world cities and the rest of the world. Today the world's largest TNCs locate in world cities and exploit the economic knock-on effects. However, the ripple effect of the global recession of 2007 illustrates that so-called multiplier effects can be both positive and negative.

Secondary cities (e.g. Chongqing)

Shanghai

Regions (e.g. south-east Asia)

Economic centres (e.g. Shenzhen Special Economic Zone)

Political decision-making
The leaders of some world cities influence trading and economic links between countries – the mayors of London and New York promote their cities globally. Cities also need the right policies to support economic growth – without resolving transport and crime issues, addressing crime or promoting enterprise and education, a city is unlikely to succeed.

Migration
As the largest economic centres, world cities also attract large numbers of migrant workers. While many workers are unqualified, the flows of capital attract the most wealthy and qualified. They are able to take advantage of the globalised flows of information and communication.

Shanghai

Shanghai is situated on the east coast of China and is a world city that benefits economically from its large size. There are a number of factors that help to explain its economic success:

- *Migration* – Shanghai has more than 100 000 graduates per year from 60 higher education institutions. Consequently, over one-quarter of the city's labour force has a college education. This pool of enterprise and skills is added to each year from overseas workers and students.

- *Production* – with the creation of several more open cities along the Yangtze River as far as Chongqing, there is now a line of production centres west of Shanghai. These cities support Shanghai's export-oriented economy as well as fast becoming export centres in their own right.

- *Political decision-making* – in 1984, the Chinese government set up Shanghai as one of 14 open cities. As a result, economic and technological development zones were established and there has been huge inward investment. In 1990, Pudong New Zone, on the eastern side of Shanghai, was created. What was once impoverished swampy farmland is now home to China's financial and commercial centre (Figure **2**).

- *Business* – with favourable terms offered to overseas companies, Shanghai attracts a disproportionate amount of overseas investment. This effect is exaggerated as risks are perceived to be less if there are already foreign companies in an area. This has resulted in Shanghai's continued growth at the expense of some of the smaller cities.

▲ **Figure 2** *Panoramic view of Pudong, Shanghai*

Synoptic thinking

Geographers are often required to think *synoptically* – in other words to make links between different themes or content. Whilst Human, Physical and Environmental Geography all have an identity, the best geographers will make connections between all three. For example, spend a little time considering how one chapter of this book relates to another – this chapter and Global systems and global governance (Chapter 1) are two good places to start!

ACTIVITIES

1 Why are world cities so important to the global economy?

2 Suggest reasons why education is so important to the future of cities, particularly in LDEs.

3 Study Figure **1**. Explain why world cities in HDEs are more likely to be economically successful even though they are not the most populated.

4 Study Figure **2** and also Figure **1** in section 2.5. Suggest reasons for the changes in land use.

5 'Big is best.' Discuss with reference to the role of world cities in global and regional economies.

STRETCH YOURSELF

Is the growth of world cities a cause or an effect of globalisation? Suggest reasons for your answer. Try and show synoptic reasoning in your response.

In this section you will learn about political, economic, technological, social and demographic processes associated with urbanisation and urban growth of Bengaluru

The growth of Bengaluru

Is it possible for a city to grow too quickly? Bengaluru (Bangalore) is one of India's fastest-growing cities. In the early 1970s, it was a relatively small, sustainable and green city. It is now the Indian equivalent of Silicon Valley with its focus on technology and the **knowledge economy**. In that time, the economic as well as physical growth of Bengaluru has been dramatic (Figure **1**). New economic opportunities have been created for a growing so-called middle class, but the city has also pulled poorer migrants in from the surrounding countryside and provided a foothold on the employment ladder.

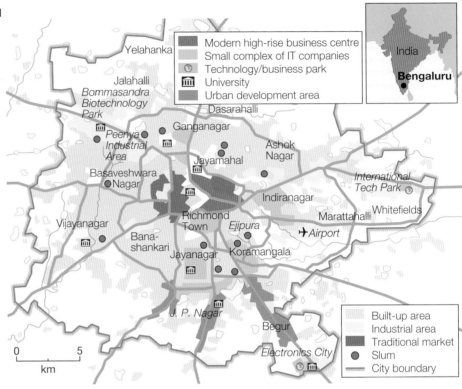

▲ **Figure 1** Land use in Bengaluru

Political processes

In the 1970s, the far-sighted state government set aside a large piece of land outside Bengaluru for a high tech business park – 'Electronics City' (Figure **2**). Texas Instruments was the first large TNC to move in, but it was not until the 1990s that the IT and associated industries rapidly expanded. The government enticed overseas firms with the promise of low taxes, relaxed licensing laws and ending of limits on currency conversions. Furthermore, by investing in education, transport infrastructure, urban development areas and new suburban housing developments, the government has been proactive in tackling some of the urban problems.

Economic processes

Bengaluru has spearheaded India's drive into a new globalised economy, an economy that relies on people and their skills rather than manufactured goods. It has been described as the largest job-creating city in India and is also the capital of aeronautical, automotive, biotechnology, electronics and defence industries.

Meeting the needs of high tech and corporate workers is the informal sector. This includes low-wage local businesses, from clothing manufacturers to teams of uniformed employees working in the private health clubs or executive homes of the elite.

▼ **Figure 2** Electronics City (or E City) was established by Karnataka State Electronics Development Corporation (Keonics) in 1978. Infosys moved to E City in 1994. Both are now hugely successful multinational companies.

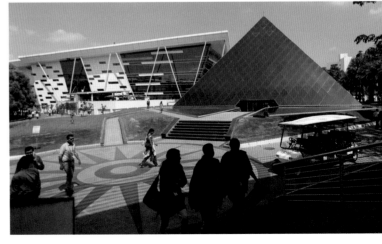

Social processes

Look at Figure **3**. The **trickle-down effects** of urbanisation have not benefited everyone. There is a growing divide between the so-called middle class and those at the bottom of the social class or *caste* system – the *Dalits*. As more skilled labour is pulled from outside Bengaluru, there is increasing pressure on housing causing a change in traditional housing patterns. Within the centre, the traditional, colonial properties, such as in Richmond Town, are being converted into exclusive homes, increasing rents and forcing inhabitants to move out. Conversely some of the wealthy migrate to the 'safe' and modern, gated suburbs such as Indiranagar. Slums are now a permanent feature of urban life, occupying marginal land such as rubbish dumps or alongside railways. Less than half of these slum dwellers have access to sanitation, clean water or electricity.

▲ **Figure 3** *Slums next to Infosys building in E City*

Technological processes

Technological development is both a cause and effect of urban growth in Bengaluru. The city's 200 engineering colleges and many universities provide a highly-skilled workforce which, since the 1960s, has supplied an educated, skilled and cheap workforce for India's defence and space research industries (average wages are 10 per cent of those in London). This has encouraged companies from overseas to invest in Bengaluru – particularly those within the knowledge economy, such as banking, finance, insurance, software development and call centres, where wages tend to be the largest cost. These companies often reinvest in their workforce, for example the upskilling of employees in IT-specific skills, which benefits the local economy. There has also been a growth in the number of home-grown IT-based companies, such as Infosys and Keonics. Furthermore, Bengaluru's airport is the third busiest in the country with an increasing number of flights to Europe and other HDEs.

Demographic processes

Bengaluru is growing at a dramatic, and some would argue, unsustainable rate. The population was 10.8 million in 2015, more than double that of 2001. Much of this increase can be explained by the influx of economic migrants from elsewhere in India and, increasingly, from foreign countries, including migrant workers from HDEs. Given the youthful age structure of Bengaluru's many migrants, population momentum will almost guarantee continued population growth for several decades. This places continuing pressures on urban infrastructure and services.

Did you know?
The built-up area of Bengaluru is now six times larger than in the 1970s. During the same period, vegetation has declined by 66 per cent and water bodies by 74 per cent.

ACTIVITIES

1 How far do the processes driving the urbanisation of Bengaluru interact with and reinforce each other? Include specific examples in your answer.

2 Study Figure **1**. To what extent is there a (spatial) relationship between location of slums and places of work? Illustrate by reference to named places.

3 Look again at the example of a megacity, Shanghai in section 3.4. Which process(es) are most associated with the growth of Shanghai?

STRETCH YOURSELF

How do we measure well-being in contrasting urban areas? Research the 'Better Life Initiative' (OECD) and comment on well-being in cities within newly emerging economies (EMEs). Are we too quick to judge 'quality of life' in cities in EMEs and LDEs?

In this section you will learn about urban change: deindustrialisation, decentralisation and the rise of the service economy

All change in the city

Are cities living systems? It may be argued that cities are a result of all the dynamic interactions of the people who live there and that they have observable and predictable characteristics – just as other living species. Furthermore, the concept of 'urban metabolism' views cities as living entities that consume energy, food, water and other raw materials and release wastes. This raises the question of what conditions are necessary for cities to thrive. How have cities changed, and how will they continue to evolve in the face of increasing human pressures?

Deindustrialisation

Deindustrialisation is the long-term decline of a country's manufacturing and heavy industries. Economic geographers have identified a number of factors common to all industrialised countries which may explain the deindustrialisation trend seen after the Second World War:

◆ *Reduced need for labour* – rapid mechanisation and application of new technologies increased productivity per worker and led to a labour surplus.

◆ *Reduced demand* – as household incomes increased, individuals tended to prefer to spend disposable income on services rather than on manufactured goods. This lowered demand for manufactured goods and, as a result, led to reduced productivity in the manufacturing sector.

◆ *Globalisation of manufacturing* (see Chapter 1) – resulted in an increasing number of multinationals outsourcing labour to LDEs, while the emergence of EMEs introduced new (and cheaper) competition to the global market place. Employing large numbers of workers in manufacturing industries in HDEs no longer made economic sense.

◆ *Increased costs* – including costs of raw materials and those associated with political decision-making, such as removal of a trade subsidy (protectionism) and tighter environmental restrictions (such as clear air acts).

Since traditional manufacturing industries were labour intensive, they tended to locate close to centres of population. Consequently, deindustrialisation adversely affected towns and cities where older industries were established. In the UK between 1951 and 1981, two million manufacturing jobs were lost in inner-city areas alone, which led to significant socio-economic and environmental problems. (see Think about).

Did you know?

The prefix 'de' is from Latin and is added to verbs (and their derivatives) to denote removal or reversal.

Think about

Why was deindustrialisation so severe in the UK?

The rapid collapse of the UK's manufacturing since the 1950s was due to:

• high levels of protectionism that did not encourage UK manufacturing industries to become more efficient

• trade unions being resistant to changes in industrial practices such as the introduction of piece rates (being paid for the output, rather than the hours worked)

• UK investors continuing to heavily invest abroad where new opportunities offered higher returns on capital investment

• outdated plants and machinery as a result of the UK's early industrialisation

• unfavourable exchange rates resulting from oil discoveries and monetary and fiscal policies.

Decentralisation

Within urban areas, competition for space combined with socio-economic and environmental push factors, are increasingly forcing businesses to move to locations beyond the CBD and inner city. In particular, the pull of the outer suburbs and rural–urban fringe and beyond offer significant advantages, such as custom-built, accessible industrial estates, business parks and out-of-town shopping developments.

On a larger scale, the negative image of cities encourages the movement of urban populations to smaller settlements further down the urban hierarchy. This process is also known as deconcentration or counterurbanisation.

Rise of service economy

Look at Figure **1**. In 1841, one-third of working people (33 per cent) in Great Britain worked in service industries. By 2011, this figure had risen to four-fifths (81 per cent) in England and Wales. Such a structural shift in employment in HDEs was prompted by:

◆ growth in corporate headquarters – a need for multinationals to coordinate their increasingly dispersed activities

◆ rise of the knowledge economy – development of producer services, such as management consultancy, legal and financial services and advertising and IT, to meet the needs of other businesses

◆ growth in research and development

◆ expanding consumer demand for services related to leisure and quality of life

◆ property development – including the redevelopment of former industrial buildings in prime locations such as on canal sides and former ports

◆ increase in tourism – resulting from greater prosperity in HDEs, cheaper air travel and increased car ownership.

The growth of such 'post-industrial' economies has not benefited all. Highly paid managerial jobs have been relatively few in number, while so-called 'back office' jobs are low-skilled, lower paid and are more likely to be part-time or temporary. In most countries, the total growth of jobs in the service sector since the 1950s has failed to compensate for the loss of manufacturing jobs.

Consequently, unemployment remains a problem in most HDE towns and cities. Furthermore, new services have tended to be pulled towards the largest cities at the top of the urban hierarchy (usually world cities such as London, Tokyo and New York). This is because multinationals favour these locations as they need access to national and international labour markets, highly skilled and educated workforces and the same specialist producer services. Consumer services are more evenly distributed through the urban hierarchy and increasingly favour suburban and edge-of-town locations in order to be near to more affluent suburban and urban fringe populations. Research and development facilities in particular seek pools of highly qualified labour, such as urban fringe locations near to universities (Figure **2**).

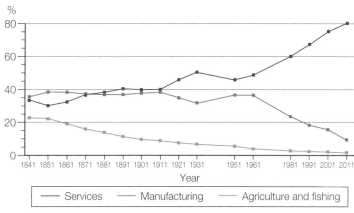

▲ **Figure 1** The structural shift in employment (1841–2011)

▼ **Figure 2** Cambridge Science Park, Cambridge, UK

ACTIVITIES

1 Summarise the effects of the three processes of economic change (deindustrialisation, decentralisation and the rise of service economy) on urban areas.

Ⓢ
2 Construct a compound (divided) bar chart for 1841, 1861, 1961 and 2011 to show the share of UK jobs in the three industries shown. Comment on the significance of the four time periods graphed.

Ⓢ 3 Look at the list of companies at www. cambridge sciencepark.co.uk/company-directory.

a Summarise the types of industry that are located at Cambridge Science Park.

b Why have these industries located in Cambridge Science Park?

GIS: planning and development

 S In this section you will learn about the role and application of GIS in urban planning and redevelopment

Geographical Information Systems (GIS)

GIS was arguably the most exciting technology of the late 20th century – applying the infinite scope and value of geography to breathtaking advances in ICT.

A GIS is, effectively, a computer database system capable of capturing, storing, analysing and displaying geographical information from a vast variety of sources. It is associated with a whole new language of geodatabases, geovisualization and geoprocessing! The geographical key is that all this data is identified according to location. A GIS, in consequence, is most often associated with maps – composite maps – because of the capacity to represent multiple layers of digital information such as satellite images, aerial photographs, census tables, graphs and written text (see Figures **1** and **2** and also section 2.8).

A GIS, therefore, stores, retrieves, manipulates, analyses and displays whatever combinations of information the specific task requires – so revealing spatial patterns and interrelationships. Furthermore, in the sampling and selection of data, followed by application of analytic rules, it can be used to create models of geographical reality that reveal important new information. This is essential to effective decision-making.

In short, a GIS makes it possible to integrate information that is difficult to associate through any other means. A particularly important component of this is its ability to produce two- and three-dimensional computer graphics to allow decision-makers to visualise and thereby understand the results of their analyses or simulations of real-world features and events. The potential of this technology is infinite and adopted increasingly by academics, governments, NGOs and businesses – researching, planning, responding to emergencies, targeting markets, and so on. Local governments use GIS in planning for, and responding to, the challenges of, and opportunities afforded by, urbanisation.

Think about

Confusing GIS with GPS

A common mistake is to confuse GIS with GPS. Global positioning systems (GPS) tell us where we are, whereas GIS capture and store information about the places around us. The difference was tragically illustrated a few years ago during a particularly hot Chicago marathon. A suburban ambulance using GPS satellite navigation picked up a victim of heat exhaustion. But the ambulance did not have a GIS database that could identify the location of the closest hospitals and the patient died before a hospital could be reached.

ACTIVITIES

1   Study Figure **2**.
    a   Identify different types of land use. Give reasons for your answer.
    b   Suggest reasons why Herbert Austin chose to locate an automobile assembly plant in Longbridge in 1905.

2   Study Figures **2** and **3**.
    a   What do you understand by the term multiple deprivation?
    b   Compare and contrast Figures **2** and **3**. (You will need to visually overlay Figure **3** on top of Figure **2** – just like a GIS).

3   Evaluate the value of GIS in urban planning and redevelopment.

4   Study Figure **1**. Construct a comparative line graph to show the percentage change in unemployment (April 2005 to January 2015) for Longbridge and Birmingham. (You may find it useful to look at 3.8 for background to the regeneration of Longbridge.)

Unemployment	Longbridge	Birmingham
April 2005 (pre-closure)	628 (3.9%)	31 690 (4.9%)
May 2005 (immediately post-closure)	1009 (6.2%)	34 136 (5.3%)
March 2010 (height of recession)	1322 (8.1%)	50 946 (7.5%)
January 2015	606 (3.7%)	31 600 (4.5%)

△ *Figure 1* *Unemployment in Longbridge and Birmingham*

STRETCH YOURSELF

John Bryson, University of Birmingham, said that in hindsight the closure of Rover was 'quite good for the city' as it spared the company 'death by a thousand cuts'. What do you think he meant by this and do you agree with him?

The use of GIS to examine the impact of the MG Rover closure, Longbridge

Look at Figures **2** and **3**. Longbridge ward is in the south-west of Birmingham. For 100 years, it was synonymous with car production. Herbert Austin bought a derelict print works at Longbridge in 1905 and several mergers later (Morris, Jaguar and Leyland), the British Leyland Motor Corporation was formed in 1968. In the face of increasing costs and overseas competition, the company continued to change ownership – owners included the UK government, British Aerospace, BMW and Honda. Finally in April 2005, the now rebranded MG Rover group collapsed and what was left of the company was sold to Chinese company Nanjing Automotive. In 2007, Nanjing Automotive's newly founded British division, MG Motor UK, restarted limited assembly at Longbridge.

◀ **Figure 2** *Longbridge ward overlaid on top of satellite image*

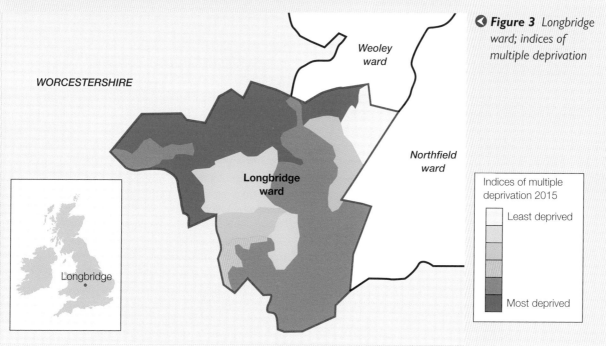

◀ **Figure 3** *Longbridge ward; indices of multiple deprivation*

3.8 Urban policy and regeneration in Britain since 1979

In this section you will learn about urban policy and regeneration in Britain since 1979

Urban regeneration

'Plan your work for today and every day, then work your plan.' (Margaret Thatcher)

Margaret Thatcher was Prime Minister of the UK between 1979 and 1990 and led the biggest shake-up in urban policy and **regeneration** in the last 50 years. Successive governments have continued to implement change, promote growth and development in our urban areas.

▼ **Figure 1**

Urban regeneration schemes 1979–2015

Phase	Year	Policy	Comment
1979–92 Conservatives assume power and Margaret Thatcher champions the role of the market and the trickle down effects of property-led regeneration (see 3.24).	1981	Urban Development Corporations (UDCs)	Thirteen UDCs established to physically, socially and economically regenerate brownfield and vacant sites. UDCs had wide-ranging powers and resources and aimed to **lever** investment from the private sector.
	1981	Enterprise Zones	Small areas of land opened up (via new planning legislation and tax incentives). The aim was to attract new growth industries.
	1986	The Inner Cities Initiative • Inner City Task Forces • City Action Teams	Consisted of small teams of civil servants who worked with the local community. The aims were to improve the economic environment and stimulate work on environmental and recreational developments.
1993–97 Change in Conservative Prime Minister prompts a fresh approach towards inner-city policy. Unelected boards were replaced with **partnerships**; which brought more targeted investment.	1991	City Challenge	Funding by competitive bids rather than those most in need. A change in emphasis away from unelected boards towards partnerships with the private sector and local communities. Thirty City Challenge Partnerships were established.
	1992	European Regional Development Fund	Liverpool was the first UK city to receive EU funds in 1993.
	1997	Single Regeneration Budget	Competitive bidding for funds continued. Disparate funding from more than twenty sources was administered by the Department for the Environment.
1997–2010 Labour government. Many of the schemes initiated by the Conservative government continue, albeit in another guise. Increasing number of **QUANGO** initiatives and a move towards more sustainable communities.	1997	Single Regeneration Budget (Challenge Fund)	Amalgamation of the two policies above. Emphasis on local priorities and involvement. Housing, crime and unemployment were particularly targeted. (Revised May 2001.)
	1997	Regional Development Agencies (RDAs)	Coordination of regional economic development and regeneration. Wide ranging powers.
	1999	English Partnerships (EPs)	National regeneration agency for England to 'support high quality sustainable growth in England'. Responsible for land acquisition and major development projects, alone or in private partnerships.
	1999	Urban Regeneration Companies (URCs)	Organised and delivered major regeneration projects in key urban locations.
		A wave of government and QUANGO initiatives emerged at the end of the twentieth century, including Education Action Zones, Health Action Zones and Millennium schemes.	
2010–15 Coalition government adopts a decentralised and localist policy agenda.	2010	Local Enterprise Partnerships (LEPs)	Based on 'real functional economic areas', 'business led' and tasked with creating jobs and driving growth. Replaced RDAs whose functions largely passed to central government.
	2010	New Homes Bonus	Part of wider approach to promote house building, including the relaxation of planning controls on home extensions. The 'bonus' is an incentive to local authorities to accept housing growth (councils receive double the council tax per home for six years).
	2010	Community Infrastructure Levy	Offers incentives at a local level to 'help communities accommodate the impacts of new development'. Neighbourhoods receive a proportion of the funds councils raise from developers.
	2011	Localism Act and Tax Increment Financing (TIF)	City leaders, alongside LEPs, can make a case to be given new powers to promote economic growth and set their own local policies (decentralisation of power). TIF gives local councils new borrowing powers to finance infrastructure and capital projects.

Note that the post-2000 policies are more important to your course than earlier policies

132

Driving forward change: the regeneration of Longbridge

St Modwen is the UK's leading regeneration specialist and was tasked in 2001 with transforming the 468-acre *brownfield* Longbridge site from one of the largest car plants in the world to a £1 billion mixed community development, the largest regeneration scheme outside London. This private company works in partnership with other stakeholders including Birmingham City Council, Greater Birmingham and Solihull LEP and Centro (responsible for the delivery of public transport in the West Midlands). By 2016, the scheme had brought about many achievements:

◆ 350 new homes – Park View and Cofton Fields housing developments are within walking distance of Longbridge rail station, and Cofton Park and junctions of the M5 and M42 are both less than three miles away

◆ Over 200 000 ft^2 of new office and industrial space

◆ The £66 million Bournville College

◆ The £100 million Longbridge Technology Park, home to more than 60 businesses, aims to make Longbridge the Midlands' leading centre for technology and innovation

◆ The £23 million Cofton Centre industrial park

◆ £8 million internet Connectivity Package – digital and physical improvements, including some of world's fastest internet speeds, to attract heavy-use operations such as data centres

◆ The Factory, a three-storey flagship youth centre including sports hall, ICT suite, workshop area and media centre

◆ Austin Park, a three-acre green park, including a newly uncovered 255 m stretch of the River Lea and public art reflecting the automobile heritage.

The £70 million town centre (Figure **2**) is home to high-street retailers, restaurants, hotel, offices, public park and car parking. Over 200 000 people, including an above-average percentage of people in the younger age groups, live within a ten-minute drive time, including affluent Barnt Green.

The regeneration scheme aims to create 10 000 new jobs across high technology, retail and leisure ventures and 2000 new homes by the completion of the project in around 2035.

Future developments include an upgraded train station, bigger park and ride scheme, new cycle link, retirement village, several contrasting residential developments and affordable apartments.

Did you know?

Annual government spending on 'core' regeneration programmes experienced a 65 per cent reduction over the two years from 2009–10.

▲ **Figure 2** *Longbridge 'town centre' – a mixed-use retail, leisure, residential and educational development*

ACTIVITIES

1 Why is any urban regeneration policy necessary, even in periods of economic growth?

2 Study Figure **1**. Suggest reasons why policy-makers increasingly tried to involve local communities in their decision-making.

3 To what extent is the regeneration of Longbridge sustainable? (Hint: consider the needs of social, economic and environmental sustainability.)

4 What do you think is the most important aspect of the regeneration of Longbridge? Justify your answer.

STRETCH YOURSELF

In general, Labour MPs represent urban, and particularly inner-city, constituencies whereas Conservative MPs represent more affluent suburban and rural constituencies. How do you think this influences Labour and Conservative approaches towards urban policy, particularly when in government?

In this section you will learn about the different urban forms of cities

It is guaranteed that you will have experienced one of the urban problems that are characteristic of every megacity around the world – traffic congestion. The congestion in some megacities, such as Jakarta in Indonesia, make that seen in the UK seem rather trivial (Figure **1**). Jakarta has a population of 10 million and drones and smartphone apps are innovative attempts to help to manage the traffic congestion, which can bring the city to a halt, particularly during the annual Muslim festival of Eid. As urban economies grow, megacities change in urban form.

What is urban form?

Urban form is the physical characteristics of built-up areas including the shape, size, density and make-up or configuration of settlements. It can be considered at different scales – from local street level to an entire county or wider region. Urban form will change over time in response to several factors:

◆ *Population:* globalisation has brought increased migration. 'Flows' of people through a city are increasingly difficult to predict and to manage. For example, lack of forward-planning or an inability to keep pace with such flows may result in shortfalls of housing, schools, health services and so on.

◆ *Environment:* for example, established physical infrastructure such as sewer systems (sewerage) or water treatment works need to keep pace with population change.

◆ *Economy:* industry has tended to locate in centres of population. As population has moved, for example, from the inner city to the suburbs and more recently to beyond the *rural–urban fringe*, industries have tended to follow.

◆ *Technology:* for example, some industries will be pulled towards the 'hardware' of wired networks or data and processing hubs where there is high-speed access to the internet and a convergence of new technologies (read again the redevelopment of Longbridge on 3.8). Elsewhere, new technologies are already supporting **teleworking** and challenging the notion of work hubs.

◆ *Policies:* government policies affecting housing, planning, transport and the economy will restrict or encourage changes at different times and in different urban locations.

Megacities are a complex 'system of systems' and it is almost inevitable that they are unstable – into these urban systems flow rapidly changing movements of capital, people, pollutants, cultures and technologies. Shocks to the system such as economic recession, mass migration, extreme weather events, resource insecurity and demographic changes place additional pressures on urban centres and the urban form they adopt.

⊘ **Figure 1** *Cities such as Jakarta, Indonesia, are now suffering from decades of underinvestment in the road network and in the public transport infrastructure.*

Did you know?
All Jakarta's rivers are polluted – 70 per cent of them are heavily polluted.

Contemporary characteristics of megacities

As a result of rapid urban growth and generally weak planning systems, many megacities include various types of urban space with diverse urban form issues. Nevertheless, most megacities share the following characteristics:

◆ *Urban sprawl:* the expansion of the urban area, usually with insufficient urban infrastructure such as streets, parks and utilities

◆ *Peripheral growth:* development of new growth poles and dispersed settlements

◆ *Edge cities:* surrounded by new forms of retail, leisure, industrial and business parks, miscellaneous warehousing and large employment buildings, ring roads and motorway interchanges

◆ *High-density living and intensification of urban centres:* vertical residential zoning such as super high-rise developments, small three-story family homes and one-room studio apartments

◆ *Residential differentiation:* different socio-economic groups likely to live geographically apart; also likely to be a housing shortage forcing the poorest into inadequate housing units

◆ *Redevelopment and conservation:* protection of historic cores; redevelopment of former industrial sites

◆ *Infrastructure:* ageing infrastructure likely to dominate with some areas poorly served

◆ *Transit-oriented development:* urban areas around railway stations and major road routes

◆ *Car dominated urban form:* led to lower-density housing estates on the edge of cities and in more rural locations making edge-city leisure, retail and employment locations – more accessible; this raises challenges as less carbon-intensive futures are sought and as fuel prices rise

◆ *Environmental problems:* including increased levels of pollution, health and waste concerns.

Within cities, much infrastructure is fixed – for example, transport networks, power stations and sewer systems may have a lifespan of more than one hundred years. However, this infrastructure needs to provide reliable and high-quality services within both relatively 'slow' changing urban forms as well as the rapidly shifting 'flows' of the twenty-first century. This is increasingly a problem for megacities that are inter-connected within the global economy and have to manage the intensification of flows of both capital and people. The problem may be further exacerbated where infrastructure is owned by many parties – including both public and private companies and individuals – resulting in decisions made on infrastructure without consideration of the impact on urban form.

Think about

Urban form and infrastructure

Infrastructure is defined as the physical and related organisational structures needed for society to operate. It includes the following sectors:

• Energy: power stations, gas pipelines, electricity distribution grid, wind turbines, photovoltaics

• Transport: roads, railway stations, airports, seaports, cycle ways, footpaths (Figure **2**)

• Water: water treatment works and pipes

• Wastewater: sewers, pumps and sewage treatment works

• Solid waste: recycling facilities, landfill sites, incinerators, waste transfer and processing sites

• ICT: wired and wireless networks (e.g. satellites), broadband cables, television masts, data cables

• Cultural and social infrastructure: schools, surgeries, hospitals, museums, community venues

• Green and blue infrastructure: the interconnected networks of land and water that support ecological systems as well as contributing to health and quality of life of urban residents such as urban parks, canals and waterways.

Figure 2 *Bus, train and cable car station of Staalden, near Zermatt. Switzerland operates an integrated public transport network. The Gotthard Base Tunnel, at 57 km, is the world's longest rail tunnel and demonstrates Switzerland's commitment to large-scale infrastructural projects.*

Comparing and contrasting different urban forms

Characteristic urban form	Reasons for urban form	Functional zones	Example
Pre-industrial cities	Largely unaffected by industrial developments and have retained much of their urban layout and characteristics. Elite groups tended to locate in the centre surrounded by the lower socio-economic groups, including artisans who worked from home and lived with other artisans sharing the same trade.	Historic buildings such as churches and castles likely to dominate the centre High class residential zone(s) near centre Less clear delineation of residential and commercial districts as today	Lincoln, Bath and York Carcassone, France
Modern (or industrial) cities	Similar activities and similar people group together. This led to homogenous areas with each area being dominated by a particular land use or social group. Arrangement of areas strongly determined by the general decline in land values outwards from the city centre.	Dominant CBD Residential zoning Industrial zone likely to be manufacturing-based	Birmingham Chicago
Post-industrial cities	Urban mosaic – more chaotic and looser structure with many smaller zones rather than one or two dominating. Post-suburban and peripheral developments with high tech corridors or zones.	Multi-nodal structure Less dominant CBD Higher degree of social polarisation Service sector-based industry that is less tied to one location	Milton Keynes (a 2nd generation new town in the 1960s) Tokyo Las Vegas Paris, La Defense (Figure **4**)
Public transport oriented (PTO) cities and motor-based cities (MBC)	PTO development takes an integrated approach towards planning. For example, minimising the walking distance between residential developments and public transport nodes and interchanges. MBC – The onset of mass motorisation from the 1950s helped increase rates of suburbanisation and the decentralisation of some economic activities.	PTO cities may develop along railway lines and main roads. More intensive development might be allowed near railway stations. MBC development will be linked to major road networks – urban freeways or motorways. Non-residential land uses, such as retailing and offices may locate in urban fringe locations.	Public transport oriented development in Hong Kong (90% all passenger trips are by public transport) Detroit is not only the birthplace of the US car industry but also demonstrates the donut effect (see 3.10) as people and economic activities have moved to edge
African cities	Many cities have grown from colonial settlements and have not experienced the industrialisation of US and European urban centres. Recent and rapid urban growth has forced changes to established and older zones as well as expansion on the periphery. Lacking resources and control, urban form may be unplanned and sometimes chaotic.	Dominant CBD that is likely to be the political and cultural centre. HQ of foreign multinationals may be present along with large hotels and historic buildings. Older industrial areas adjacent to transport routes, such as railways Newer peripheral middle-class housing served by road network Informal housing developments on marginal land	Nairobi, Kenya Capetown, South Africa
Socialist cities	Followed principle of a classless city. This meant that everyone should live in a same type of housing block irrespective of the location in the city. Housing blocks were located close to local services to encourage walking. The city centre was large and an administrative and political centre rather than a commercial centre.	Homogenous blocks throughout the city Four micro-neighbourhoods (15 000) made up a Residential District (60 000 residents). Neighbourhoods had local services; districts had higher order shops and entertainment services. City centre had prestige buildings and a central square for socialist rallies.	Prague

Figure 3 *Comparison of different urban forms*

Figure 4 *La Défense to the west of Paris is Europe's largest purpose-built business district in Europe*

Processes of urbanisation, suburbanisation, counterurbanisation and urban resurgence (see 3.2) continue to shape and change urban forms globally. The consequences of such dramatic urban change continue to influence and dictate where we choose to live, work and socialise. Throughout history, **residential differentiation** has attracted particular attention from sociologists, urban planners and particularly, geographers. As towns and cities grow, so their functionality may change, providing new opportunities for some, but creating difficulties for others. For example, the East End of London was a religious centre in the Middle Ages, a notorious Victorian deprived area in which Jack the Ripper preyed, and is now a hub for financial services centred on Docklands' Canary Wharf and Canada Place. Yet as **gentrification** and urban renewal continues in the London Docklands, some would argue this area is still home to an urban underclass. With such contrasts, how can we compare different urban forms?

One approach to the study of urban forms is to consider different types of functional zones in different types of environments.

ACTIVITIES

1 Why does every city have a unique urban form? (Hint: what factors influence urban form and how do these change over time?)

2 Suggest reasons why megacities in LDEs may develop a different urban form than cities in HDEs.

3 Study Figure **3**. Do the urban forms listed share any similarities in either the reasons or patterns of their urban form?

STRETCH YOURSELF

The World Bank and national governments in most developed nations support the development of contained, high density, mixed use cities and towns. How is this particular high density and mixed urban form of benefit to people and to the environment? Is smaller, more manageable growth better than high density?

In this section you will learn about urban characteristics of Los Angeles and Mumbai

In 2010, Hollywood (Los Angeles) and Bollywood (Mumbai) signed an agreement to boost cooperation for producing and distributing their films. However, the cities have more in common than just their film industries.

Los Angeles

Land use

Los Angeles grew out of the Gold Rush of northern California in the mid-nineteenth century when it became known as 'Queen of the Cow Counties' – so-called because it supplied beef to the miners. Subsequently, largely white and middle classes moved to Los Angeles to escape the problems of the big industrial cities such as Chicago, Detroit or Pittsburgh. Their vision was one of clean, orderly and organised residential communities of local neighbourhoods. While many of these inhabitants of 100 years ago moved to the city to retire, others were independently wealthy and increased car ownership led to suburbanisation. Today, Los Angeles typifies modern **urban sprawl** (Figure **1**) – Greater Los Angeles incorporates 88 cities and is largely made up of low density housing.

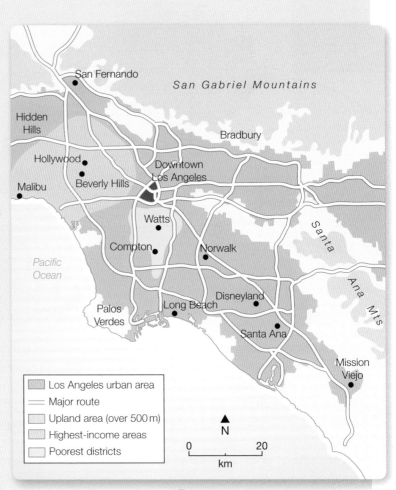

◆ **Figure 1** *Greater Los Angeles sprawls for 115 km east to west and covers an area of 1166 km²*

Economic inequality

As businesses and services followed workers to cheaper and larger sites in the suburbs, the CBD or downtown Los Angeles suffered economically. Described as a 'donut city', downtown Los Angeles is dominated by headquarters of TNCs that offer few employment opportunities for the lower-skilled local population. Other employers, even including those offering lower-paid service jobs such as retail work, have moved to the **exurbs** or **edge cities** such as Anaheim (a two-hour drive away from the downtown area and home to Disneyland).

Cultural diversity

Mexican Hispanics are the largest of many ethnic groups in Los Angeles. Distinctive ethnic enclaves, many within and around the downtown area, are testament to the present-day multicultural feel of the city – 'Chinatown', 'Little Italy' and the African-American dominated South Central and Watts districts. These groups have been left behind by the flight to the suburbs and suffer from deprivation such as unemployment, high crime rates and poor access to schools and health.

Mumbai

Situated on the west coast in the state of Maharashtra (Figure **2**), Mumbai is the largest, most populous, richest, arguably most modern and rapidly growing city in India. Metropolitan Mumbai today crams more than 12 million people into an area of just 438 km². The large harbour meant the city became known as 'Gateway of India', which is just as apt today as the city has gained a status as India's commercial, economic, transportation and cultural centre and continues to attract much foreign (inward) investment.

Land use

Mumbai was originally founded on seven islands – the land has been drained to form one of the most crowded cities in the world (27 000 people per km²). Skyscrapers dominate the Fort area or CBD. Shopping areas, such as the Colaba Causeway Corner, meet the needs of a growing middle class. High tech industries have moved in to take advantage of cheap, skilled labour. However, in stark contrast to the luxury of residential areas such as Malabar Hill (Figure **3**), around half of the city's 20 million people live on the streets or in shanty towns (*bustees*), such as the slum district of Dharavi to the north.

Economic inequality

Mumbai is a city of contrasts. Huge numbers live in poverty and the city's growth is largely unsustainable. While there is much foreign investment and wealth, low taxes and a large informal economy do not provide the necessary funds to allow much-needed investment in urban infrastructure. For many, daily life is difficult and full of risks – around five million have no access to clean drinking water and several thousand die annually from road and rail accidents.

Cultural diversity

Each day Mumbai receives around 1000 new migrants who contribute to the cultural diversity of the city. Several languages are spoken, including Marathi (the mother-tongue of local Maharashtrians), English and 'Bambaiya Hindi' – a blend of two languages reflecting the mix of dialects. Many faiths are practised in the city, including Buddhism, Hinduism, Islam and Christianity, and a host of festivals are celebrated, including Diwali, Christmas and even a gay pride parade.

Figure 2 The location of Mumbai, India

Figure 3 Luxury living on Malabar Hill, Mumbai

STRETCH YOURSELF

What do you think the future holds for world cities like Los Angeles, Mumbai and Bengaluru? Consider problems (such as issues of sustainability) as well as the benefits (such as economic growth) of living in these fast-changing cities.

ACTIVITIES

1 a Study Figure **1**. Describe the urban sprawl of Los Angeles. (Hint: use the scale and make good use of place references.)

 b How does Los Angeles compare to the urban sprawl of Mumbai?

2 a Copy and complete the summary descriptive table for Los Angeles and Mumbai:

	Los Angeles	Mumbai
Land-use patterns e.g. industry, residential		
Economic inequality e.g. employment		
Cultural diversity e.g. contrasting ethnic groups		

 b To what extent do Los Angeles and Mumbai share similar urban characteristics/form?

3 Summarise the factors that influence land-use patterns, economic inequalities and cultural diversity. How far is in-migration a cause of change?

In this section you will learn about town centre mixed-use developments, cultural and heritage quarters, fortress developments, gentrified areas and edge cities

Have you ever become lost when driving through a city? Cities can be difficult places to find your way around – regeneration and redevelopment mean that maps become quickly out-of-date (and confusing even the best SatNav system!). New urban landscapes are emerging in locations throughout cities and causing localised changes to traditional urban forms.

Town centre mixed-use development

Mixed-use development blends residential, commercial, cultural, institutional and industrial uses, which are interconnected physically and also functionally. They may be safely and easily accessed by pedestrians and include multiple functions within a short walk (including even within the same building). For example, the newly created Longbridge Village (3.8) and BedZed (3.22).

Cultural and heritage quarters

Many cities across the UK have developed a cultural quarter in order to encourage growth and revitalise the local economy in the arts and creative industries.

The Dr Who Experience, the Welsh Assembly and one of Europe's largest civil engineering projects are all located in Cardiff Bay – a hugely successful cultural and heritage quarter formerly known as Cardiff docklands (Figure **1**). Completed in 1999, a 2 km² freshwater lake with 13 km of waterfront continues to stimulate new commercial, tourist and leisure developments, including Techniquest Science Discovery Centre, Craft in the Bay, Butetown History and Arts Centre and the Wales Millennium Centre.

Fortress developments

Urban planners and architects are increasingly considering the needs of defensible spaces and so-called fortress cities (see Extension). This planned and physical use of space creates strong boundaries and, some may argue, an increasing polarisation of society. Examples of fortress developments include gated residential developments, shopping centres patrolled by private security guards and new technologies used to monitor and survey public spaces. They do not encourage social mixing as only similar higher income groups will tend to live in the same defended spaces.

⬣ **Figure 1** *Cardiff Bay with the Wales Millennium Centre in the centre*

⬣ **Figure 2** *Oscar Pistorius leaving court in 2014; he was found guilty of murdering his girlfriend Reeva Steenkamp in 2015*

In South Africa rates of crime, particularly violent crime, are high which has led to a national anxiety about becoming a victim of crime. The government has been criticised for doing little to stop crime and this has caused a growing number of middle classes to live in gated communities and to use private armed security companies. This anxiety was brought into the global spotlight, when Paralympic and Olympic sprinter Oscar Pistorius fatally shot his girlfriend. He said he had mistaken her for an intruder hiding in the bathroom of his home, which was located within a gated community (Figure **2**).

Gentrified areas

Gentrification is a form of inner-city regeneration. It usually involves the movement of affluent, usually young, middle-class people into traditionally run-down (and cheaper) areas of an inner city. As money is invested in these older homes, property values tend to increase and a positive multiplier effect is felt in the wider neighbourhood. Local services may be upgraded to better cater for the needs of this new and upwardly mobile population – pubs may become wine bars and corner shops may be turned into coffee shops. However, the trickle-down effects of gentrification are not always felt by all and the established (often so-called working-class) population can feel alienated and priced out of the local housing market.

Oxford is one of the least affordable cities in the UK (house prices are 11 times the average local salary). Consequently, gentrification is a popular route to establishing a foothold on the property ladder. This has helped to further widen the wealth gap. Oxford City Council has responded by investing in schemes that bring people together – new community centres, a visiting health bus and community events.

Edge cities

Has suburbia ended? Some US geographers believe that such has been the pace of suburbanisation and the subsequent decentralisation of people and economic activities, that suburbs have now matured into centres with city-like qualities. So-called edge cities are characterised by mixed office, residential and leisure spaces and tend to be located in the outer suburbs near to motorway or main road junctions. In the UK, even established smaller cities include edge city developments. Look at Figure **3**. Bunkers Hill to the east of Lincoln (a city of less than 100 000 residents) is typical of edge city development. The site is a mixed-use modern development and includes 500 estate homes, primary school, supermarket, dentist, pharmacy, coffee shops and numerous retail units.

ACTIVITIES

1 Study again the new urban developments described in this section. To what extent are they sustainable? (Hint: consider contrasting needs of social, economic and environmental sustainability.)

2 Think about your nearest urban centre. List examples of recent changes (e.g. new retail, leisure or residential developments and infrastructural improvements, such as roads). Suggest reasons why these changes were necessary and which stakeholders are most likely to have benefited.

EXTENSION

Defensible space and 'fortress cities'

Medieval castles perfectly demonstrated an area or space that could be protected or defended. Oscar Newman (1972) first used the term 'defensible space' and his philosophy has subsequently been adopted by urban planners and architects. The British Crime Survey has established that crime, for example burglaries, is not random but particularly associated with certain types of built environment (the proximity of a housing estate to the inner city or the type of housing – flat, terraced, detached).

Look back at 3.10, Mike Davis (1992) argued that new luxury developments in Los Angeles, such as Hidden Hills, Bradbury and Palos Verdes (Figure **1**, 3.10), are 'fortress cities' that use walls, guarded entrances and security firms to defend their elitist spaces. In contrast, he argued that less safe public spaces (such as toilets and playgrounds) had all but disappeared as a result of Central Hall planning decisions; even shopping malls were forced to guarantee security for retailers following the Watts riots of 1965 and South Central riots of 1992. Davis said that this all led to strict policing of who was allowed into these places, exclusion of the poor and a reduction in the amount of freely shared space.

Did you know?

Edge cities have also been previously referred to as superburbia, technoburb, galactic cities and even pepperoni-pizza centres!

▶ *Figure 3* *Bunkers Hill, Lincoln – an example of a mixed edge development*

In this section you will learn about the characteristics of the postmodern western city

Some films end with many unanswered questions in the audience's mind. Ridley Scott's 1981 film *Blade Runner* (Figure **1**) has no clear-cut plot and leaves the audience with contrasting interpretations. It is an example of **postmodernism** and portrays an eerily recognisable Los Angeles in 2019 that is in crisis – it has no identity, no centre and is made up of a patchwork of styles.

The rise of the postmodern western city is very real and such cities represent a significant change to the modernist cities that they preceded.

What is a postmodern city?

The postmodern city is an urban form associated with changes in urban structure, architectural design and planning, and reflects the changed social and economic conditions of the late twentieth century in some western cities.

Change associated with urban form	Characteristic
Urban structure	• Chaotic multimodal structure • High tech corridors • Post-suburban developments
Urban architecture and landscape	• Unusual mix of different styles and shapes that incorporate meaning and symbolism • Celebration of the past through historical references
Urban government	• Encouragement of mobile international capital • Services provided by the market (rather than public services) • Public and private sectors work in partnership
Urban economy	• Service (and especially quaternary) sector dominated • Globalised • Consumption oriented
Planning	• Spatial 'fragments' designed for aesthetic rather than social ends • Attempts to incorporate the views of many stakeholders
Culture and race	• High levels of social polarisation • Highly fragmented

⊘ *Figure 2* *Characteristics of the postmodern city*

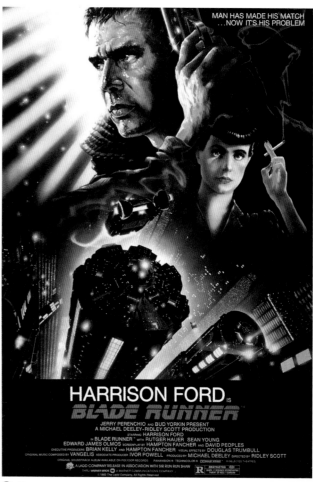

⊘ *Figure 1* Blade Runner – *a postmodern film blending hyperreality (where images of Los Angeles seem more real than the city today) and science fiction in the postmodern landscape of a future Los Angeles*

What is postmodernism?

Postmodernism started as a critique to modernism, albeit any precise definition is shrouded in uncertainty (one dictionary definition states that postmodernism 'has no meaning. Use it as often as possible')! Postmodernism is a so-called philosophical movement that applies a particular viewpoint to the world in which we live. Postmodernists see western world society as an outdated lifestyle that is impersonal and faceless, often associated with difference, scepticism and plurality. For example, postmodernists celebrate ethnic diversity and spectacular 'different' forms of architecture.

Las Vegas – a postmodern city

In 1905, the railroad arrived in Las Vegas in the state of Nevada, connecting the city with the Pacific Ocean and the country's main rail networks. In 1931, construction began on the massive Boulder Dam (later renamed the Hoover Dam), drawing thousands of workers to a site just east of the city. Casinos and entertainment venues opened up just outside the city's jurisdiction in order to meet the needs of this workforce (Figure **3**). The Strip, as it became known, rapidly expanded as a result of the cheap HEP and funding from both gangsters as well as more reputable entrepreneurs. In the last 30 years, the Strip has undergone a postmodern transformation with the construction of massive complexes that take inspiration from global landmarks. Today, Las Vegas displays the characteristics of a postmodern western city.

Why is Las Vegas a postmodern city?

- Urban Structure: post-suburban development is perhaps epitomized by Lake Las Vegas, a 30-minute drive into the Nevada desert. This is essentially a gated community with pseudo Mediterranean villas on the shores of a human-made 'Italian' lake. In the 1970s, everyone drove in Las Vegas, whereas now, the Strip has become a promenade for the tens of thousands who walk down it each day.

- Urban architecture and landscape: epitome of over-the-top consumerism of American architecture (Figure **3**).

- Urban government: local government work in partnership with private corporations (e.g. High-Impact Projects team) in planning for the infrastructure necessary to accommodate the millions of tourists each year. However, the powerful gaming/casino industry dominates a fragmented and weak local government that draws from a limited tax base.

▲ *Figure 3* *The Las Vegas Strip – an eclectic and often surreal mix of postmodern architectural styles*

For example, MGM's CityCenter opened in 2009 and was the most expensive private development in the country, at a cost of almost US$10 billion.

- Urban economy: service-dominated tourism drives the economy.

- Planning : it is a fragmented city where builders are given free rein to build what and where they want, particularly in the four-mile 'Strip'. CityCenter is a mixed-use complex that has its own (free) tram service between Monte Carlo Resort to the south and the Bellagio to the north.

- Ecology of fear: Nevada is consistently ranked amongst the highest in the US for violent crime.

ACTIVITIES

1. Study Figure **2**. Why do you think that African cities are not described as postmodern?

S
2. Construct a suitable chart to show the data in Figure **4**. Comment on your results.

3. Read again the example of Los Angeles in section 3.10. To what extent does the city show characteristics of postmodernism? Are there any similarities with postmodern Las Vegas? You are encouraged to complete further research.

Race	Population	% of total
White	277 000	45.9
Hispanic or Latino	200 359	33.2
African American	63 970	10.6
Asian	37 417	6.2
Two or more races	18 104	3.0
Other	6 638	1.1
Total	603 488	100.0

▲ *Figure 4* *Multicultural Las Vegas (2013 census)*

In this section you will learn about multiculturalism and issues associated with, and strategies to manage, cultural diversity

Social and economic issues associated with urbanisation

With half of its population born outside Canada and from 200 ethnic groups, Toronto is often referred to as the most multicultural city in the world. However, most of the world's cities are multicultural. More than 300 languages are spoken in London, 20 per cent of Parisians were born outside France, 16 major languages are spoken in Mumbai and about 45 per cent of New Yorkers speak a language other than English at home.

Immigration into Britain

During the 2016 EU referendum, immigration was one of the most discussed topics. The Prime Minister, David Cameron, supported Britain remaining in the EU. Nevertheless, his views towards immigration have not always been consistent (Figure **2**).

⊙ *Figure 2 David Cameron quotes regarding immigration*

Year	Quote
2010	'We will take immigration back to the levels of the 1990s – tens of thousands a year, not hundreds of thousands.'
2011	'When it comes to immigration to our country, it's the numbers from outside the EU that really matter.'
2013	'Our membership of the EU allows British people free movement to travel, live and work in other European countries… The same freedom of movement is true for EU nationals coming to Britain.'
2014	'We want to create the toughest system in the EU for dealing with abuse of free movement … But freedom of movement has never been an unqualified right, and we now need to allow it to operate on a more sustainable basis.'
2015	'Britain is one of the most successful multiracial democracies in the world. I am so proud of that. But to sustain that success, immigration needs to be controlled… Our approach will be tougher, fairer and faster.'

Issues associated with cultural diversity

◆ *Economic* – migrants may meet labour shortages particularly in the service (e.g. the NHS) and manufacturing sectors. However, the perception of 'jobs for migrants' may cause resentment and racial intolerance, particularly during times of economic recession where employment opportunities are fewer. Nevertheless, with a younger age profile, ethnic minority groups are projected to comprise a growing share of UK employment. These jobs will remain disproportionately in the trade, accommodation and transport sectors that are associated with lower-than-average pay. In the business and other services sector, there will most likely remain an under-representation where pay is generally higher than average.

⊙ *Figure 1 In 2005, Royal Mail issued a set of stamps showing, by means of ethnic stereotypes, 'the diversity of British cuisine in today's multicultural society'*

The same or different?

Multiculturalism refers to a society that recognises values and promotes the contributions of diverse cultural heritages and ancestries of the various groups within it (Figure **1**). It is often a sensitive and emotive issue and any attempt to define should consider choice of words carefully. Evidence for a multicultural society includes contrasting places of worship, celebration of different religious festivals, speciality shops and services, and specialist language newspapers and media services (TV, radio, and internet channels).

◆ *Housing* – new migrants generally are poor upon arrival in a country. Consequently, multiple occupancy in rented accommodation, usually within the poorer inner city, is widespread. Furthermore, ethnic minorities have traditionally been less successful in securing mortgage loans, a situation that has been worsened by a credit squeeze since the financial crisis of 2008. In a process known as *residential succession*, established ethnic groups may move to suburban locations leaving the housing empty for newly arrived migrant groups to subsequently occupy.

◆ *Education* – children, particularly of primary age, will usually attend their nearest school. Such is the concentration of ethnic groups in some parts of an urban area that it has led to schools becoming dominated by one ethnic group. This has an impact on the curriculum as, for example, additional English lessons may be necessary and special religious provision may be requested by parents. In some cases, specialist schools may open, such as faith schools, as either academies, free or independent schools. There is also significant variation in educational attainment of different ethnic groups. A report published in 2015 by the Institute for Fiscal Studies reported that all ethnic minority groups were more likely to go to university than their White British peers (Figure **3**). The report suggested that this is, in part, a result of higher aspirations by ethnic minority students and their families.

◆ *Health* – as many ethnic minority groups continue to live in inner-city areas, there tends to be a close association with poorer levels of health. However, this tends to be more a result of the poorer quality of the built environment than the underlying poor health of the population.

◆ *Religion* – migrants may wish to follow their own religious calendars. This could lead to friction with employers and local communities. Misunderstanding of religious practices such as wearing traditional clothing is also a potential cause of conflict.

◆ *Gender* – women from ethnic minorities might adopt different cultural preferences, traditions and norms. For example, there may be a tendency for women to be the primary carers and to look after family dependents (old and young). As a result, women from ethnic groups may be disadvantaged in terms of possible employment opportunities as they require more flexible working, which also tends to be more poorly paid.

◆ *Food culture, pop music and sport* – all help to support what is described as interculturalism or a blending of cultures. In a recent survey, around 75 per cent of Europeans considered sport as a means of integration, for example, by supporting a local or national sporting team or through competing in sport. In the English Premier League, domestic players accounted for less than a third of playing time in the 2015–16 season (Figure **4**). This compares with 69 per cent 20 years ago!

◆ *Language* – if the host country language is not adopted quickly, it will act as a barrier to integration and restrict employment and educational opportunities.

⬥ **Figure 4** *Against all odds, Leicester City won the English Premier League in 2015–16; Anthony Knockaert (France), Chris Wood (New Zealand), Luke Moore (England) and Riyad Mahrez (Algeria) were all part of their squad*

⬥ **Figure 3** *Ethnic minority groups can have higher aspirations than their White British peers*

Background to cultural diversity in Batley, near Bradford

Batley is situated in the Kirklees District of West Yorkshire, 11 km from the city of Bradford (see Figure **5**). From the 1800s, migrants from Western Europe, notably Germany and Italy, arrived in the region looking for work in the expanding wool industries. In Bradford, this resulted in new ethnic enclaves such as 'Little Germany' and 'Little Italy'. After the Second World War, the now industrialised valleys outside Bradford welcomed migrants from Pakistan and India and also the Caribbean. In Batley, cheap migrant labour worked in the local textile industries, particularly the *shoddy* industry (using recycled or reclaimed material). Today, people of south Asian origin make up about one-third of the population of Batley. In Batley East the figure rises to over 50 per cent. (see Figure **7**).

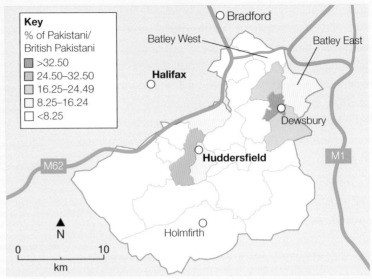

Figure 5 *Percentage of Pakistani/British Pakistani Kirklees District, West Yorkshire*

Cultural diversity in Batley

♦ *Ethnically mixed areas* – look at Figure **6**. Whilst ethnic minority groups have grown in Batley and Kirklees, the separation of religious groups has decreased. For example, Batley East is an ethnically diverse ward and has a Simpson's Diversity Index score of 11.7 (see Skills box).

♦ *Ethnic diversity within families* – the proportion of households with two or more people that contain more than one ethnic group has grown and is now around 10 per cent of all households. These families are now having children, raising the percentage that has a mixed ethnic identity to around 2.5 per cent of the population in Kirklees.

♦ *Language* – Gujarati, Punjabi and Urdu are widely spoken, as well as English, within the community. This represents a challenge for local government and schools in the communication and sharing of information.

♦ *Education* – around one in three schools in Kirklees is a faith school – this is both in line with the national average and also reflects an increase across all faiths (including Church of England). The Al Hashim Academy opened in 2014 as a dedicated secondary school for the Muslim community.

♦ *Poverty* – Batley is a deprived community and despite targeted government funds, a significant proportion of the population suffers from multiple deprivation. However, poverty largely exists because of the prevailing difficulties of the built environment rather than being a particular feature of minority ethnic groups. Increasingly, ethnic minority groups are taking ownership of businesses and a large mosque on Purwell Lane was recently built from money generated in the local community.

♦ *Community action* – is a strong feature of the town and the annual Batley Festival is a celebration of multiculturalism. The Indian Muslim Welfare Society is one of several local charities that work with individuals of all faiths.

♦ *Housing* – the local housing stock includes high-density Victorian terraced housing as well as post-war council housing – both of which are in need of renewal and redevelopment. The Sadeh Lok Housing Group is one example of a social housing provider that builds affordable new homes.

ⓢ Using indices

An index is a statistical measure of change. Indices are used frequently in geography to aggregate/represent changes in many individual indicators – such as the Human Development Index which is used to measure quality of life.

The most commonly used measure of segregation is the Index of Dissimilarity (also known as the Segregation Index), which calculates a summary measure of the spread of a group across space compared with the spread of the rest of the population. The Index is calculated by comparing the percentage of a group's total population that live in an area (usually an Output Area or OA) with the percentage of the rest of the population that live in the same area. The absolute difference is added up across all the OAs, and then halved so that the index is between 0 and 100. Zero per cent indicates a completely even spread of the population and 100 per cent means complete separation.

Ethnic diversity can be measured using Simpson's Diversity Index. This takes account of the number of ethnic groups present as well as the relative abundance of each ethnic group. The index may be standardised by stretching it to be always within the range 0 to 100. The index is greatest when there are equal numbers in each group, when it is equal to 100, and lowest when there is only one group in an area, when it is equal to 0. The average across Kirklees is 3.9 – for England & Wales it is 3.6.

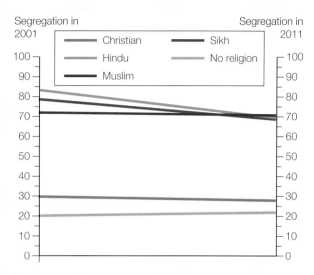

Δ **Figure 6** *Change in religious group segregation in Kirklees (2001–11)*

▽ **Figure 7** *Ethnic diversity in Batley East*

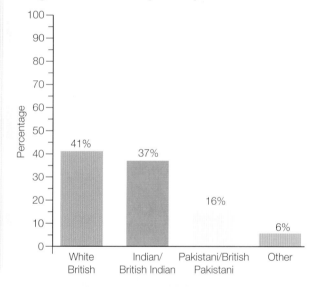

ACTIVITIES

1 Study Figure **1**. The originally published title for this set of stamps was 'A Celebration of Food', but it was later changed to 'Changing Tastes in Britain'. Why do you think the title was changed?

2 Read again the views of David Cameron (Figure **2**).
  a Suggest reasons why immigration was such a discussed topic within the EU Referendum debate.
  b Do you think David Cameron supports immigration and a more multicultural Britain? Explain your answer.

3 Explain why causes of cultural diversity are interconnected.

4 'Multiculturalism presents many challenges.' Discuss this viewpoint and include comment on the extent to which Batley has met this 'challenge'?

STRETCH YOURSELF

Complete further research on recent international migration into the UK (e.g. economic migrants from eastern Europe or refugees from western Asia). Suggest how such flows of immigrants into the UK might add to cultural diversity as well as presenting new challenges.

In this section you will learn about:
◆ the impact of urban forms and processes on local climate and weather
◆ urban temperatures and the heat island effect

Urban microclimates

Have you ever visited a central city park early in March and wondered why the blossom was so full and the daffodils so bright – yet your garden still dull from winter? Or turned a city corner and been rocked by an unexpected wind? These examples illustrate how urban areas create their own climate and weather – their own **microclimates**.

Some geographers talk of cities creating their own *climatic dome* with distinctive:

◆ temperature ranges
◆ precipitation generation and patterns

◆ (relative) humidity
◆ wind speeds, turbulence and eddies
◆ (reduced) visibility.

Within this dome, there are two levels – an urban canopy below roof level, where processes act in the spaces between buildings, and the urban boundary layer above. Prevailing winds extend the climate dome downwind as a *plume* into the surrounding countryside for tens of kilometres (Figure **1**). Consequently, urban patterns of precipitation and air quality, for example, can be spread to the immediate rural area downwind.

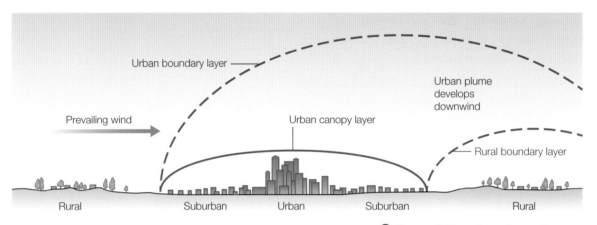

Look at Figure **2**. It is the structure of the built-up area and the activities within it that cause these changes to climate and weather. For example, buildings and streets interfere with airflow, and high levels of pollution play a part too. However, the best known microclimatic impact of built-up areas is the *urban heat island* effect.

🔼 **Figure 1** The urban climate dome

🔽 **Figure 2** Characteristics of urban microclimates. Note that relative humidity is the amount of water vapour that is in the air as a percentage of the total amount that the air can hold at that temperature.

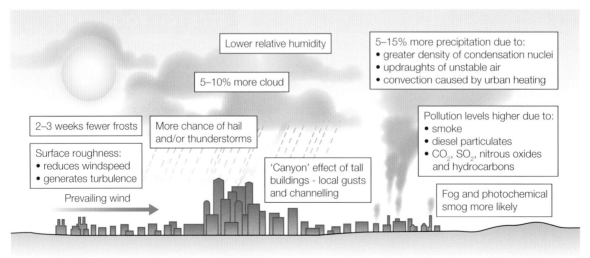

The urban heat island

The urban heat island defines the urban area as a significantly warmer 'island' surrounded by a rural 'sea' of cooler temperatures. There are several reasons for this:

◆ Urban areas have lower **albedo**. Extensive dark surfaces such as tarmac and roofs absorb heat during the day and release it slowly at night.

◆ Large expanses of glass and steel reflect heat into surrounding streets.

◆ Although the vegetation in urban parks and gardens provides moisture through evapotranspiration, and urban lakes and ponds evaporate, there is far less vegetation, so less evapotranspiration. Furthermore, drains and sewers remove surface water quickly so the amount of moisture in the air (humidity) is reduced. Consequently less heat energy is lost in evaporating it.

◆ Buildings 'leak' heat through poor insulation in winter and air conditioners pump hot air into the streets in summer.

◆ Power stations, industries, vehicles and the inhabitants themselves all generate their own heat.

◆ Rising heat, water vapour from power stations and industry, and condensation nuclei from air pollution provide the conditions necessary for precipitation. Consequently, heavier and more frequent convectional thunderstorms occur in urban areas.

The temperature decline from urban centre to rural–urban fringe is known as the *thermal gradient*. This difference can be more than 6 °C in late summer and around 2 °C in winter. Furthermore, it is usually strongest at night with additional urban cloud and dust acting as a blanket to reduce long-wave radiation losses – this is effectively a locally-enhanced *greenhouse effect* (see 5.18)

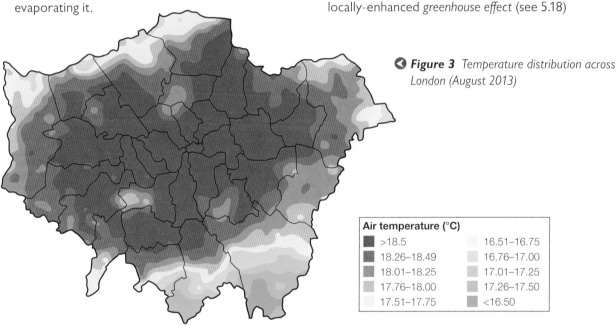

◀ **Figure 3** *Temperature distribution across London (August 2013)*

Air temperature (°C)

>18.5	16.51–16.75
18.26–18.49	16.76–17.00
18.01–18.25	17.01–17.25
17.76–18.00	17.26–17.50
17.51–17.75	<16.50

ACTIVITIES

1 Study Figure **3**.

  a Describe and suggest reasons for the temperature variations across London.

  b What might account for the linear extensions to the north, west and east, and the islands of lower anomalies elsewhere?

2 Describe and explain how microclimate might differ in a city in an LDE compared to a city in a HDE.

Ⓢ **3** Alternative techniques to show these temperature changes include line and bar graphs drawn as transects (cross-sections across the city).

  a Using the figures from the choropleth key, construct a north–south line graph and an east–west bar graph.

  b Comment on the comparative advantages and disadvantages of all three techniques.

  c Suggest an alternative technique to show these temperature changes. Compare the two techniques.

In this section you will learn about precipitation and wind in urban environments

Urban precipitation

Look again at 3.14, Figure **2**. Notice that urban areas have 5–15 per cent more precipitation than rural areas, in both quantity and also in number of days of rainfall. This may seem counterintuitive given that relative humidity levels are up to 6 per cent lower because of less urban vegetation and surface water. However, urban air pollution results in more condensation nuclei – the tiny particles essential for cloud droplets to form – which helps in the formation of clouds. Furthermore, air is warmer in cities, all leading to higher precipitation.

Remember, in order for precipitation to form, air has to rise, cool below the **dew point** and allow water vapour to condense. Both air turbulence amongst buildings of varying heights and also heat island-related convection promote this uplift. Dust and pollution from industry and vehicles increase the density of condensation nuclei and 'seed' the cloud droplets.

Precipitation falling as snow tends to lie on the ground for less time in urban areas. Both the generally higher urban temperatures and heat-retaining darker road and roof surfaces encourage more rapid melting than in the countryside.

Fog

Fog, and mist if less dense, is effectively cloud at ground level. Again, the higher concentration of condensation nuclei over cities encourages their formation, especially during cooler overnight temperatures. Fog and mist tend to be thicker and persist longer in anticyclonic (high-pressure) conditions where winds are too weak to blow them away.

The relationship of fog to condensation nuclei is emphasised further when weather records are compared to industrialisation. During the height of the Industrial Revolution in the nineteenth century, the centres of manufacturing across Britain experienced an increase in the number of days of winter fog – especially associated with the burning of coal. Today, Beijing and New Delhi suffer similarly although vehicle emissions, as well as coal burning, are responsible for the thick fog that can cause chaos (Figure **1**).

△ **Figure 1** *Fog hangs over New Delhi, India; fog is a threat to health, not only by causing breathing difficulties but also by increasing the risk of road accidents*

▽ **Figure 2** *Thunderstorm over Beirut, Lebanon*

Thunderstorms

Look at Figure **2**. While fog and mist are most associated with winter, urban convection is especially powerful in summer. The rising heat (including water vapour and condensation nuclei from power stations, industry and, to a lesser extent, vehicles) can trigger heavier and more frequent late afternoon and early evening thunderstorms. The more intense the heating, the more violent the storm. This is because updraughts of hot, humid air can rise higher in the atmosphere – and more quickly. The air cools and condenses rapidly, forming water droplets, hail and ice which charge the thundercloud and discharge as lightning.

Wind

Urban structures greatly interfere with wind by slowing, redirecting and generally disturbing the overall airflow (Figure **3**). For example, buildings create friction and act as windbreaks resulting in urban mean annual velocities up to 30 per cent lower than in rural areas. Indeed, as a general rule, the effects of buildings extend downwind by ten-times the height of the building. In short, the heights, shapes, orientation and layout of buildings, as well as connecting roads and transport infrastructure, all have microclimatic impacts. Indeed, as building designs become ever more dramatic and ambitious, so architects and structural engineers have to consider the potential impacts of airflow, turbulence and emissions (pollution). Two key effects, *urban canyons* and *venturi effect* are of particular significance:

- Urban canyons – relatively narrow streets bordered by high-rise buildings funneling and so concentrating winds

- Venturi effect – a particularly violent form of gusting caused in particularly narrow gaps by air rushing to replace low pressure vortices beyond structures (in the lee of buildings).

Did you know?

Major urban areas experience 25 per cent more thunderstorms than non-urban areas and hailstorms are 400 per cent more likely!

Figure 3 *Tall buildings interfere with airflow in urban areas*

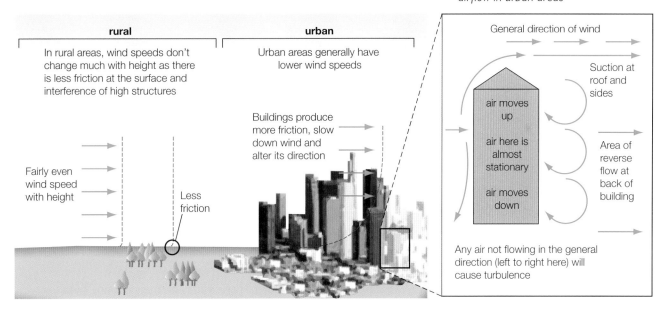

ACTIVITIES Ⓢ

Any discussion on urban microclimates is likely to involve interpretation of actual/located statistics given. Anticipation and understanding of urban rural contrasts is therefore important.

Referring to all information in sections 3.14 and 3.15 summarise microclimatic contrasts between urban and rural areas. You might want to present these in a table:

Microclimatic contrast	Urban compared to rural areas
Annual mean temperatures	
Occurrence of frosts	
Precipitation amount and frequency	
Occurrence of fog and mist	

STRETCH YOURSELF

'Statement architecture' is an increasingly apparent feature of megacities and world cities in LDEs and EMEs. Research a small selection of recent, ambitious building designs. Locate them and explain what parallels can be drawn between urban microclimates and earthquake-proof design (*AQA Geography A Level & AS Physical Geography, 5.10*).

In this section you will learn about air quality and pollution reduction

'A dense, green-yellow fog choked the streets. Cars edged forwards with passengers sitting on the bonnets shouting instructions. From behind the wheel, drivers could not even see as far as their own headlights.' (*Daily Mail*)

The Great Smog of late November/early December 1952 was the worst of London's infamous winter fogs – 'pea-soupers' (Figure **1**). Such was the death toll (12 000 in just 4 days) due to respiratory problems and, to a lesser extent, traffic accidents that it contributed to the most life-changing legislation ever passed in Britain – the 1956 Clean Air Act.

This Act, and subsequent legislation, consigned such horrors to British history. However, many cities throughout the world still have seriously high and dangerous levels of air pollution. It is one of the great urban issues of our time.

Urban air quality

Air quality in urban areas is invariably poorer than in the surrounding countryside. This results in damaging impacts upon our environment and health. For example:

- carbon monoxide causes heart problems, headaches and tiredness
- twice as much carbon dioxide enhances the greenhouse effect (see 5.18)
- ten times more nitrous oxides causes haze, respiratory problems and acid rain (which weathers buildings)
- two hundred times more sulphur dioxides than in rural areas causes haze, respiratory problems, acid rain and damage to plants.

In addition, photochemical oxidants cause eye irritation, headaches and coughs, and **particulates** from power stations and vehicle emissions cause respiratory problems and dirty buildings. Both cause smog.

▲ *Figure 1* A London 'pea-souper'

Smog

Smog is a mixture of smoke and fog. It occurs when smoke particulates and sulphur dioxide from burning coal mix with fog. The London smogs were caused by anticyclones trapping these pollutants in a pollution dome. During these conditions, the cold air, unusually, lies below warmer air. However, such temperature inversion is more commonly associated with cold air sinking to the floor of sheltered valleys and displacing the warmer air above it (Figure **2**). Smog today is far more likely to be photochemical – caused by sunlight reacting chemically with industrial and vehicle emissions to form a nasty cocktail of secondary gases. Although present in all modern cities, it is most common in those with warm, dry, sunny climates such as Los Angeles (Figure **3**).

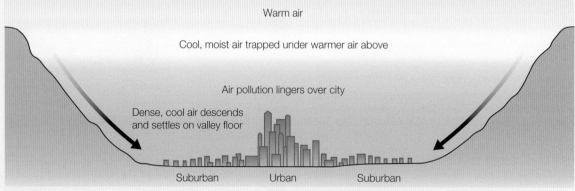

Warm air

Cool, moist air trapped under warmer air above

Air pollution lingers over city

Dense, cool air descends and settles on valley floor

Suburban   Urban   Suburban

▲ *Figure 2* Temperature inversion traps urban pollution in calm weather conditions

Think about

Anticyclones

Anticyclones are areas of high air pressure where descending air 'piles up' on the surface. They are slower moving than areas of low pressure (depressions), which bring cloud, precipitation and wind. Although infrequent, once established, anticyclones are likely to last for many days or even weeks. Descending air warms as it falls and picks up moisture. Consequently, condensation is unlikely and so clouds are rare. However, early morning dew and mist is likely in summer with frost and fog in winter. Gentle winds blow clockwise and outwards from the high pressure centre. The clear skies ensure calm, sunny weather in summer, and bright, but cold conditions in winter.

⬤ *Figure 3* *A photochemical smog hangs over Los Angeles, California, USA*

Pollution reduction

Reducing air pollution is a major challenge in all urban areas, particularly in rapidly developing LDEs and EMEs. Strategies involve a mixture of legislation, vehicle restrictions and technical innovations.

Legislation

Many countries set strict controls on emissions, including smoke-free zones. Clean Air Acts throughout HDEs have been hugely influential, but there is never room for complacency. For example, there is further contemporary research on the hazardous effects of diesel particulates. In the UK, air quality targets are set and monitored by the Department for Environment, Food and Rural Affairs (DEFRA), which has created a simple banding system to allow the publication of specific, targeted pollution warnings during, for example, weather forecasts.

Vehicle restrictions

Simply reducing the number of vehicles in central urban areas is achieved through various means. For example, pedestrianisation of city centres and Park and Ride schemes are common throughout the UK. Selective vehicle bans (on specified days determined by their number plates) operate in Mexico City. Switzerland even goes so far as banning all but emergency, utility and electric vehicles throughout some resorts (Figure **4**). Congestion charging in London, Singapore, Stockholm and Oslo has proven to be effective, but still contentious. Such is the rate of development of GPS vehicle tracking and driver monitoring systems, that the management and convenience of vehicle restriction systems can only improve.

Technical innovations

Industrial pollution controls have forced factories to reduce gas and particulate emissions, which can be as simple as using filters. Lead-free petrol and catalytic converters on vehicle exhausts have long been established and the current pace of development of lean-burn engines is remarkable. However, it is the development of hybrid electric (HEVs), 'plug-in' hybrids (PHEVs) and fully electric vehicles (EVs) that is most exciting new technology, although this raises new concerns about battery toxins. Boserup's 'necessity is the mother of invention' (see 4.26) is particularly well illustrated in all these innovations.

⬤ *Figure 4* *Car-free Zermatt, Switzerland*

ACTIVITIES

1 a What are the main sources and types of urban pollution?

   b What is the difference between smog and photochemical smog?

2 New Delhi plans to introduce congestion charging, but New York rejected similar plans for Manhattan. Outline the environmental, economic, social and political arguments for and against managing urban traffic in this way.

STRETCH YOURSELF

Switzerland lies at the geographical heart of Europe yet enjoys remarkable air quality. Research the Swiss government's environmentally-friendly policies to combat air pollution and promote 'ecological forms of mobility'. What lessons can the rest of Europe, including the UK, learn from these policies?

In this section you will learn about:
- urban precipitation, surfaces and catchment (drainage basin) characteristics
- impacts on drainage basin storage areas
- the urban water cycle: water movement through urban catchments as measured by hydrographs

Urban precipitation, surfaces and catchment characteristics

We have seen in 3.14 that urban areas have 5–15 per cent more precipitation than rural areas mainly because:

- warmer air in cities can hold more moisture
- dust and pollution make more condensation nuclei.

However, less vegetation and therefore less evapotranspiration reduces moisture in the air (humidity). Less vegetation also means less interception and more precipitation landing on hard, urban surfaces. The slate, tile, concrete and tarmac used in urban areas are impermeable and so the urban catchment is dominated by surface runoff (overland flow), particularly in city centres (Figure 1). Both shallow and deep infiltration is significantly reduced and drains are therefore needed to remove surface water quickly.

Impacts on drainage basin storage areas

Drainage basins act as systems of inputs, transfers, outputs and stores (see *AQA Geography A Level & AS Physical Geography*, 1.1). The relationship between these variables is dynamic – they will change with circumstances. Urbanisation is especially significant in altering storage. For example:

- Urban rivers are primarily the exit for water transferred through the drainage basin, but they are also important stores. Management of river channels by dredging, embanking and **channelisation** will increase their storage capacity.
- Reservoirs, lakes, ponds and swimming pools are permanent stores, but vulnerable to evaporation.
- Depression storage, such as surface puddles, following rain is temporary.
- Interception storage is reduced owing to the replacement of vegetation by impermeable structures such as buildings, roads and pavements engineered to drain the water rapidly into the nearest river.
- Soil moisture storage will vary according to ground conditions. For example, clay soils retain more water than sandy ones. However, there is usually less soil storage capacity as urban development reduces exposed surfaces and vegetated areas.

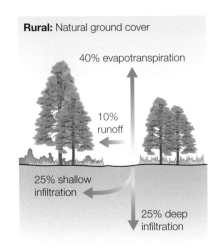

Rural: Natural ground cover
40% evapotranspiration
10% runoff
25% shallow infiltration
25% deep infiltration

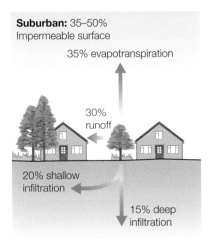

Suburban: 35–50% Impermeable surface
35% evapotranspiration
30% runoff
20% shallow infiltration
15% deep infiltration

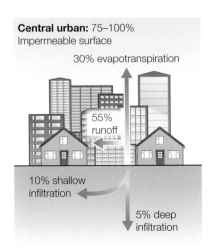

Central urban: 75–100% Impermeable surface
30% evapotranspiration
55% runoff
10% shallow infiltration
5% deep infiltration

▶ *Figure 1 Catchment characteristics change depending on the nature of the surface; rural, suburban and central urban land uses vary markedly in their percentage losses*

The urban water cycle: flood hydrographs

Look again at *AQA Geography A Level & AS Physical Geography*, 1.6. A flood hydrograph shows how river discharge responds to storm events. Measuring, recording and understanding the relationship between discharge and precipitation is essential in urban areas because of the increased flood risk caused by:

◆ the higher proportion of urban precipitation making its way into urban river channels

◆ the speed that this happens (shown by reduced lag times).

Look at Figure **2**. Urban hydrographs are 'flashy' – they show a rapid rise in discharge over a short period of time which produces a 'peaky' graph. This is because water is mainly entering the river via surface runoff.

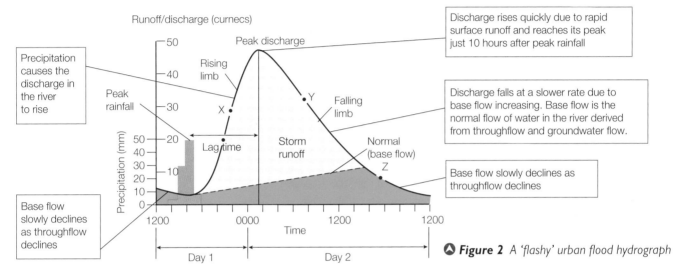

▲ **Figure 2** A 'flashy' urban flood hydrograph

Urban areas are not solely responsible for flooding – a town or city is likely to represent only a small area of the drainage basin as a whole. But, the need to build on floodplains to meet the increased demand for housing and the high proportion of impermeable surfaces certainly increases the threat. On occasions, therefore, heavy rainfall in urban areas can cause flash flooding as water flows rapidly along roads and pavements. Drains may be unable to cope with the sheer volume of water, and culverts may become blocked with debris. Furthermore, urban rivers themselves can get blocked by debris, particularly where bridges narrow the channel and so constrict the flow (Figure **3**).

▲ **Figure 3** *Torrential rain and hail caused the River Wey to burst in banks in Guildford, December 2013*

ACTIVITIES

1 Study Figure **2**.
  a Copy and complete the following. State reasons for each one:
    • The lag time is the …
    • At point X, surface runoff and discharge are …
    • At point Y, discharge is …
    • At point Z, there is no longer an input from surface runoff and throughflow and base flow are both …
  b If the channel capacity at the measuring point is 30 cumecs, what is inevitable, and when?
2 Study Figure **1**. Comment on the relationships between vegetation cover and the impermeable surfaces associated with urban areas.

STRETCH YOURSELF

Sketch a flood hydrograph in a rural area. The discharge will not rise so high. The lag time will be extended over a longer period of time because more water is entering the river via throughflow and groundwater flow, and less via surface runoff.

In this section you will learn about:
- management of drainage within urban catchment areas
- Sustainable Drainage Systems (SuDS)

Management of drainage within urban catchment areas

The rapid removal of water from urban surfaces is essential if day-to-day city life is to continue normally. But, efficient drainage into nearby rivers is only viable if they can cope with the huge volumes of water entering them. Otherwise, flooding is inevitable.

Hard and soft engineering

Urban drainage has long been established as an engineering challenge. Sewerage and water treatment plants will always be required for the safe management of human waste and industrial effluent. However, similar *hard-engineering* approaches to surface water drainage are no longer assumed to be the only solution. Simply transferring water to the nearest river necessitates management if its capacity is not to be overwhelmed (section 3.17). Hard-engineering projects invariably involve high costs of construction and also ongoing maintenance. They also require long periods of planning – not least to determine their social, environmental and economic impacts. So *soft-engineering* approaches, working with, rather than against, natural processes are increasingly adopted as more affordable, sustainable solutions (Figure **1**). Also see *AQA Geography A Level & AS Physical Geography*, 3.9.

Did you know?

Green roofs have existed for centuries. The Vikings used grass sods on the roofs of some of their buildings. Modern buildings use a more high tech approach, with impermeable base layers, and drainage and irrigation systems.

Hard engineering	Soft engineering
River straightening involves cutting through meanders to create a straight channel. This increases the gradient and speed of flow which may increase flood ris k further downstream. In some places the straightened sections are lined with concrete (see *channelisation* below).	*Afforestation* (planting trees to establish woodland or a forest). Trees increase interception and reduce throughflow and surface runoff because they take up water to grow. In short, evapotranspiration from both leaves and branches dissipates water that would otherwise end up in the river channel.
Natural *levées* can be made higher, so increasing capacity. *Embankments* are raised riverbanks using concrete walls, blocks of stone or material dredged from the river bed. The latter is arguably a more sustainable, environmentally friendly option and looks more natural than the concrete walling more common in urban areas.	*Riverbank conservation* by planting bushes and trees reduces lateral erosion, bank collapse and so silting-up of the channel. This is because their roots stabilise the banks by binding the loose material/sediments together.
Diversion spillways (flood relief channels) by-pass the main channel. They can be for emergency use only (controlled by sluice gates) when high flow levels threaten flooding or a permanent feature enhancing the environment by creating new wetlands and recreational opportunities.	*Floodplain zoning* restricts different land uses to certain locations on the floodplain (e.g. nearest the channel may only be used for pasture or for recreational use). Natural floodplains act as a natural soakaway, so protecting them from development and reducing surface runoff into the channel.
River *channelisation* involves lining straightened channels with concrete. This reduces friction, improves the rate of flow and reduces the build-up of silt because it prevents the banks from collapsing. But, channelisation looks unsightly and damages local ecosystems.	*River restoration* involves a return of the channel to its natural course and so reversal of artificial drainage management 'solutions' adopted in the past. This return to nature is discussed in 3.19.

⬆ *Figure 1* Contrasting drainage management approaches

Sustainable Drainage Systems (SuDS)

Sustainable Drainage Systems (SuDS) represent the ultimate in realistic, yet environmentally friendly, replication of natural drainage systems within any built environment. (The acronym has stuck even though the 'urban' has been removed because the techniques can be adopted in all urban or rural developments whether housing, industrial, commercial or transport.) They hold back and slow surface runoff from any development and allow natural processes to break down pollutants.

Techniques

SuDS techniques include:

◆ swales – wide, shallow drainage channels that are normally dry (Figure **2**)

◆ permeable road and pavement surfaces – use of porous block paving and concrete (Figure **3**)

◆ infiltration trenches – gravel filled drains and filter strips

◆ bioretention basins – gravel and/or sand filtration layers beneath reed beds and other wetland habitats to collect, store and filter dirty water (and provide a habitat for wildlife)

◆ detention basins – excavated basins to act as holding ponds for water storage during flood events

◆ rain-gardens – shallow landscape depressions planted with flowers and shrubs

◆ green roofs – super-insulating wildflower habitats with minimal runoff to gutters.

Benefits

The benefits of SuDS are remarkable in:

◆ slowing down surface water runoff and reducing the risk of flooding

◆ reducing the risk of sewer flooding during heavy rain

◆ preventing water pollution

◆ recharging groundwater to help prevent drought

◆ providing valuable habitats for wildlife in urban areas

◆ creating green spaces for people in urban areas (Figure **2**).

▲ **Figure 2** *Swales can be landscaped as attractive community spaces.*

▼ **Figure 3** *Permeable through-draining car park. The blocks have a 5 mm gap filled with grit allowing water to soak away into the ground below.*

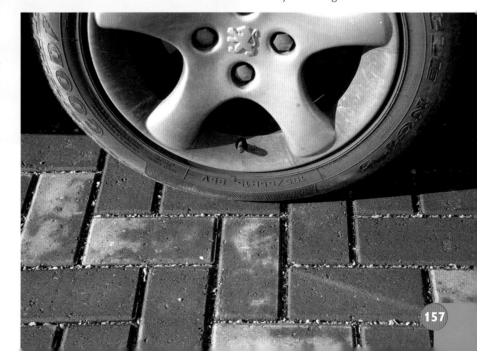

Lamb Drove, Cambourne

Look at Figures **4** and **5**. Lamb Drove in Cambourne, west of Cambridge, is an award-winning one hectare Cambridge Housing Society development of 35 'affordable' homes. Cambridgeshire is a relatively low-lying county where flooding in river valleys and urban watercourses is a major concern. The project was part of a European-funded programme (FLOWS), which featured 40 projects throughout Germany, the Netherlands, Norway, Sweden and the UK.

The original aim of the Lamb Drove SuDS scheme was to:

▲ *Figure 4* *Detention basin, Lamb Drove, Cambourne*

◆ showcase practical and innovative sustainable water management techniques within new residential developments

◆ demonstrate that SuDS are a viable and attractive alternative to more traditional forms of drainage and to deliver practical solutions for new housing areas.

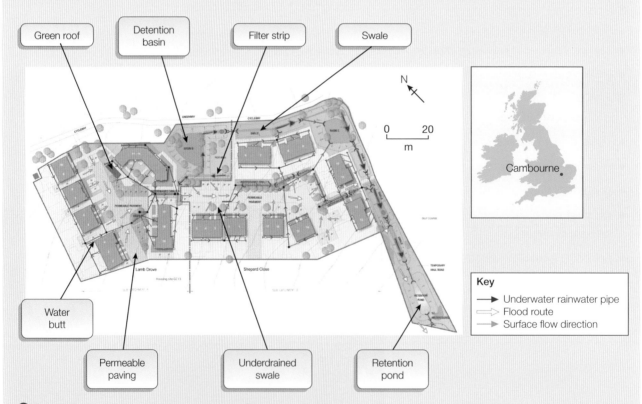

▲ *Figure 5* *Final design of the Lamb Drove, Cambourne SuDS scheme*

A range of SuDS components were used to demonstrate different available techniques and the application of a management train (see below) from prevention to site control and regional control components. The measures include:

◆ water butts to collect roof water for garden irrigation

◆ permeable paving allowing water to enter porous storage zones and to filter out pollutants.

◆ a green sedum roof to reduce and treat runoff

◆ swales (shallow open channels) to collect all excess water from the site, further slowing the flow and continuing the water treatment process

◆ creation of detention basins and wetlands in open spaces to slow down the runoff rate and store water on a temporary short-term basis during extreme (flood) events (Figure **4**)

◆ a retention pond for final storage of water before being released to a drainage ditch beyond the development site.

The developers adopted the concept of a *management train* at the site. This uses simple, natural and visible drainage components in series to improve the water quality. A management train also controls the quantity of runoff incrementally by reducing flow rates and volumes. In short, water management was considered from the point at which it falls on land and buildings to the point at which it leaves the site – so mimicking as much as possible the natural pattern of drainage prior to development. Roofwater that is not collected in water butts flows directly to grass swales or under-drained swales where pollutants (including heavy metals) are filtered. Rain falling on roads or paths passes through the permeable block paving, again where it is filtered and stored in the permeable layer of crushed rock below.

As such, the management train has water travelling downstream through a series of swales, detention basins and wetlands until it reaches the final retention pond.

The project is a proven success having been monitored and appraised ever since its completion in 2006. Concerns that standing water might prove to be a hazard have proved unfounded. Furthermore, it is cost-effective – both construction and ongoing maintenance costs have been 10 per cent less than conventional pipe drainage systems. There has also been a substantial improvement in the biodiversity, ecology and subsequent quality of life at Lamb Drove compared to typical residential developments. For example, the sculptured swales and detention basins have resulted in a visually enhanced and attractive landscape providing an increased amenity and social value to both residents and the local community. The use of SuDS has also resulted in an improved quality of water leaving the site compared with traditional piped drainage systems.

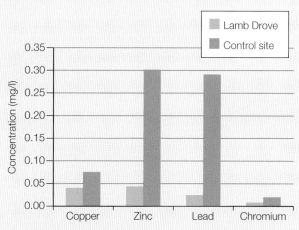

Figure 6 *A comparison of heavy metal concentrations at Lamb Drove, Cambourne and a nearby control site*

ACTIVITIES

1 Study Figure **1**. Critically assess hard- and soft-engineering solutions to urban drainage management.

2 Study Figures **5** and **6**.

Explain why the Lamb Drove development is now cited as 'best practice' for all planners nationally. Consider social, economic and environmental costs and benefits.

STRETCH YOURSELF

SuDS could be described as a hybrid of hard- and soft-engineering drainage techniques. Discuss this point of view.

Think about

It is worth considering whether or not SuDS systems are designed to reduce the impact that the surface water drainage system of one site has on other sites downstream, rather than reducing flooding on the development site itself. In fact would it be appropriate to go as far to suggest that it is a misconception of SuDS systems that they reduce flooding on the development site?

In this section you will learn about restoration and conservation in damaged urban catchments

River restoration and conservation

River restoration involves removing all hard engineering adaptations to restore meanders, wetlands and floodplains. Drainage and therefore flood management is returned to nature.

River Skerne, Darlington

A good UK example of river restoration is the River Skerne in Darlington (Figure **1**). The River Skerne is a tributary of the River Tees in north-east England. Between 1850 and 1945, the river was straightened and channelised (with the river corridor markedly narrowed) to accommodate industrialisation and urbanisation. Subsequent widening and deepening was undertaken in the 1950s and the 1970s to further improve drainage and reduce flood risk, and new housing developments were built along the entire north side of the river. The area's rich industrial heritage included ironworks and the first railway company, Stockton and Darlington.

Deindustrialisation had left it a polluted wasteland, with much of the floodplain raised by old industrial waste tipping. Gas and sewer pipes running alongside the river severely constrained the scope of restoration opportunities. A project adopting the use of soft revetments, such as willow mattress, on the outer banks of the restored meanders to protect the pipelines from erosion was initiated in July 1995. The River Restoration Centre has since restored a 2 km stretch of the river into an attractive wetland environment thriving with wildlife.

🔺 **Figure 1** *The restoration of the River Skerne, Darlington*

Such a complete back-to-nature approach is often impractical in urban areas because the built environment is so established. Urban river restoration has to be partial, therefore, with a greater emphasis on conservation. Indeed, such an approach is likely to widen the benefits markedly by:

- maintaining drainage and managing flood risk
- improving the appearance of brownfield sites
- stimulating investment to encourage the return of industry to inner-city areas
- providing routeways such as cycle paths
- promoting educational activities such as nature walks and wildlife gardening
- providing recreational opportunities for anglers, bird watchers and ramblers
- maintaining rich biodiversity by protecting native species and restricting or reducing non-native wildlife.

ACTIVITIES

1 Summarise the benefits of river restoration and conservation in urban areas.
2 With reference to the Blue Loop, outline the challenges to its successful and sustainable management.

STRETCH YOURSELF

Assess the priorities of each stakeholder involved in the Blue Loop Community Project with particular reference to potential conflicts of interest.

Sheffield city centre's Blue Loop

To walk the Blue Loop is to follow Sheffield's industrial past (Figure **3**). During the 1800s, the River Don was essential in providing water for industrial cooling and processing, but it proved difficult to navigate. The opening of the much anticipated Sheffield and Tinsley Canal in 1819 provided the infrastructure for the mass export of coal, steel and manufactured goods. But, the arrival of the railway to Sheffield in 1848 led to a marked decline in the prosperity of the canal. While some commercial use continued until the early 1970s, the canal and associated industrial buildings fell into increasing disrepair and dereliction. It was not until the injection of funding and management of the Sheffield Development Corporation (1992), helped by British Waterways, that restoration and conservation of both the polluted, lifeless River Don and the canal could start.

Additional funding by Natural England and the National Lottery has allowed the regeneration to evolve into an admirable multi-agency and multi-purpose conservation area. The Blue Loop Community Project is:

◆ rejuvenating much of a 13 km waterfront for cyclists, walkers and runners

◆ encouraging biodiversity by restoring natural ecosystems, and protecting wildlife species and habitats – song thrushes, water fowl, herons and kingfishers now thrive

▲ **Figure 2** *The local community is much engaged with recreational opportunities associated with the Blue Loop Community Project. Family anglers along the River Don near the Salmon Pastures Nature Reserve.*

◆ clearing non-native species of buddleia and Japanese knotweed (see *AQA Geography A Level & AS Physical Geography*, 6.11)

◆ engaging local communities through volunteering, and promoting family events, festivals and recreation (Figure **2**)

◆ reducing flood risk by increasing wetland areas and vegetation on the floodplain, and adopting SuDS techniques (see 3.18). For example, businesses are educated in sustainable living – installing green roofs to help conserve wildlife, improve the appearance of buildings and reduce the risk of flooding.

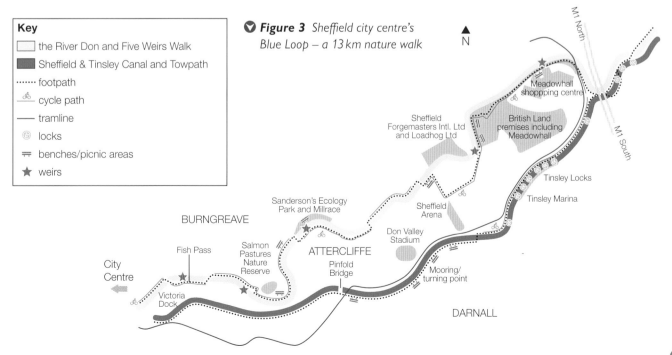

Key
- ☐ the River Don and Five Weirs Walk
- ▬ Sheffield & Tinsley Canal and Towpath
- ⋯⋯ footpath
- 🚲 cycle path
- — tramline
- Ⓛ locks
- = benches/picnic areas
- ★ weirs

▼ **Figure 3** *Sheffield city centre's Blue Loop – a 13 km nature walk*

N

In this section you will learn about:

◆ sources of urban waste
◆ the relation of waste components and waste streams to economic characteristics, lifestyles and attitudes
◆ the environmental impacts of alternative approaches to waste disposal

What a load of rubbish!

Look at Figure **1**. Up to one-fifth of all waste generated globally is likely to be from an urban area (municipal), although the proportions are estimates (measuring, let alone classifying waste is notoriously difficult). However, from the statistics available, it would appear that:

◆ on average, people in HDEs produce 10 to 30 times more waste than those in LDEs

◆ waste generation globally is growing exponentially.

Urban waste is not simply domestic rubbish, but also comes from industrial and commercial activity (Figure **2**).

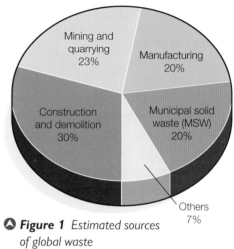

△ **Figure 1** *Estimated sources of global waste*

◯ **Figure 2** *All sources of urban waste*

Source and type of waste	Related issues
Domestic (residential) waste is generated as a consequence of household activities such as cooking, cleaning, repairs, hobbies and redecoration. It includes empty containers, packaging, clothing, old books, paper and old furnishings.	This waste is categorised as biodegradable, recyclable or inert.
Municipal waste results from municipal activities and services such as street cleaning. It includes dead animals, market wastes and abandoned vehicles.	The term is more commonly applied in a wider sense to include solid domestic and commercial wastes – hence municipal solid waste (MSW) becoming synonymous with urban waste (see Figure **3**).
Commercial waste from offices, wholesale and retail stores, restaurants, hotels, markets, warehouses and other commercial establishments.	Again, this waste is categorised as biodegradable, recyclable or inert.
Animal and vegetable waste resulting from the handling, storage, sale, cooking and serving of food.	This waste quickly becomes putrid, producing strong smells and therefore attracting rats, flies and other vermin. It requires immediate attention in its storage, handling and disposal.
Institutional waste from institutions such as schools, universities, hospitals and research institutes. This includes wastes that are considered to be hazardous to public health and the environment.	Hospital waste is categorised as Risk or Non-Risk Waste. The latter includes 'sharps', in addition to infectious, pharmaceutical, chemical and radioactive waste. All require specialist disposal including incineration.
Ashes are the residues from the burning of wood, coal, charcoal, coke and other combustible materials for cooking and heating in houses, institutions and small industrial establishments. Ashes consist of a fine powdery residue, cinders and clinker, often mixed with small pieces of metal and glass.	When produced in large quantities at power generation plants and factories, these wastes are classified as industrial wastes.
Bulky waste include domestic furniture and 'white goods', commercial packaging and containers, and industrial crates, pallets and metal banding**.**	This waste require special collection.
Street sweeping waste includes paper, cardboard, plastic, dirt, dust, leaves and other vegetable matter.	The mixed nature of this waste usually makes separation of the biodegradable, recyclable or inert uneconomic.

Source and type of waste	Related issues
Dead animals, both large and small, that die naturally or are killed accidentally.	If not collected promptly, dead animals are particularly offensive and a threat to public health because they attract flies and other vermin as they rot. Carcass and animal parts from slaughterhouses are regarded as industrial wastes.
Construction and demolition waste consists mainly of earth, stones, concrete, bricks, timber, roofing and plumbing materials, heating systems and electrical wires.	If not recycled, this waste will make up a significant proportion of landfill.
Industrial waste covers a vast range of substances which are unique to each industry. Major generators of industrial solid waste include thermal power plants (coal ash), integrated iron and steel works (slag) and pulp and paper industries (lime).	

Municipal solid waste (MSW)

Municipal solid waste (MSW) is known commonly in the UK as refuse or rubbish (Figure **3**). Most definitions do not include industrial waste, agricultural waste, medical waste, radioactive waste or sewage sludge. Furthermore, the composition of MSW varies greatly from country to country and changes significantly through time. For example, at the start of the twentieth century, the majority of domestic waste (53 per cent) in the UK consisted of coal ash from open fires. Nowadays, in HDEs and EMEs (without significant recycling activity), MSW predominantly includes food wastes, market and street wastes, plastic containers and product packaging materials, and other miscellaneous solid wastes from residential, commercial and institutional sources.

◉ **Figure 4** *Assorted hazardous waste material awaiting recycling*

Type of waste	Examples
Biodegradable waste	Food and kitchen waste, green waste and newspaper (most can be recycled although some plant material may be excluded if difficult to compost)
Recyclable materials	Paper, cardboard, glass, bottles, jars, tin cans, aluminum cans, aluminum foil, metals, certain plastics, fabrics, clothes, tyres and batteries
Inert waste	Construction and demolition waste, dirt, rocks and debris
Electrical and electronic waste	Electrical appliances, light bulbs, washing machines and other 'white goods', TVs, computers, mobile phones, alarm clocks and watches
Composite wastes	Waste clothing, drinks cartons and waste plastics such as toys
Hazardous waste	Most paints, chemicals, tyres, batteries, light bulbs, electrical appliances, fluorescent lamps, aerosol spray cans and fertilisers (Figure **4**)
Toxic waste	Pesticides, herbicides and fungicides
Biomedical waste	Expired prescription (pharmaceutical) drugs

⊙ **Figure 3** *A typical classification of MSW*

Waste streams

Both the nature of waste and its 'journey' from source to disposal will vary according to the economic characteristics, lifestyles and attitudes prevalent within any particular society. The complete flow of waste from its domestic, commercial or industrial source, through to recovery, recycling or final disposal is known as a **waste stream**. In HDEs, this is increasingly regulated and managed. However, in most LDEs and EMEs indiscriminate and improper dumping of MSW without treatment is particularly common. This raises several serious environmental issues including:

◆ loss of recyclable resources such as metals, plastic and glass

◆ loss of potential resources such as compost from organic waste, and energy from controlled incineration

◆ contamination of land and water bodies (from **leachates**)

◆ air pollution due to emissions from burning and the release of methane from decomposition

◆ multiple risks to human health (including respiratory problems, skin and other diseases).

Did you know?

Ninety per cent of all products bought become waste within six months of purchase!

Global waste trade

The **global waste trade** is the international trade of waste between countries for its disposal, recycling or further treatment. The trade is predominantly from HDEs of the North to EMEs and LDEs of the South. A notable exception is the THORP reprocessing plant at Sellafield in Cumbria. It imports used nuclear (power) fuel rods from all over the world in order to extract reusable uranium and plutonium (see 5.19).

Critics of the global waste trade claim that inadequate regulation has allowed many EMEs and LDEs to become toxic dumps for hazardous waste. Most of the world's most dangerous and toxic wastes are produced by western countries (such as the USA and in Europe), yet people in countries that produce little or no toxic waste suffer negative health effects. EMEs and LDEs do not always have safe recycling processes or facilities, and workers process the toxic waste with their bare hands, leading to both illness and death (Figure **5**).

Also, hazardous wastes are often not disposed of properly or treated, leading to poisoning of the surrounding environment and disastrous effects upon natural ecosystems. For example, an estimated 50 million tonnes of waste electrical and electronic equipment (WEEE) are produced each year, the majority of which comes from the United States and Europe. Most WEEE is shipped to EMEs and LDEs in Asia and Africa to be processed and recycled. Heavy metals, toxins and chemicals leak from these discarded products into surrounding waterways and groundwater, poisoning the local people. Workers in the dumps, local children searching for items to sell and people living in the surrounding communities are all exposed to dangerous health risks as a consequence.

⊙ Figure 5 Workers in New Delhi, India dismantle obsolete computers and extract valuable materials such as nickel and copper with their bare hands

Manila, the Philippines

Manila in the Philippines is typical of an EME facing major waste-disposal problems. It is estimated that only around 10 per cent of Manila's waste is (officially) recycled or composted, leaving thousands of tonnes of MSW generated daily to be disposed of. Four-fifths is collected and transported to vast landfill sites – the rest is burned or dumped illegally.

Look at Figure **6**. Payatas is the largest of Manila's landfill sites – six mountains of rubbish tens of metres high covering 200 hectares! Opened in 1973, it remains in use despite being officially closed following a collapse during a rainstorm in July 2000 which killed over 200 people. More than 80 000 slum dwellers live around Payatas – their lives blighted by the stench, and drinking water contaminated with heavy metals, lubricants and solvents. Over 4000 waste-pickers face severe health problems including typhoid, hepatitis, cholera and other infectious diseases.

Figure 6 *Waste disposal, Manila, Philippines. Over 4000 waste pickers, including children as young as five, sort through rubbish at Payatas landfill site. They sell what they find to dealers who resell to manufacturers to produce new products from the waste.*

Alternative approaches to waste-disposal

Given the potentially harmful environmental impacts of the phenomenal quantities of MSW produced in urban areas, effective management is essential. Disposal methods vary globally according to location, economic and political circumstances.

EU and UK government legislation and targets relating to submarine dumping, landfill, incineration and recycling are both strict and regularly reviewed. For example, in the UK, submarine dumping of sewage sludge has been prohibited since 1998 and radioactive waste since 1999. The latter is buried in steel-clad or concrete and lead-lined vitrified glass containers (see 5.19). Only fish wastes and inert material of natural origin such as rock and mining wastes are now dumped at sea. Indeed, over 99 per cent of submarine dumping is locally-generated sediment resulting from the dredging of harbours and their approaches to ensure they are navigable. (In order to minimise ecological impacts on the sea bed, most dredged material is dumped at established sites. It is also used for beach nourishment or land reclamation.)

Nowadays, incineration and recycling tend to be the first thoughts in waste management. Yet burial in landfill remains the most usual fate of MSW whether directly or following extraction of recyclable waste in materials recovery facilities (MRFs). But landfill sites in HDEs are not the vast, hazardous waste mountains typified by Manila's Payatas in the Philippines (Figure **6**). A modern sanitary landfill is not a dump, but more usually a well protected, engineered facility distant from built-up areas. Most commonly, after the waste is dumped, it is compacted by large machines and then sealed with plastic sheeting before burial under top soil. Health, safety and environmental controls are strict and so associated pollution minimised. The arguments for and against landfill, incineration and recycling are nothing if not powerful! (See Figure **7**.)

Finally, composting, resource and energy recovery targets feature explicitly in many new initiatives given ambitions for a more sustainable future (Figure **8**).

Landfill

For

- Makes good use of abandoned quarries
- Easily managed
- Methane can be vented and used as a fuel
- On reaching capacity can be sealed, top-soiled and landscaped for recreational use
- Cost-effective and relatively safe if managed efficiently

Against

- Attracts vermin, flies and scavenging birds
- Wind-blown material becomes unsightly litter
- Burying organic waste leads to anaerobic decay
- Subsidence is common as the waste degrades
- Produces methane – a powerful greenhouse gas
- Leachates percolating into groundwater can be toxic
- Heavy, dirty lorry traffic is generated
- Smell is unpleasant, particularly in hot weather

Incineration

For

- Produces energy from burning MSW
- Heat, steam and ash produced are valuable resources
- Requires far less land than landfill sites
- Long life span
- Cost-effective once constructed and operational
- Safe disposal of hazardous waste such as medical waste

Against

- Particulate emissions require managing
- Chimney emissions can be toxic if not managed
- Carbon dioxide emissions are a greenhouse gas
- Not all MSW is combustible

Recycling

For

- Doorstep 'wheelie-bin' collection sanitary and safe
- 'Single-stream recycling' – all recyclable materials in one bin popular and convenient
- Organic waste can be composted and sold to enrich garden soil
- 'Resource recovery' implicit given reprocessing of recycled materials into new products
- Supports associated niche markets such as architectural salvage
- More recycling means less landfill

Against

- Public collection points, such as bottle banks, can generate litter
- Public recycling facilities require expensive, safe operation to avoid hazardous leakage – acid in batteries and CFC gases in old fridges
- Public resistance if separately charged
- Public separation of paper, (washed) plastic, metals, garden waste and landfill inconvenient and prone to error
- Electrical and electronic waste (WEEE), such as computers, contain toxic components including lead, cadmium and beryllium, so careful dismantling is essential

▲ **Figure 7** *Arguments for and against landfill, incineration and recycling*

Energy from Waste (EfW), Lincoln

Lincolnshire's £125 million Energy from Waste facility started operating in July 2013. The site processes 462 tonnes of MSW every day that would otherwise be dumped in landfill. Eleven MW of electricity is generated, enough to power 15 000 homes each year! The 24-hour facility employs 33 people and the visitor centre is a popular educational resource. All ash is recycled and air pollution is minimal and strictly controlled (Figure **8**).

Waste management in Bristol

Bristol is the largest city in the south-west of England with a population expected to reach 500 000 by 2029. The UK's first city to be awarded European Green Capital status (2015), it is committed to a programme of radical economic, social and environmental changes, so improving quality of life markedly for its citizens. Urban regeneration, a changing economic structure, improved (integrated) transport infrastructure, new green spaces, and cultural and sporting facilities are transforming this 'core city.' Increasingly efficient waste management is an integral element of this overall strategy (Figure **9**).

Bristol City Council committed to:

◆ reducing the amount of waste generated per household by 15 per cent

◆ reducing the amount of MSW sent to landfill

◆ increasing waste recycling to 50 per cent.

Initiatives include specialised kerbside collections of recyclable waste, exacting targets for waste management contractors and recycling education in schools. Furthermore, the Avonmouth waste treatment plant processes 200 000 tonnes of MSW per year, incinerating enough non-recyclable waste to generate electricity for nearly 25 000 homes.

▲ *Figure 8* *Lincolnshire's Energy from Waste facility, North Hykeham, Lincoln*

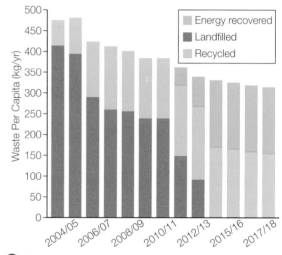

▲ *Figure 9* *Existing and projected changes in Bristol's waste management*

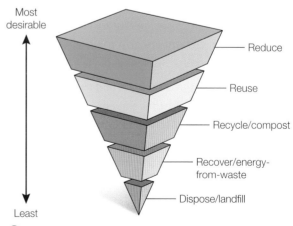

▲ *Figure 10* *The 3Rs – reduce, reuse and recycle*

ACTIVITIES

1 Suggest reasons for the following statements:
   a On average, people in HDEs produce 10 to 30 times more waste than those in LDEs.
   b Waste generation globally is growing exponentially.
2 Outline the political, economic, social and environmental arguments for reducing landfill, but increasing recycling and incineration.
3 Look at Figure **10** and again at Figure **7**. Discuss the following statement.
   'The key to sustainable MSW management is the *3Rs* – reduce, reuse and recycle.'
4 Study Figure **9**. Describe the existing and projected changes in Bristol's waste management.

STRETCH YOURSELF

Read section 4.26 on the *circular economy*.
To what extent could the circular economy be a way forward in addressing environmental issues associated with urbanisation?

In this section you will learn about environmental problems in contrasting urban areas and strategies to manage these problems

Atmospheric pollution

Concentrated energy use and vehicle emissions in urban areas lead to greater air pollution with significant impact on human health. Strategies to manage these domestic, industrial and vehicle emissions are examined in detail in section 3.16. Clean air legislation, vehicle restrictions and technical innovations are improving air quality markedly in many cities, particularly in HDEs, but widespread progress is not yet global.

Water pollution

Urban water pollution comes from:

◆ domestic waste water from sinks, washing machines, bathrooms and toilets (sewage)

◆ effluent from industries

◆ leachates from illegal dumping and poorly managed landfill

◆ rainwater runoff from roads, pavements and roofs.

It is possible to identify point sources such as pipes from factories discharging effluent into a river and non-point (diffuse) sources such as surface-water drainage, urban runoff from brownfield sites and roads. All of these are contaminating and will significantly harm groundwater and therefore rivers. They also pose a significant risk to health. For example, sewage contains pathogens which spread infectious diseases leading to diarrhoeal disease and increasing child mortality, especially in poorer countries. Urban surface runoff alone contains oils and traces of heavy metals – the latter persisting for a long time in river sediments. Leachates and effluent are also toxic. Urban waste water collection and treatment is therefore essential (Figure **1**).

Fortunately, sewage treatment and rigorous legislation on the treatment of industrial effluent is commonplace in HDEs. For example, the EU's 1991 Urban Waste Water Treatment Directive (revised 1998) makes sewage and effluent treatment obligatory, with measurable improvements to river and ocean water quality in consequence.

In LDEs and EMEs progress is inevitably slower, but organisations such as the UN and World Bank are investing huge resources into addressing both water quality and supply issues. Treatment plants, dam and reservoir schemes, and aquifer identification and exploitation are just some examples. Add to this the work of NGOs such as the British charity Water Aid, and the remarkable Bill and Melinda Gates Foundation and there is considerable hope for the future (Figure **2**).

▼ *Figure 1 Modern urban waste water treatment*

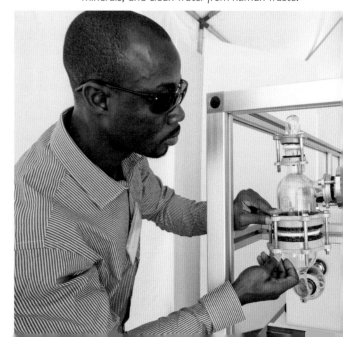

▼ *Figure 2 The Bill and Melinda Gates Foundation has raised more than US$75 million to build thousands of freshwater wells and develop appropriate technologies to address water pollution. This prototype toilet (designed by Loughborough University researchers) extracts biological charcoal, minerals, and clean water from human waste.*

Urban waste water treatment in the UK

Before waste water can be treated it needs to be collected. Every day in the UK, over 600 000 km of sewers collect over 11 billion litres of waste water from homes, municipal, commercial and industrial premises, and rainwater runoff from roads and other impermeable surfaces. There are three main types of collection system:

- surface-water drainage that collects rainwater runoff from roads and urban areas and discharges direct to local waters
- combined sewerage that collects rainwater runoff and waste water from domestic, industrial, commercial and other premises
- foul drainage that collects domestic waste water from premises.

Both surface water and foul drainage may eventually connect to combined sewerage where there are no local environmental waters to which surface-water drainage can discharge. Combined sewerage systems are not uncommon in the UK and elsewhere in Europe. A basic requirement of combined sewerage systems is that they need to cater for all normal local climatic conditions including storm water from peak seasonal wet weather. However, there may be times when heavy continuous rainfall will temporarily exceed the capacity of combined sewerage systems and so 'combined sewer overflows' are designed and built as an integral part of combined sewerage systems. These allow excess waste water to be discharged to local rivers to avoid sewers being overwhelmed and waste water 'backing up' along sewers and flooding streets and properties, or overwhelming waste water treatment plants.

Look at Figure **3**. Around 9000 waste water treatment plants usually process combined sewerage in four stages:

- preliminary treatment – to remove grit, gravel and screen large solids
- primary treatment – to settle larger suspended, generally organic, matter
- secondary treatment – to biologically break down and reduce residual organic matter under controlled conditions
- tertiary (advanced) treatment – tailored to address specified pollutants using different treatment processes.

An emerging issue is linked to pharmaceuticals and other chemicals such as antibiotics, birth control pills and chemotherapy agents. Tertiary treatment is therefore increasingly challenged and will have to evolve from the established:

- reduction of nutrients (phosphorus or nitrogen compounds) to protect waters from eutrophication (an excess of nutrients in water)
- reduction of nitrates and pathogens to protect bathing or shellfish waters
- reduction of ammonia to protect freshwater fisheries.

▶ **Figure 3** *The basic principles of waste water treatment*

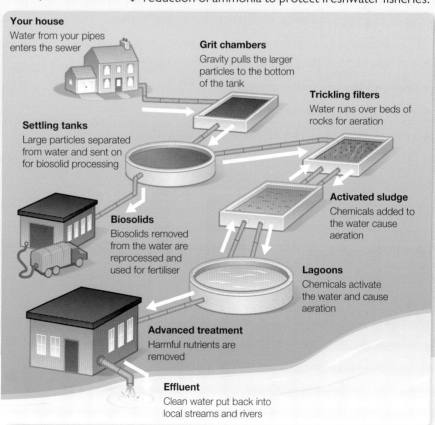

Your house
Water from your pipes enters the sewer

Grit chambers
Gravity pulls the larger particles to the bottom of the tank

Trickling filters
Water runs over beds of rocks for aeration

Settling tanks
Large particles separated from water and sent on for biosolid processing

Biosolids
Biosolids removed from the water are reprocessed and used for fertiliser

Activated sludge
Chemicals added to the water cause aeration

Lagoons
Chemicals activate the water and cause aeration

Advanced treatment
Harmful nutrients are removed

Effluent
Clean water put back into local streams and rivers

Dereliction

Derelict (neglected or abandoned) urban land results from:

◆ the inevitable ageing and decay of buildings with the passage of time (causing maintenance costs to spiral)

◆ the movement of urban activities to better and more profitable locations

◆ changes in an urban economy brought about by deindustrialisation (Figure **4**).

Brownfield site development

Redevelopment of brownfield sites is a key feature to the development of sustainable cities (see 3.22). But there are difficulties with building on them:

◆ Thousands of brownfield sites have been contaminated by previous industrial uses and may present significant risks to human health and to the wider environment. Decontamination is both time-consuming and very expensive.

◆ Not all brownfield sites have the physical access necessary for residential development.

◆ The neighbouring land may still be used for industrial purposes making the brownfield site unsuitable for new homes, for example, a sewage treatment works or a heavy industrial plant.

In the UK these unsightly brownfield sites represent a valuable urban resource (Figure **5**) – despite the potential high costs of clearance and decontamination. For example, their redevelopment for housing:

◆ starts to address the pressing need for more homes

◆ improves the urban environment

◆ reduces urban sprawl and so protects green-belts

◆ reduces demand on car use (commuting from suburbs or further afield).

▲ **Figure 4** *Urban dereliction can be hazardous and also create eyesores*

Did you know?
Friends of the Earth argue that there are already 750 000 unoccupied houses in British cities which could be upgraded, while a further 1 250 000 could be created by either subdividing large houses or using empty space above shops and offices.

◀ **Figure 5** *New housing development at Lysaght Village on the site of the former steelworks, Newport, Gwent*

Brownfield regeneration of Bristol's river docks area

Between 2006 and 2013, 94 per cent of new housing in Bristol was built on brownfield sites and it is estimated that a further 30000 new homes will be needed by 2026. The city council is confident that this can be achieved without using any greenfield sites.

Look at Figure **6**. In the 1960s, when cargo ships became too large to come up the River Avon, the docks area went into decline and dependent industries such as tobacco factories closed. The subsequent regeneration and redevelopment of the area has required cooperation between the city council, landowners, private developers and the South West Regional Development Agency.

Much has been achieved:

◆ Decaying industrial buildings have been redeveloped for residential purposes.

◆ Several listed buildings have been preserved.

◆ Derelict land has been cleared and decontaminated for new housing, businesses, and cultural and leisure facilities.

Such an approach is now commonplace where deindustrialisation has blighted waterfront and dockland areas in some of the UK's largest cities. For example, both planned and completed projects are to be found in London, Liverpool, Glasgow, Cardiff and Belfast.

⬆ **Figure 6** *Brownfield site regeneration of Bristol's river docks area*

ACTIVITIES

1 Create a list of ten environmental problems facing urban areas today. Rank them from the most to the least pressing and justify your decisions.

2 Outline the primary sources, impacts and treatments of urban waste water.

3 Explain why building new homes on brownfield sites is more expensive than developing greenfield locations.

4 Whether for example the Bristol river docks area, London Docklands (3.24) or Cardiff Bay (3.11), all such schemes have transformed areas of urban dereliction. With reference to any one of these, or a similar named project in your area, identify and outline five key advantages and five key disadvantages associated with the regeneration. (For both advantages and disadvantages, think of social, economic and environmental aspects.)

In this section you will learn about:
- the environmental impact and ecological footprint of major urban areas
- dimensions of sustainability: natural, physical, social and economic
- the nature and features of sustainable cities
- contemporary opportunities and challenges in developing more sustainable cities

The impact of urban areas on local and global environments

Throughout this chapter you will have learnt that urban areas have a marked impact on both local and global environments. High densities of people and buildings compete for space and consume vast quantities of water, energy and other resources. There are also problems of air pollution, traffic congestion and waste disposal. By developing the **ecological footprint** concept (the area of land needed to provide the necessary resources and absorb the wastes generated by a community), the impact of cities on the environment can be highlighted dramatically (Figure **1**). For example, in the first such analysis (2002) of a major world city, London's ecological footprint turned out to be 120 times the area of the conurbation itself! Indeed, such are the needs and demands of urban populations in HDEs that their ecological footprints are more than ten-times those of people living in similarly sized urban areas in LDEs.. In short, urbanisation nowadays is so important and dominant that any hopes of approaching global sustainability depend on developing more sustainable cities.

Figure 1 *The urban ecological footprint*

Sustainable cities and liveability

Sustainability simply means meeting the needs and hopes of today without messing up the future! In effect, making the best use of resources, protecting the environment, controlling growth and waste – all of which have an acute relevance to urban lifestyles. But, it is the concept of **liveability** that best describes the natural, physical, social and economic dimensions of sustainability in an urban context. It has been a buzz word in city development for some time – it encapsulates that understandable urban ideal of collectively improving everyone's quality of life both now and in the future (Figure **2**). The truly liveable, sustainable city might well prove to be an impossible dream, but careful planning and management can certainly take us a long way towards it.

Figure 2 *What does 'liveability' mean to you? The size of each word is proportional to the number of times it occurred in all responses to a Twitter competition to define liveability run by the San Francisco County Transportation Authority.*

Did you know?
The ecological footprint of the Tokyo metropolitan area has been calculated to be almost three times the land area of Japan as a whole!

The nature and features of sustainable cities

◆ Greener built environments: using energy and water more efficiently, reducing MSW and managing it better (see 3.19 and 3.20).

◆ Improved transport: developing infrastructure, networks and **modes** to meet demand without increasing congestion and pollution. Everything from excellent public transport to vehicle restrictions and technical innovations can be considered (see 3.16). The new term 'livable communities' stems from this ideal – what the US Department of Transport defines as 'transportation, housing and commercial development investments … coordinated so that people have access to adequate, affordable and environmentally sustainable travel options'.

◆ Planned expansion: encouraging 'compact cities' rather than uncontrolled and unrestricted urban sprawl. Developing brownfield sites is hugely significant (see 3.21).

◆ Conserving buildings and open spaces to be used and enjoyed by everyone: restoring important historic buildings, brownfield clearance to create new green spaces and protection of existing open spaces. Improving biodiversity within urban river systems and ecosystems is important in this respect (see 3.19).

◆ Carbon-neutral development: building structures such as houses that generate as much energy as they use – so reducing pollution. Beddington Zero Energy Development (BedZED) in Hackbridge, London is an example of this (Figure **3**).

▲ *Figure 3* *BedZED consists of 82 'affordable' homes, 18 workplaces, retail and leisure areas. It includes a children's nursery, medical centre, sport pitch, exhibition centre, offices and meeting rooms. The scheme includes a biomass-combined heat and power plant, an onsite sewage treatment and rainwater recycling system, and natural wind driven ventilation.*

Contemporary opportunities and challenges in developing more sustainable cities

All cities are unique. All offer enormous potential for positive change. But as more people live and work in them, prosperity increases and better transport is demanded, and more energy, goods and food are consumed. This generates more pollution and waste unless the opportunities to develop more sustainable cites are embraced.

This inevitably presents challenges:

◆ Political will: there needs to be long-term strategic planning, 'joined-up' thinking involving all relevant **stakeholders** including governmental departments. Given the short-term thinking that dominates most government – particularly in democracies – this is not a straightforward as it seems.

◆ Globalisation: the inter-connectedness of cities within the global economy has increased the power and influence of TNCs. These large and powerful companies must, therefore, also embrace the need for change.

◆ Economic gains for all: there must be economic incentives for both the wealthiest and poorest if positive change is to happen. Sustainable cities must be inclusive.

◆ Climate change: sustainable cities must stimulate economic growth without increasing greenhouse gas emissions.

Did you know?

In 2010, the San Francisco County Transportation Authority held a competition on Twitter to answer the question: What does 'liveability' mean to you? The elegant, rhyming six-word winning definition was 'accessible places, natural spaces, minimal traces'.

Strategies for developing more sustainable cities

This is an exciting age to be a Geographer, not least for those involved in urban planning. Throughout the world, there are examples of innovative, positive management of sustainable urban change. We now live in an information age, where the internet and globalisation spread knowledge efficiently, so the best strategies for achieving greater urban sustainability can be understood, evaluated, copied and adapted according to need (Figure **4**).

Figure 4 *Globally influential examples of strategies promoting urban sustainability*

Strategy to promote urban sustainability	Example
Greener built environments – to create greener built environments that use water and energy efficiently, reduce urban waste and increase recycling	*Quezon City, Philippines* An inclusive approach to the management of MSW has been adopted including raising awareness of recycling in local communities and schools. This has led to a 38 per cent reduction of MSW deposited in landfill
Improved transport – to expand and develop existing transport infrastructure and networks to meet demand	*Vitoria-Gasteiz, Spain* Bicycle lanes, an expanded public transport network, car sharing and charging points for electric vehicles and have all been introduced (Figure **5**)
Planned expansion – to encourage 'compact cities' and planned expansion, rather than uncontrolled and unrestricted urban sprawl	*Nantes, France* This was the first city to successfully reintroduce electric trams as part of its public transport system. This has helped to decrease commuter traffic on the roads and encourage reurbanisation (urban resurgence)
Economic opportunities – to provide a range of local economic opportunities, including new opportunities in a 'green economy'	*San Antonio, Texas, USA* 125 'green jobs' created as a result of the success of the city's recycling programme
Conserving buildings and open spaces – to protect existing, and create new, green spaces that are valued and made use of by all members of the community. And to support high levels of biodiversity within urban ecosystems	*Queen Elizabeth II Olympic Park, Stratford, London* As part of the sporting complex built for the 2012 Summer Olympics and Paralympics, this brownfield regeneration includes the creation of 100 hectares of new urban park along the River Lea. The new *wildlife corridor*, links Hackney Marshes to the north with the Thames – otters are now able to move freely for the first time in a century
Carbon-neutral development – to remove as much carbon dioxide from the atmosphere as was put into it in construction – a zero carbon footprint	*BedZED, Hackbridge, London* Beddington Zero Energy Development (BedZED) was completed in 2002 and built on a brownfield site. It is the largest zero-energy and carbon-neutral urban village development in the UK (Figure **3**)

Freiburg: a sustainable city

The German city of Freiburg (Figure **6**) set itself the goal of urban sustainability as early as 1970. Natural, physical, social and economic dimensions of sustainability were central to the plans. For example:

◆ Green spaces are both protected and enhanced. Forty per cent of the city is forested with over half of these woodlands protected as nature conservation areas. Only native Black Forest trees and shrubs are planted. Furthermore, excepting flood retention basins, the River Dreisam is unmanaged and so provides natural habitats for flora and fauna.

◆ A sustainable water supply harnesses both rainwater harvesting and wastewater recycling. Groundwater is the city's most important source of drinking water and so has to be protected from pollution. In consequence, Sustainable Drainage Systems (SuDS) including green roofs, permeable road and pavement surfaces, and bioretention basins are widespread (see 3.18).

Figure 5 *Charging points for HEVs, PHEVs and EVs are increasingly common, especially in city centre car parks*

◆ MSW is reused and recycled as much as possible. There are 350 community collection points for recycling, a biogas digester processes all food and garden waste, and energy for 28 000 homes is produced by incineration.

◆ Social dimensions include the provision of 'affordable' energy-saving homes and locals are encouraged to invest in renewable energy resources. In addition to the financial dividends, investors receive free football season tickets!

◆ Economic dimensions include the creation of 10 000 jobs in 1500 environmental businesses. Over 1000 people are employed in solar technologies – in research, development and manufacture of solar energy systems including solar panels (Figure **7**).

⬢ Figure 6 *The location of Freiburg, Germany*

⬢ Figure 7 *Freiburg in the Black Forest has around 400 photovoltaic installations; this truly 'green city' is a leading solar energy capital*

ACTIVITIES

1 Approximately 80 per cent of the population in HDEs live in urban areas (compared with around 50 per cent in LDEs). To what extent is the aim of achieving sustainable cities less of a challenge for HDEs? Explain your answer fully.

2 Suggest reasons why achieving the goal of a sustainable city necessitates long-term planning solutions to economic, social and environmental problems.

3 Examine to what extent 'liveability' is synonymous with 'urban sustainability'.

4 Calculate your ecological footprint in order to understand which areas of your lifestyle have most impact on the planet. Which behaviours will you (realistically) be able to change? Internet search engines have many examples of interactive quiz activities that will enable you to do this.

STRETCH YOURSELF

Evaluate the claim that Freiburg is a truly sustainable city.

In this section you will learn about an EME Olympic city – Rio de Janeiro

Rio de Janeiro – a city of contrasts

Do you want a great party in February? Then get to Brazil because there's none bigger, better or brighter than Rio de Janeiro's world famous annual carnival! Join the 2 million revellers who hit the streets for five days and nights of parades, music and to dance the incomparable samba (Figure **1**).

Rio is one of the most breathtaking cities in the world. As the host city of the FIFA World Cup (2014) and Olympic and Paralympic Games (2016), its global recognition is assured. Twelve million inhabitants are squeezed between miles of sandy beaches, coastal lagoons and steep, forested mountains. Overlooking it all are the rearing granite Sugar Loaf Mountain and the Corcovado with its 40 m high statue of Christ the Redeemer (Figure **2**).

Rio's magnificent scenery, hot, sunny (tropical) climate, Copacabana and Ipanema beaches, and rich cultural and political history make it a magnet for tourism. However, Rio is infamous for high unemployment and crime, traffic congestion, pollution, fatal landslides and extraordinary extremes of wealth.

Figure 1 *Rio's world famous carnival – dancing the night away*

The geography of Rio

Rio was founded by the Portuguese in 1565 as an important port, trading centre and gateway to the rest of South America. It became the colony's capital and Brazil's largest and most important city. Although the industrial megacity of Sao Paulo has overtaken it in size and population, and Brasilia is the new capital, Rio's role in both the regional and global economy is still very important. It is:

◆ South America's top tourist destination

◆ a major centre for banking, finance and insurance

◆ a manufacturing centre for chemicals, pharmaceuticals, clothing, furniture and processed foods

◆ a major port exporting coffee, sugar and iron ore to all parts of the world.

Figure 2 *The locals call Rio 'Cidade Maravilhosa' which means 'Marvellous City'.*

Look at Figure **3**. Spatial patterns of land use, economic inequality and cultural diversity are all illustrated in Rio's division into four main areas – Centro (including the Central Business District, CBD), and North, West and South Zones.

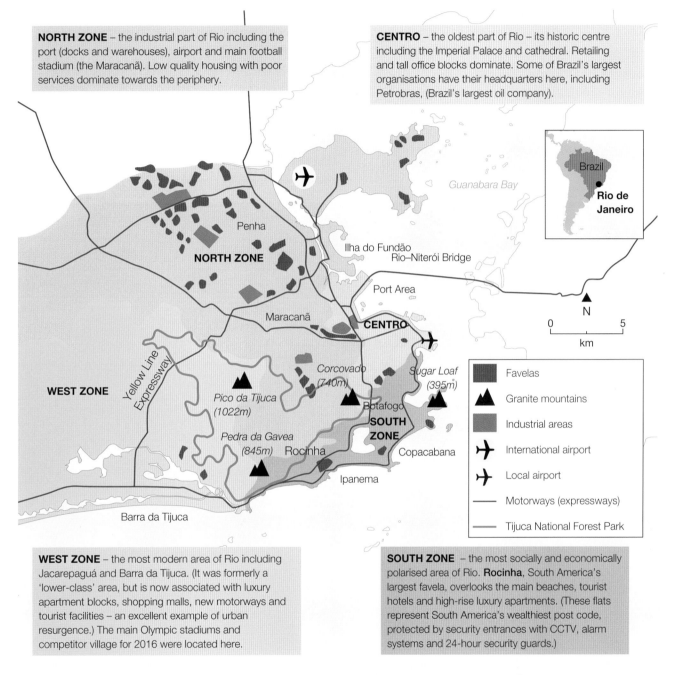

NORTH ZONE – the industrial part of Rio including the port (docks and warehouses), airport and main football stadium (the Maracanã). Low quality housing with poor services dominate towards the periphery.

CENTRO – the oldest part of Rio – its historic centre including the Imperial Palace and cathedral. Retailing and tall office blocks dominate. Some of Brazil's largest organisations have their headquarters here, including Petrobras, (Brazil's largest oil company).

WEST ZONE – the most modern area of Rio including Jacarepaguá and Barra da Tijuca. (It was formerly a 'lower-class' area, but is now associated with luxury apartment blocks, shopping malls, new motorways and tourist facilities – an excellent example of urban resurgence.) The main Olympic stadiums and competitor village for 2016 were located here.

SOUTH ZONE – the most socially and economically polarised area of Rio. **Rocinha**, South America's largest favela, overlooks the main beaches, tourist hotels and high-rise luxury apartments. (These flats represent South America's wealthiest post code, protected by security entrances with CCTV, alarm systems and 24-hour security guards.)

▲ **Figure 3** *Location map of Rio de Janeiro*

Favelas – common misconceptions

A *favela* is often described as a slum or a shanty town. This is a generalisation that is often not the case. Residents take great pride in their homes which are usually brick-built with electricity, running water, sewerage systems and even internet access. The settlements have retail facilities including video rental outlets, bars, travel agents and fast food such as McDonald's. Nor are they ghettos – they have a religious and cultural mix of groups from all over Brazil and are full of life and activity. Residents tend to have low to average incomes but they are far from destitute.

Urban challenges and solutions in Rio

Rapid population growth caused by both natural increase and in-migration puts enormous strains on this densely populated city. Problems of insufficient and inadequate housing, health and services as well as unemployment, congestion, pollution and crime dominate local planning.

Housing, health and services

With over 100000 in-migrants arriving each year, Rio faces acute housing problems (Figure **4**). The result is that many squat on public or private land – often in hazardous environments, such as swamps, hills or beside roads or railways. The remainder rent multiple-occupancy housing.

One-third of informal inhabitants have no access to sewers or electricity and there is not yet piped water for all. It is therefore almost inevitable that infant mortality is high at above 50 per thousand and life expectancy is in the 50s. However, government initiatives are proving to be successful. The Favela-Bairro Project has integrated over 250000 residents in over 140 neighbourhoods. Many houses in Rocinha, for example, were provided with basic sanitation, plumbing and electricity, new leisure, health care and education facilities, secured hillsides to prevent landslides and paved and formally named roads.

Unemployment, underemployment and crime

Unemployment rates are high (around 20 per cent) with the poorly paid and irregular informal service sector (black market) dominating. Encouragement of new economic activities in, for example, ICT and public transport is helping, as is strengthening of traditional industries. But, poverty still dominates with the poorest 50 per cent earning only 13 per cent of Rio's income – virtually the same as the richest 1 per cent! Corruption is widespread, with some favelas controlled by criminal gangs involved in drugs, gun crime, robberies, kidnapping and murder – 6000 people are killed every year. Community policing, by Pacifying Police Units (UPPs) was introduced in 2008 and has restored control in nearly 200 favelas.

⊙ *Figure 4 Rocinha is sited on a very steep hillside in Rio's South Zone, overlooking the beach districts of Ipanema and Copacabana. It is Rio's largest and oldest favela with an estimated population of 150000.*

Congestion, air and water pollution

Forty per cent of inhabitants live in Rio's suburbs resulting in 4 million cars jamming the roads daily. Major road improvement projects such as the 14 km-long Rio-Niterói Bridge and 21 km Yellow Line Expressway (complete with tunnels under the mountains of the Tijuca National Park) have reduced congestion and air pollution significantly. Cable cars link otherwise inaccessible favelas and the Metro system is being extended to five lines totalling 75 km of track (Figures **5** and **6**). All such investment proved to be essential in the build-up to the World Cup and the Olympic Games. Furthermore, tighter environmental laws and controls are reducing raw sewage, industrial effluent, oil and landfill leachates in Guanabara Bay.

▲ **Figure 5** The cable car system, Teleférico do Alemão, services the group of hillside favelas, Complexo do Alemão; it is the first true mass transit system in any slum in Rio

◄ **Figure 6** The Rio de Janeiro Metro was opened in 1979 and has been steadily extended since then; the latest extension was into the West Zone in preparation for the 2016 Olympics

ACTIVITIES

1 Study Figures **3** and **4**.
  a Describe Rocinha's site and situation.
  b What evidence demonstrates that Rocinha:
    (i) has its origins as an informal squatter settlement?
    (ii) has benefited from community improvement schemes?
    (iii) has access and security problems?
2 a Outline the arguments for self-help schemes in the improvement of all of Rio's favelas.
  b Why is demolition of illegal communities, or any shanty, rarely an option? Is it ever an option?

STRETCH YOURSELF

Jardim Gramacho was one of the world's largest landfill sites and a major source of leachates into Guanabara Bay. It closed in 2012 after 34 years of operation. But, every two months, Rio still produces enough MSW to fill the Maracanã stadium! Suggest and justify sustainable solutions to this waste-disposal problem.

In this section you will learn about an HDE Olympic city – London

London – back from the brink

Rio de Janeiro is certainly breathtaking, but London is the most visited city in the world with over 16 million tourists, spending more than £3.3 billion a year. The UK's thriving capital enjoys world city status (see 3.4) although its global economic importance is disproportionate to its total population (8.6 million in 2015) – it is, in fact, the sixth richest place on Earth! Yet London's population declined from 8.6 million in 1939 to 6.8 million in 1981 because of East End counter-urbanisation, deindustrialisation and closure of the docks, which were no longer able to compete with new container ports such as Tilbury and Felixstowe. However, since 1981 its population has been on the increase.

By the early 1980s, huge swathes of London were in terminal decline with male unemployment in some East End communities as high as 60 per cent! This prompted the government under Margaret Thatcher to plan for a *post-industrial economy* by directing investment towards:

◆ a knowledge economy of high-value footloose business, focusing on expertise, management, consultancy, IT, media, finance and banking

◆ the world's fastest-growing industry, tourism, resulting from greater prosperity in HDEs, cheaper air travel and increased car ownership

◆ property, not least brownfield development for housing, office and retail land uses.

So started the world's single largest urban regeneration – 750 hectares of derelict land was reclaimed in London Docklands, 200000 trees planted and 130 hectares of open space created amongst a spatially diverse multi-purpose urban environment of financial, media and high tech industry, radically improved transport infrastructure and over 20000 new homes. The 25 years of redevelopment included gentrification without which, arguably, the 2012 Olympic and Paralympic Games might never have been awarded to London by the International Olympic Committee in 2005. This sparked a similar explosion of investment and redevelopment of the area chosen for the Olympic Park in East London.

Urban regeneration of the East End

The revival of the East End was first met by economic and **property-led regeneration** led by the London Docklands Development Corporation (LDDC, 1981) and Enterprise Zone (1982) (see Figure **1**, 3.8). When the LDDC finished in 1998, the whole Docklands area had seen spectacular redevelopment including the establishment of a secondary financial district. At its heart was, what was then, the UK's tallest building – 1 Canada Square, Canary Wharf (Figure **1**). Improved transport infrastructure included the Docklands Light Railway, Jubilee Line extension of the underground, and City Airport. However, the regeneration was criticised for failing to meet the social needs of the East End, particularly in the provision of affordable housing and appropriate employment.

From 2004 to early 2013, the London Thames Gateway Development Corporation (LTGDC) (see Figure **1**, 3.8) was central to continued East End regeneration. In contrast to property-led regeneration, the LTGDC worked alongside and coordinated existing organisations, such as the Mayor's London Development Agency. This large-scale **community-focused regeneration** was characterised by its social aims and focus on sustainability. By the end of 2012, LTGDC had put in place the programmes, partnerships, projects and sites to deliver a further 10500 new homes (35 per cent of them 'affordable'), 5000 jobs, £1.25 billion private sector investment, 225000 m^2 of commercial floor space, nearly 15000 m^2 of new education space and enhancement of nearly 300 hectares of open space.

▶ *Figure 1* *Canary Wharf financial district*

The Olympic legacy

In July 2005, the International Olympic Committee announced that the 20012 Games would come to London. This proved to be the perfect opportunity to continue the transformation of the East End. Jack Straw, then foreign secretary, promised that the Olympics would be 'a force for regeneration' and the 'greenest' and 'most sustainable' Games in history. The aim was to create a lasting legacy of jobs, homes and environmental improvements for the East End.

Look at Figures **2** and **3**. Delivering this **sports-led regeneration** was not without difficulties, but it was successful overall. An Olympic Park Legacy Company has been set up to manage a sustainable future for the sporting venues. From 2016, for example, the Stadium became the new home of West Ham United FC and the National Competition Centre for athletics in the UK, as well as a major venue for other sporting events – it was one of the stadia used in the 2015 Rugby World Cup. It will be open all year round for arts and cultural events, conferences, visitor tours and music concerts.

▲ **Figure 2** *Access to and organisation of the Games was admired the world over and the stadium, athletes' village and River Lea waterside park all have a sustainable future*

Focus of regeneration	Aim	Legacy
Housing	• Derive 4000 homes from the athletes' village including 3000 'affordable' homes.	• 2800 homes have been derived from the athletes' village but of these only 1200 are 'affordable'.
Employment	• Create 3000 new jobs. After the Games, the International Broadcast Centre (IBC), described as the best connected building in London, intended to spearhead a 'London Silicon Valley' in which media and technology companies will locate.	• From 2014, the former IBC will be home to iCITY – a new sustainable, high tech commercial business district, with the most advanced digital infrastructure in Europe. 4600 jobs will be created on the site and 2000 jobs in the resulting supply chain. But prior to redevelopment, 380 companies employing 11 000 people were forced to relocate – and not all stayed locally.
Transport	• Improve transport infrastructure including the establishment of an international station on the Kent Link (connecting to Channel Tunnel Rail Link).	• New transport infrastructure includes roads and the impressive Stratford International rail station. Once there is sufficient demand, Eurostar services to the continent will stop here.
Environment	• Create the largest urban park in Europe from a clean-up of 200 ha of contaminated brownfield sites. • Clean up polluted River Lea and canal network. • Establish new wildlife habitats.	• 4000 trees have been planted and new habitats, supported by breeding programmes, have been created. • A waterside park has been created from the clean-up of the River Lea and adjoining canals. • 30 000 tonnes of rubbish have been cleared and towpaths opened up to walkers and cyclists.

▲ **Figure 3** *Sports-led regeneration – London Olympic aims and legacy*

ACTIVITIES

1 Outline the advantages and disadvantages of a knowledge economy as a replacement for manufacturing industries and docks.

2 Evaluate the Olympic legacy in economic, social and environmental terms.

STRETCH YOURSELF

In any democracy political aspirations are rightly ambitious but subject to the economic constraints of the time. That they are rarely achieved in full is no criticism. However, urban regeneration of London's East End since the 1980s is nothing if not remarkable. Discuss.

Now practise...

The following are sample practice questions for the Contemporary urban environments chapter.

They have been written to reflect the assessment objectives of Component 2: Human geography Section C of your A Level, and Component 1: Physical geography and people and the environment Section B of your AS Level.

These questions will test your ability to:

◆ demonstrate knowledge and understanding of places, environments, concepts, processes, interactions and change, at a variety of scales [AO1]

◆ apply knowledge and understanding in different contexts to interpret, analyse and evaluate geographical information and issues [AO2]

◆ use a variety of quantitative and qualitative and fieldwork skills [AO3].

---

1   Distinguish between a mega-city and a world city. (4)

2   Compare the process of suburbanisation in HDEs in the twentieth century with that occurring in selected LDEs in the twenty-first century. (9)

3   For a specific project of river restoration and conservation that you have studied, evaluate its outcomes. (9)

---

4   Explain how urban areas affect local patterns of precipitation. (6)

5   Evaluate the success of **one** urban policy designed to promote regeneration in the UK since 1979. (9)

6   Study Figure **1**, a comment published in the medical journal *The Lancet*. To what extent do you agree that 'poor sanitation is still an urgent worldwide problem'? (9)

◉ **Figure 1** *Concerns raised over water quality at the Rio Olympics*

A recent investigation by Associated Press on water quality at aquatic venues for the 2016 Olympic Games in Rio de Janeiro, Brazil, has raised concerns about the risk to the health of athletes who will compete next year.

Nearly 1400 of the more than 10000 athletes competing at the games will be directly exposed to the contaminants in this water as they engage in sailing, swimming, canoeing, rowing, etc. However, despite the very real worries about the health of competitors, the story largely glosses over the broader and more important issue of how water quality affects the people of Brazil, and misses the opportunity to, even briefly, reflect on how poor sanitation is still an urgent worldwide problem.

(*The Lancet*, September 2015)

7 How and why does the generation of urban physical waste vary around the world? (6)

8 With reference to Figures **2** and **3**, assess the importance of transport planning in promoting sustainable urban development. (9)

🔺 **Figure 2** *Visualisation of proposed segregated two-way cycle track on Tower Hill, central London*

9 As an urban form, is the postmodern western city a success or a disaster? Justify your answer. (20)

'A network of Quietways and Cycle Superhighways will – when completed – make up 100 km of safer cycle routes known as the Central London Grid.

◆ Quietways are signposted cycle routes, run on quieter back streets to provide for those cyclists who want to travel at a more relaxed pace.

◆ Cycle Superhighways are on main roads and often segregate cyclists from other traffic.

The Central London Grid is funded by the Mayor's Vision for Cycling, a 10-year plan to deliver cycling improvements across London with spending set to total £913 million by 2022.'

(Transport for London, 2016)

🔺 **Figure 3** *Central London Grid*

10 Explain the concept of liveability with reference to sustainable urban development. (6)

11 For one or more urban areas, assess the value of incineration versus landfill as approaches to waste disposal. (9)

12 'Despite the recent economic growth in cities across many LDEs, the deterioration in air and water quality threatens to stifle further growth.' Assess this statement. (9)

13 Study Figure **4**. Using these figures and your own knowledge, analyse some of the issues associated with economic inequality and social segregation in urban areas of the UK today. (20)

	British Muslims	**UK population**
% unemployed	12.8	5.4

	British Muslim women	**British women**
% of unemployed women who say they can't work because they have to look after the home	44.0	16.0

	British Pakistani women	**White British women**
% of women who are asked about family life in job interviews	12.5	3.3

🔻 **Figure 4**

(Source: House of Commons all-party Women and Inequalities committee, August 2016)

183

Alternative protein sources will be needed for humans and livestock to reduce land and energy use. How could insects play a part in ensuring future food security?

Your exam

'Population and the environment' is an optional topic. You must answer one question in Section C of Component 2: Human geography, from a choice of three: Contemporary urban environments or Population and the environment or Resource security. Component 2 makes up 40% of your A Level.

Your key skills in this chapter

In order to become a good geographer you need to develop key geographical skills. In this chapter, whenever you see the skills icon you will practise a range of quantitative and relevant qualitative skills, within the theme of 'people and the environment'. Examples of these skills are:

◆ Drawing and annotating diagrams or models of human systems and human processes 4.3, 4.18

◆ Using atlases and other map sources 4.1, 4.2, 4.4, 4.7, 4.15, 4.16, 4.20, 4.22, 4.27, 4.28, 4.30

◆ Use of electronic databases, and innovative sources of data such as crowdsourcing and 'big data' 4.30

◆ Presenting data and interpreting graphs 4.7, 4.8, 4.10, 4.12, 4.13, 4.19, 4.21, 4.23, 4.25, 4.29

◆ Analysing quantitative and geospatial data, including applying statistical skills 4.9, 4.14, 4.17, 4.30

Fieldwork opportunities

1 Investigating the relationship between soils and human activities: A farm visit

There are several organisations that link up school groups with local farms, including LEAF (Linking Environment and Farming). Try to contact the organisation or the farmer directly before the trip, to discuss the specifics of your investigation – you will then get more out of the site visit. How does the physical environment (both the local climate and soil) affect the decisions taken by the farmer on a daily, monthly and annual basis? Take photographs and draw sketch maps highlighting any areas of the farm experiencing soil problems such as waterlogging or soil erosion. How does the farmer use technology to overcome physical constraints and keep costs to a minimum? How does he or she use GIS on the farm?

2 Investigating perceptions of place and well-being

Suggest a number of fieldwork techniques that you could use to collect and analyse qualitative data about the attitudes of local people towards the place in which they live, their well-being and any links between the two. Try to be specific, for example, naming study locations, describing sampling techniques and methods of data collection. How might you get access to the people you'd like to talk to? Once you have a plan, you can begin to approach potential respondents. A small pilot of your fieldwork may be beneficial.

3 Using online databases to inform your work

As suggested in 4.9, the Environment and Health Atlas of the UK is an excellent resource that includes mapped data about specific health conditions, as well as environmental agents. The atlas draws upon a range of resources produced by organisations as diverse as the UK Meteorological Office, the Office for National Statistics and the UK's water companies. The Index of Multiple Deprivation is another excellent example of free, online geospatial data presented simply for ease of use. In addition, of course, the ONS's own census data is easily interrogated if you need to gather quantitative data at the level of a ward or parish, health authority or local authority.

In this section you will learn about physical factors affecting population, key population parameters, the key role of development processes, and global patterns of population growth

How is the world's population changing?

In July 2015, the world's population topped 7.3 billion – 7 300 000 000! The UN predicts that it will reach 8 billion by 2025, 9.2 billion by 2050 and may reach 11 billion by 2100!

Look at Figure **1**. Notice how the rate of population growth (shown by the steepness of the curve) increases from the 1800s and particularly from about 1950. This is called **exponential growth**. What will be the impact of this growth on people's quality of life? Is it sustainable? Will we exceed the **carrying capacity** of the planet? These are some of the themes examined in this chapter.

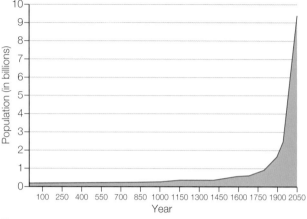

△ **Figure 1** *Growth of the world's population*

What is the pattern of population change?

Look at Figure **2**. It is a map showing estimated growth rates of population around the world. Notice the following patterns:

◆ Population is growing in most countries of the world. There are only a few countries, mainly in Eastern Europe, where the population is actually decreasing.

◆ The population of much of Africa is growing rapidly (over 2 per cent per year). Notably, growth is fastest in LDEs.

◆ Elsewhere, population is increasing between 0.5 and 1.5 per cent per year.

What factors affect population change?

How many children (if any) will you have when you are older? Do you think your answer would be the same if you were asked the same question in five or ten years time? What factors might influence your decision-making – and also the decision-making of other young adults worldwide? Soil, climate, resource availability, development processes and human behaviour are amongst the factors that affect population numbers – there is no single factor. When taken collectively, it might mean that in reality for some of the world's population there are a limited number of choices when 'deciding' family size.

◁ **Figure 2** *Estimates of population growth rates (2013)*

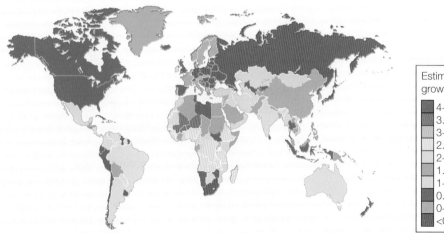

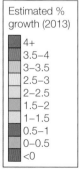

Estimated % growth (2013)

- 4+
- 3.5–4
- 3–3.5
- 2.5–3
- 2–2.5
- 1.5–2
- 1–1.5
- 0.5–1
- 0–0.5
- <0

Niger, North Africa

Look at Figure **3**. According to the UN's Human Development Index (2015), Niger, in the Sahel region of Africa (*AQA Geography A Level & AS Physical Geography*, 2.13), is the poorest country on the planet. The country also has the highest total fertility rate (at nearly seven births per woman). It would be too simplistic to regard poverty alone as the explanation for the high population growth rate.

Northern areas of Niger typically receive only around 200 mm of rain each year. Rainfall increases in the south-west, with the majority of the rain (600 mm) falling between May and September. Over the last 50 years, rainfall has been declining and droughts (and subsequent food shortages) are common. Furthermore, underdeveloped infrastructure means that during times of crises, food aid cannot be distributed without difficulties.

▲ **Figure 3** *The location of Niger, Africa*

In the absence of real political will to invest in irrigation systems, planting of crops tends to be limited to the more fertile soils of the south bordering Lake Chad and the River Niger. Elsewhere, subsistence farming dominates on the dry, dusty and nutrient-deficient soils, particularly the nomadic herding of cattle, sheep and goats. However, overgrazing and loss of cattle during drought periods have forced farmers and their families to give up this traditional method of farming and to move to the towns to find work. Larger family sizes are both a cultural norm (polygamy is common and competition exists between some wives to have larger families) and also considered as an economic necessity for subsistence farmers as children can work on the land at a young age and earn money from jobs as they get older. Niger cannot feed itself alone – around 2.5 million people have no guaranteed source of food. They are food-insecure.

ACTIVITIES

1 Study Figure **1**. How confident can we be about predicting future world population size? Give reasons for your answer.

2 Study Figure **2**.

   a Using an atlas and referring to individual countries, as well as physical regions (such as climatic or vegetation zones), describe the pattern of population growth rate.

   b Suggest reasons for the patterns described.

3 How is the strength of an economy likely to impact on natural population change?

STRETCH YOURSELF

Using the most recent table of the Human Development Index and its components (http://hdr.undp.org) and also your own research using the internet, complete the following table for five countries with significantly different HDI ranks. Comment on the completed table.

HDI rank	HDI value	Physical environment characteristics (e.g. climate/ soil/resources)	Population parameters (e.g. distribution/ density/total/ change)	Level of development (e.g. GNI per capita)

In this section you will learn about global and regional patterns of food production and consumption

Food for thought

If great quotations often illuminate what should be obvious, then Feuerbach's 'a man is what he eats' is just as pertinent today as it was when written in 1850. Regardless of race, creed or wealth, nutrition is central to human functioning. Nutrition is fundamental in determining both our capacity to work and also our quality of life, not least our susceptibility to illness and capacity for recovery. Consequently, food consumption is as paramount today as it was in Feuerbach's time when food production was described as the most important of all economic activities.

Yet geographically, patterns of food production and consumption demonstrate remarkable contrasts. For example, the FAO (United Nations Food and Agricultural Organisation) estimated that in 2014–16 one in nine (780 million) of the global population were suffering from **chronic undernourishment** – and that almost all of these lived in developing countries (Figure **1**). Two-thirds of this total live in Asia, yet sub-Saharan Africa is the region with the highest percentage – one in four is undernourished.

Look at Figure **2**. *Undernourishment* is the most common expression of **malnutrition** in the form of *undernutrition* – both protein-energy malnutrition (a lack of calories and protein) and also *micronutrient* (vitamin and mineral) deficiency. Malnutrition also includes *overnutrition* and, globally, eating too much is now a more serious health risk than eating poorly. Indeed, in 2014, more than 1.9 billion adults, were overweight – over 600 million were **obese**.

Feeding the world

Europe, North America and Australasia have enough farmland to provide the food they need, with significant surpluses to export. In contrast, half of all LDEs lack sufficient farmland and technology to be self-sufficient and are too poor to import. Enough food is currently produced to feed the world and the 2015 **Millennium Development Goal** to halve the proportion of hungry people was almost met. However, one-third of all food produced worldwide is still wasted, so one in four calories produced for consumption is never eaten! Undernutrition will persist until food is more evenly distributed and waste reduced.

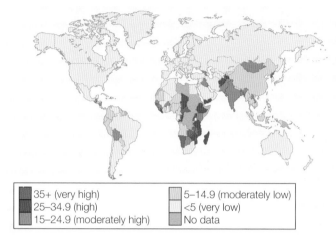

35+ (very high)	5–14.9 (moderately low)
25–34.9 (high)	<5 (very low)
15–24.9 (moderately high)	No data

⬥ **Figure 1** *Percentage of undernourishment in the population, (2014–16)*

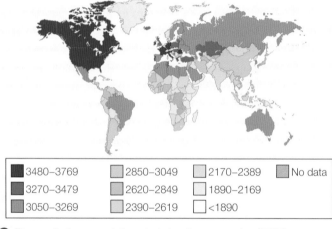

3480–3769	2850–3049	2170–2389	No data
3270–3479	2620–2849	1890–2169	
3050–3269	2390–2619	<1890	

⬥ **Figure 2** *Average daily calorie intake per capita (2011)*

Did you know?

Every year in the UK alone, 18 million tonnes of food, worth £23 billion, ends up in landfill – approximately one-third from producers and the supply chain, one-third from retailers and the remaining one-third from households.

Too large to handle – an oversize epidemic?

Look at Figure **3**. Worldwide obesity has more than doubled since 1980 – 39 per cent of adults were overweight in 2014 (13 per cent were obese) and 41 million children under the age of five were either overweight or obese. Yet obesity is preventable.

The fundamental cause of overweight and obesity is eating too many calories and not getting sufficient physical activity. Globally, there has been:

◆ an increased intake of energy-dense foods, particularly those high in fat, sugar and salt

◆ an increase in physical inactivity due to the increasingly sedentary nature of many forms of work and leisure activities, changing modes of transportation and increasing urbanisation.

The health consequences of being overweight, particularly obesity, are often discussed in the context of 'diseases of affluence' associated with over-indulgence and the pressured way of life in HDEs. But many obese people are low-income earners, with imbalanced diets dominated by cheap carbohydrates and/or convenient micronutrient-poor processed foods. It is therefore inadvisable, and insensitive, to generalise. Yet facts are facts and it is easy to quote obesity in the USA, for example, as 'a plague of the twenty-first century … reaching epidemic proportions'. However, the socio-economic, age and even race variables associated with overnutrition in the USA, let alone the health and economic consequences, are found worldwide. Furthermore, many LDEs and EMEs continue to deal with the problems of infectious diseases and undernourishment, while they are also experiencing a rapid upsurge in non-communicable disease risk factors related to obesity and overweight, particularly in urban areas. These include coronary heart, liver and gallbladder disease, type 2 diabetes, cancers, high blood pressure and stroke.

Indeed, it is not uncommon to find undernourishment and obesity co-existing within the same country, community and even household.

Combating overweight and obesity involves shaping people's choices by making the choice of healthier foods and regular physical activity easier – more accessible, available and affordable. But individual responsibility can only have its full effect where people have access to regular physical activity and healthier dietary choices. Policies such as the Jamie Oliver-backed sugary drinks tax, to be introduced in the UK in 2018, will help, and particularly governmental and societal pressure on the food industry to:

◆ reduce the non-saturated fat, sugar and salt content of processed foods

◆ ensure that healthy and nutritious choices such as fruit, vegetables, legumes (peas, beans), whole grains and nuts are available and affordable to all consumers

◆ restrict marketing of foods that are high in fats, sugars and salt, especially those aimed at children

◆ support regular physical activity practice in the workplace.

⊗ **Figure 3** *Obesity has consequences beyond debilitating and life-threatening health issues; everything from airline seats to clothing to crematoria furnaces need to be reconfigured*

ACTIVITIES

1 Study Figures **1** and **2**.
  a Identify those countries and regions with the highest and lowest average daily calorie intakes.
  b Suggest reasons for the patterns identified.
  c Describe and explain any correlation between patterns of average daily calorie intake and hunger (Figure **1**, 4.7 may also help).

2 'Something must be done'. But by whom? Rhetoric is easy to preach regarding obesity. Working in small groups, using Jamie Oliver as your inspiration, identify how social media might be adopted productively and sensitively to tackle overweight and obesity.

STRETCH YOURSELF

Geographically, patterns of food production and consumption demonstrate huge variations, with dramatic consequences both now and for the future. For example, world population is projected to be 9.2 billion by 2050. Working out how to feed all these people – while also advancing economic development, reducing greenhouse gas emissions, and protecting crucial ecosystems – is arguably one of the greatest challenges of our era. Evaluate and discuss this statement.

In this section you will learn about:
- agricultural systems and agricultural productivity
- the relationship with key physical environmental variables – climate and soils

Around 2 billion of the global population (28 per cent) are employed directly or indirectly in food production. We have already established that, despite huge geographical variations in food production, consumption and waste, enough food is currently produced to feed everyone. But what is produced, and also how and where, varies greatly and will continue to do so as world population grows and diets change. For example, the world's per capita meat and milk consumption is growing, especially in China and India, and is expected to remain high in Europe, North America, Brazil and Russia. These foods are more resource-intensive to produce than plant-based diets.

Agricultural systems and agricultural productivity

Agriculture, like any other economic activity, involves inputs, processes and outputs (Figure **1**):

- *Inputs* are the physical, human and economic factors that determine the type of farming in an area, i.e. what goes in to make the farm.

- *Processes* (farming methods) are the activities carried out to turn inputs to outputs. They vary depending on the inputs and also by the level of technology available.

- *Outputs* are the products from the farm – the crops cultivated and animals reared.

Again, like any other system there are also feedbacks (see *AQA Geography A Level & AS Physical Geography*, 1.1), such as the reinvestment of profits into buying additional land and equipment, the use of manure as fertiliser and the use of fodder crops produced for feeding livestock. There will also be changes to the system – either physical environmental such as the impacts of droughts and floods, or human such as changes in demand, market price or political policies.

Look at Figures **3** to **9**. Agricultural systems vary according to physical environmental variables, such as relief, but primarily climate and soils. Human factors are also important. Socio-economic, behavioural, cultural, scientific and even political influences have a significant bearing on agricultural land use and productivity also. It is important therefore to appreciate that farming practices will demonstrate great variability and frequently contain elements of more than one type. We certainly cannot simply associate particular agricultural systems with rich and poor countries. For example, the poorest LDEs may well be dominated by subsistence and near-subsistence farming, but will often have some large-scale commercial enterprises, such as plantations, too. Likewise, intensive and extensive systems occur in both rich and poor countries.

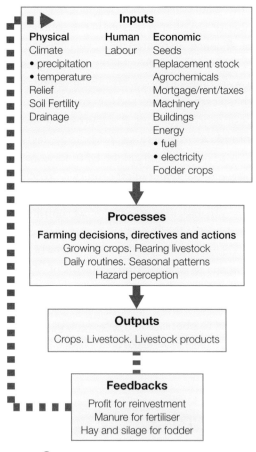

Figure 1 *Agriculture as a system*

Physical environmental factors influencing agriculture

Of all the physical environmental factors affecting agricultural productivity, climate, soils and relief are the most important. For example:

◆ *Temperature* dictates the length of the growing season. In temperate climates, such as the UK's, a growing season exceeding three months (with mean annual temperatures above 6 °C) is required.

◆ *Precipitation* is an important factor in determining the water supply. The relevance and effectiveness of mean annual rainfall totals depends on temperatures and therefore rates of evapotranspiration. But seasonal distribution of rainfall is more important to agriculture than annual totals. In the UK, we take rain for granted, but a failure of the monsoon in south-east Asia, or an extended drought in Africa's Sahel, can result in famine.

◆ *Wind* and storm frequency restricts cultivation of grain crops. However, winds can also be beneficial, such as the warm *chinook* ('snow-eater'), which melts snow on the North American Prairies increasing the length of the growing season.

◆ *Soil quality* is fundamental to successful agriculture and determined by factors such as depth, texture, structure, mineral content, pH, aeration, capacity to retain water and vulnerability to **leaching** (see 4.5, 4.6). Soils influence crop production by their supply or deficit of soil moisture, together with their type and stock of available nutrients. In the UK, for example, potatoes fail if soil acidity falls to less than pH4.

◆ *Relief variables* interrelate with the climate and soil factors above. They include altitude, angle of slope and aspect (Figure **2**). In the UK, for example, the upper limit for hay and potatoes is 300 m and slopes of more than 11 degrees become impractical for safe ploughing.

Figure 2 *The direction in which a place faces is its aspect which influences local temperature variations. The south-facing aspect of these vineyards in the Rhône Valley, Valais, Switzerland, maximises their insolation. The north-facing slopes, in contrast, are markedly cooler in the shade, and woodland extends to the valley floor.*

Did you know?
It takes 300 potatoes to make one 750 ml bottle of vodka. The potatoes are mixed with water and fermented, before the liquid is then distilled.

 Working out the mean

The *mean* value is the sum of all the values of each observation divided by the number of observations. It is useful to have a single value that represents the centre of distribution of a set of data. This straightforward calculation allows every value to be considered and comparisons made across **continuous** and **discrete** data sets. But, as every value is included in the calculation, the mean is less reliable when the data set has a large *range*.

Agricultural systems and examples

Note that these groups can be linked, for example, commercial arable farming.

Arable farming is the farming of cereal and root crops, usually on flatter land where soils are of a higher quality. Arable farming can be subsistence such as slash-and-burn shifting cultivation in Latin America, Africa and south-east Asia, or commercial such as potato cultivation in the UK.

The humble potato has long been the staple root crop ideally suited to the UK's temperate maritime climate. A Hertfordshire farmer, William Chase, was so frustrated that supermarkets would reject potatoes that weren't 'cosmetically perfect' that he founded a hand-fried brand of crisps, Tyrrells. Having since sold the business for £40 million, he now produces premium vodka.

◀ **Figure 3** *Former farmer William Chase, founder of Tyrrells potato crisps*

Pastoral farming involves livestock rearing and can be subsistence, such as nomadic pastoralism (herding of cattle, sheep, goats and camels) in semi-desert regions of West Africa, or commercial, such as sustainable beef cattle ranching on the South American Pampas.

Argentina's Pampas thrives in the temperate climate and is one of the richest areas of grassland biodiversity in the world. Towards the end of the twentieth century, all of its world-renowned beef cattle grazed freely. Since then, many cattle ranchers have sold their land for more profitable soy, wheat and maize production. Only 20 per cent of Argentinean beef today is reared sustainably on grass – the 26 000 ha Estancia Ranch (with 20 000 cattle) is one of the few remaining traditional extensive pasture-based ranches in Argentina.

◀ **Figure 4**
Cattle around El Calafate on Argentina's Pampas

Mixed farming is, as the name implies, the production of both arable crops and livestock. Commercially sensible in allowing flexibility (including diversification into farm shops and leisure activites such as 'glamping'), it is particularly suited for spreading risk. Mixed farming is the most common form of agriculture in the UK.

Mixed farming in Fife takes advantage of some of the most productive, easily worked soils in Scotland. This 160 hectare self-contained organic livestock and arable farm (Balcanquhal) also includes several areas of amenity woodland. It has been managed along with three other farms, reducing costs through economies of scale. Diversification across the group includes pig production and an established game (pheasant and partridge) shoot.

▲ **Figure 5** *Balcanquhal Farm and Lomond Hill, Fife*

Intensive farming involves high investment in labour and/or capital such as machinery, glasshouses and irrigation systems. It produces high yields per hectare from often small areas of land. Fruit, flower and vegetable production (horticulture) in south-west England and the Netherlands are examples.

Horticulture in Cornwall uses both polytunnels and glasshouses but the mild climate gives the region an advantage over other areas in the production of early spring vegetables and flowers. Scientific innovation is particularly notable in horticulture, such as with the increasing use of hydroponics (growing plants in mineral nutrient solutions without soil).

▶ **Figure 6**
Tending seedlings in a polytunnel, Truro, Cornwall

Commercial farming typically involves farmers and **agribusinesses** maximising profits by specialising in single crops (monoculture) or raising one type of animal. Commercial farming will often involve high investment of capital into land, contractors, machinery, agrochemicals (fertilisers, pesticides, fungicides and herbicides) and animal welfare. Examples are grain cultivation in North America, tea plantations in East Africa and cattle ranching in South America.

The Canadian winter wheat harvest represents large-scale commercial agribusiness at its most efficient. High productivity is achieved through inputs, processing and marketing being run by the company and reduced labour costs through using specialist, highly mechanised contractors at all stages including ploughing, agrochemical spraying and harvesting.

◄ **Figure 7** *A field of winter wheat, Manitoba, Canada*

Extensive farming uses low inputs of labour, machinery and capital but usually involves large areas of land; yields per hectare are consequently low. It is therefore the opposite of intensive farming. Hill sheep farming in the upland regions of the UK is a good example.

Much of the character of beautiful upland areas like Snowdonia, the Lake District and Yorkshire Dales is maintained through sheep farming. Family farms producing lamb and wool are often run at the margins of profitability. The rugged relief, severe climate and thin, leached soils prohibit any other form of agriculture, although diversification into tourism-related ventures such as holiday lets may be an option.

▲ **Figure 8** *Snowdonia National Park, North Wales*

Subsistence farming involves the direct production of sufficient food to feed the family or community involved, with any excess produce sold or bartered. Nomadic pastoralism in West Africa, slash-and-burn shifting cultivation in Latin America, Africa and south-east Asia are examples of the few purely subsistence farming systems left. However, 'mainly' subsistence is still important.

Amerindian tribes in the Guiana Highlands of Venezuela clear a small area of tropical equatorial rainforest, burn the dried vegetation to provide fertile ash and cultivate the plot for 3–5 years. Growth of manioc (for cassava bread), yams, peppers, beans and maize is rapid in the hot, wet conditions. Continual weeding is essential before the plot loses fertility. Tribes then clear another plot, returning only when the original vegetation has regenerated naturally.

▲ **Figure 9** *Clearing forest for cultivation by slash and burn, Venezuela*

ACTIVITIES

S

1 Study Figures **1** and **3–9**.

  a Choose any one of the agricultural system examples described in Figures **3–9** and draw a systems diagram specific to it.

  b Describe and explain how outputs also form inputs in this agricultural system.

  c Consider one physical change and one human change to your chosen system and outline how these changes will affect the agricultural system.

2 Whereas grass-fed cows may take three to five years to be ready to sell, a South American farmer can turn around a soy or maize crop in a matter of months. Eighty per cent of Argentinean beef is now factory-reared on corn and soymeal. Outline the commercial and sustainability issues raised by these facts.

3 Summarise the key physical environmental variables influencing agricultural systems and agricultural productivity.

In this section you will learn about:
- ◆ the characteristics and distribution of two climatic types and their influence on human activities and numbers
- ◆ climate change as it affects agriculture

The tropical monsoon climate

If importance was measured solely by the numbers of people affected, then the tropical monsoon climate would be the most important climatic type in the world. Over half of the world's population live in over 21 Asian countries affected by seasonal monsoon winds. This includes six of the world's most populated nations – China, India, Bangladesh, Pakistan, Japan and Indonesia.

Look at Figure **1**. The tropical monsoon climate is distinguished by wet and dry seasons. The summer wet season is from May to October when the overhead sun heats the land intensively, causing a low pressure system to form where the air above rises. Moist air from the sea is sucked into the interior, bringing heavy rain (especially where forced to rise over high plateaus or mountain areas such as the Himalayas). By November the overhead sun has again moved southwards causing the sea to get hotter than the land, with associated low air pressure, and the wind directions are reversed. Consequently, in winter, cool, dry winds blow from the Asian interior to the sea, with any moisture having been lost over the Himalayas before descending to the coast.

For many subsistence farmers, their very survival depends upon the seasonal nature of the tropical monsoon climate – this is because rice is cultivated during the monsoon season. Rich in carbohydrate, fibre and vegetable protein, the importance of this staple food should not be underestimated (Figure **2**). Nothing is wasted; whether it is the 'wet' rice varieties grown in, for example, the flat, fertile, alluvial flood plains of the Ganges Valley, or 'dry' rice varieties cultivated in the irrigated hillside terraces of Indonesia. The rice is threshed and winnowed (to separate the rice from straw and other chaff) and the 'waste' is used as fodder for animals, kindling for fires and even woven into hats, mats, screens and baskets. As long as there is sufficient water for irrigation, even the paddies can be reused during the dry season for second rice crops, or for beans, lentils and wheat.

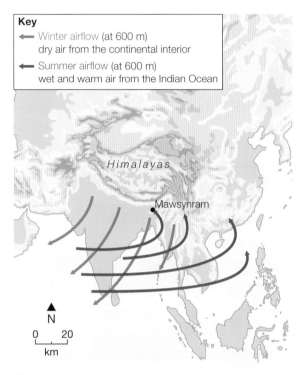

Key
- ← Winter airflow (at 600 m) dry air from the continental interior
- ← Summer airflow (at 600 m) wet and warm air from the Indian Ocean

Himalayas

Mawsynram

▲ N

0   20
km

⊘ **Figure 1** *Airflow associated with the tropical monsoon climate in south-east Asia*

Did you know?

The wettest place in the world is Mawsynram in India. This 1500 m plateau, in the Khasi Hills overlooking the plains of Bangladesh, has an astonishing mean annual rainfall of 11 871 mm. The wettest month ever recorded was in July 1861 in Cherrapunji (10 miles east of Mawsynram) when 9300 mm fell. Since the previous August, there had been a record-breaking annual total of 26 470 mm.

◁ **Figure 2** *Indian women transplanting rice seedlings into a padi field in Tamil Nadu illustrates just how labour-intensive this agricultural system is; everything from preparation of both nursery seed beds and padi fields (for flooding), to transplanting, weeding, harvesting, threshing and winnowing, demands a high input of labour*

The polar tundra climate

We have established that the importance of the tropical monsoon climate is measured by the majority of the world's people living with it. In contrast, the polar tundra climate supports a fraction of these numbers, yet tundra and ice cap regions of polar climates cover more than 20 per cent of the Earth!

The tundra climate is cold – very cold. At such high latitudes, hours of summer sunshine may be long, but average temperatures rising above 0 °C are measured in weeks rather than months. Consequently, human activities are restricted and characterised by fishing, adventure tourism and particularly mineral exploitation, rather than land-based agriculture (see 5.23).

However, the harsh climate and associated tundra vegetation have supported **indigenous** people sustainably, albeit at subsistence levels, for thousands of years. For example, the Inuit in northern Canada and Greenland have hunted caribou and seals in winter, and fished in summer, but always sustainably. This is because of their relatively low numbers in relation to the vast area covered as a result of their transient lifestyle. Likewise the Sami of northern Europe have followed the seasonal movements of reindeer northwards to the treeless tundra in summer and southwards to the boreal (coniferous) forests in winter. Their hunting has long provided most of their food and material needs – sustainably because of their low population density.

ACTIVITIES

1 In what ways, and for what reasons, do indigenous people manage their environment sustainably?

2 Using an atlas and blank world outline map:
   a Shade the tropical monsoon and polar tundra climatic regions in contrasting colours. (Don't forget to add a key.)
   b Annotate the map with the key physical environmental and human issues threatening the long-term sustainability of human activities in these regions.

3 **Extension**. Refer to 'Extension' text, your map for activity 2 and section 5.18. Explain the significance of this positive climate change feedback.

Climate change as it affects agriculture

Traditional subsistence rice cultivation is both water- and labour-intensive. The fear that climate changes will trigger far less predictable weather conditions, and in particular variations in precipitation that jeopardise traditional rice production, has led to research into less water-intensive methods of cultivation, in which the grain is sown directly into the soil. A Norwegian project called *ClimaRice*, which assesses variations in climate and the impacts of the availability of water on rice production, has run trials in Andhra Pradesh and Tamil Nadu, India, where water shortages are already evident. These trials demonstrated:

◆ increased yields
◆ lower labour costs (transplanting seedlings currently amounts to a third of the work effort in rice production)
◆ deeper root systems making the uptake of nutrients more efficient
◆ markedly reduced methane emissions generated from microbe activity in padi fields.

A notable disadvantage with the alternative method is that weeds pose more of a problem, necessitating greater use of expensive herbicides during the initial growth phase.

For more than 30 years, northern high latitudes have been showing signs of climate change with significant warming of up to 1 °C per decade on average. Furthermore, climate modelling predictions project a continuation of these trends leading to a warmer, wetter climate, resulting in further widespread thawing of permafrost and glacial retreat, a shorter snow season and reduced sea ice. Consequently, coastal erosion, permafrost and slope instabilities will get worse. Changing migrations patterns of wildlife, such as caribou, will have significant effects on the indigenous population, threatening the sustainability of existing settlements, infrastructure and lifestyles. However, sea transport, tourism and mineral exploitation could benefit, and fishing, forestry and agriculture will inevitably change.

EXTENSION

Permafrost melting is often thought of as the most dangerous feedback associated with climate change. About half of all tundra regions will be warm enough by 2050 to replace sparse shrub land with dense boreal (coniferous) woodland. Given the latter's darker surfaces and lower albedo, the sunlight reflected back into the atmosphere will decrease and increase warming. Higher temperatures will also lessen the reflectivity of snow cover.

In this section you will learn about the characteristics and distribution of two key zonal soils and their relationship with human activities

We are all guilty of taking natural resources for granted – the air we breathe, the water we drink – yet some are so fundamental that we almost feel guilt when we're forced to consider their significance. Soils fall into this category because they represent the link between a lifeless world of rock and the living world of plants, animals and humanity. In short, soils integrate inorganic and organic elements of the environment. They may even be viewed as the Earth's most important natural resource – effectively irreplaceable if degraded or lost to erosion.

Tropical red latosol

Five degrees either side of the Equator is the equatorial climate belt, best associated with the tropical equatorial rainforest biome, which supports half of all living organisms on Earth! Yet paradoxically, the zonal soil of this region – tropical red latosol – is inherently infertile, hence the inadvisability of forest clearance for agricultural use. How can this be the case?

Look at Figure **1**. The world's most predictable climate – hot, wet, humid and without seasons (until you move a few more degrees from the Equator) – promotes the perfect growing conditions for the world's most luxuriant vegetation and unparalleled biodiversity. The year-round growing season means that deciduous tropical equatorial rainforest trees can shed their leaves at any time of the year. This provides a constant supply of leaf litter, which decomposes with other biota rapidly into humus, supplying nutrients to support sustainable new growth very quickly. The fact that this nutrient cycling is so rapid means that if the tropical equatorial rainforest is cleared, the ready supply of new humus is halted and the tropical red latosol becomes quickly exhausted of stored nutrients. It then becomes exposed to excessive leaching of nutrients, and to erosion by gulleying during the heavy, daily convectional rainstorms (see 4.6).

For generations, humans have lived sustainably within tropical equatorial rainforests (see 4.3), but recent decades have seen deforestation on a massive scale, by felling, bulldozing and burning. Population growth and economic development are cited as reasons for this deforestation, specifically the need for:

◆ land for settlement and associated infrastructure

◆ land for ranching, cash-cropping and plantations

◆ hardwood timbers

◆ access for mineral exploitation.

Classification of soils

Classification is necessary if we are to organise our knowledge in a systematic way. By classifying, we can come to understand broad generalisations in addition to detailed individual descriptions or explanations. Classifications do not add to knowledge but they organise knowledge more efficiently, despite the occasional appearance of anomalies.

The most common classification of soil types is the *zonal* system. This subdivides the world's soils into three major categories:

◆ zonal – mature soils reflecting climatic conditions and associated vegetation

◆ intrazonal – reflecting the dominance of other factors, such as the characteristics of the parent rock

◆ azonal – generally immature and skeletal, with poorly developed profiles.

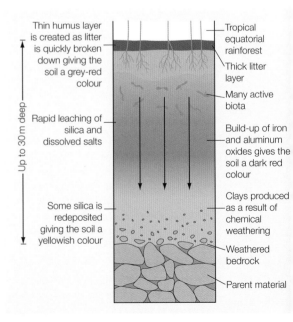

Thin humus layer is created as litter is quickly broken down giving the soil a grey-red colour

Rapid leaching of silica and dissolved salts

Up to 30 m deep

Some silica is redeposited giving the soil a yellowish colour

Tropical equatorial rainforest

Thick litter layer

Many active biota

Build-up of iron and aluminum oxides gives the soil a dark red colour

Clays produced as a result of chemical weathering

Weathered bedrock

Parent material

🔺 *Figure 1* *Profile of iron-rich tropical red latosol*

Podsol

This is the zonal soil of the taiga, the vast continuous belt of subarctic climate across North America and Eurasia between the tundra to the north and temperate grasslands to the south. Prolonged harsh winters and cool summers restrict the vegetation to boreal coniferous forest. Podsols are also found in other areas with cool climates, such as on heathland and moors of the UK and in areas of sandy soil such as fluvioglacial outwash plains.

Look at Figure **2**. Podsols are often associated with coniferous evergreen trees such as spruce, fir and pine. Areas of boreal forest do not receive particularly heavy rainfall but the podsolisation process requires a general downward movement of water through the soil. However, precipitation exceeds evapotranspiration because of the low temperatures. Also, forest vegetation influences the soil moisture balance – coniferous trees shelter the ground from drying winds. Consequently, moderate precipitation can provide a surplus and so allow downward infiltration and percolation.

The podsol has a very poor nutrient cycle. Coniferous evergreens do not take up elements such as calcium, magnesium and potassium and so these nutrients are not returned to the soil when the leaves (needles) fall – hence a poor mor (acid) humus. Particularly notable characteristics of the podsol are the:

♦ accumulation of a hard pan of iron beneath the zone of leaching and marking the highest point of the water table

♦ clear differentiation of horizons indicating, in contrast to tropical red latosols, that there may be fewer mixing agents such as earthworms and ants.

In the UK, podsols are most associated with upland sheep farming and heather moorland managed with controlled burning for the breeding of grouse for shooting. Small areas of surface heather are burnt off in a 10–15 year rotation – each square kilometre can have around six burning patches, resulting in a patchwork quilt pattern of heather at various stages of regrowth. This provides feeding areas for grouse, with unburnt nesting cover nearby. Given that grouse shooting employs 2500 people and generates £150 million annually, its importance to the UK's rural economy cannot be overstated.

Most podsols lie beneath the North American and Eurasian taiga, where hunting of moose, caribou and brown bear is no longer sustainable without legal protection due to the diminishing of wildlife habitats through deforestation. Logging in North America is now managed but, since the fall of the Soviet Union, commercially lucrative logging has been encouraged – the taiga is slowly disappearing.

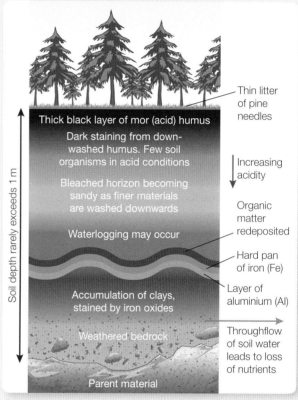

▲ **Figure 2** *Profile of podsol*

ACTIVITIES

1 Outline the physical environmental and human similarities between tropical red latosols and podsols.

2 How does climate account for the distribution and character of tropical red latosols and podsols?

3 Why is grouse shooting so controversial? Research and then outline the arguments for and against this highly profitable industry.

STRETCH YOURSELF

Agriculture, tropical red latosols and podsols just don't mix. Discuss.

In this section you will learn about soil problems and their management as they relate to agriculture

Soil problems

It usually takes thousands of years for soils to become sufficiently deep and mature for agriculture. Furthermore, whilst there will always be natural loss through erosion, leaching and mass movement, there is usually an equilibrium between the rate of soil formation and degradation. However, human mismanagement of soil is increasingly evident – and that natural balance can no longer be assumed.

At the World Economic Forum in 2012, it was reported that soil is being lost at between 10 and 40 times the rate at which it can be naturally replenished. It was also stated that 40 per cent of soil used for agriculture around the world is classed as either degraded or seriously degraded. Without soil management, food production would decline by 30 per cent over the next 30–50 years.

Did you know?

Around half of the world's population are at risk from malaria, with pregnant women and children under five particularly vulnerable. Ninety per cent of all malaria deaths occur in Africa.

◉ Figure 1 *A ranch in Oklahoma about to be engulfed in a giant dust cloud, 15 April 1935*

Soil erosion

Soil erosion is often blamed on the removal of natural vegetation cover, leaving the ground directly exposed to the elements. Replacement of this vegetation by planned land use, such as well-managed agriculture, can and does preserve soils albeit in altered states. Erosion is inevitable where poor land management, often provoked by population pressure, has allowed overgrazing, overcultivation, reduced fallow periods and deforestation. In many LDEs and EMEs, erosion relating to these factors can be alarming but HDEs are not immune. Deep ploughing up and down slopes and also monoculture aid the removal of fertile topsoil by wind and water. This reduces soil thickness and hence room for roots.

During the 1930s, the American Mid-West was hit by giant dust storms caused by catastrophic wind erosion. This erosion was not solely the result of sequential drought, but also due to the removal of the natural grassland vegetation – a practice that left the soil exposed for long periods between crops (Figure **1**).

Hedgerows have been removed in south-east England to increase the cultivable area and also to increase mechanical efficiency. This has had the effect of making the land more prone to wind erosion.

Waterlogging, salinisation and structural deterioration

Soil degradation as a result of water might reasonably be assumed to be caused by rain splash, sheet wash, rill and gulley erosion (Figure **2**). However, waterlogging, salinisation and structural deterioration are more of a problem.

Waterlogging

Waterlogging occurs whenever the water table rises to the point of soil saturation and there is insufficient oxygen in the pore spaces for plant roots to respire adequately. This *anaerobic* environment (lack of oxygen) causes root tissues to decompose and, if not addressed, the crops will die. Many farmers do not realise that a site is waterlogged until surface water appears (Figure **3**), by which time the plant roots may already be damaged and the potential yield severely affected.

In hot climates, waterlogged soils provide excellent breeding grounds for mosquitoes, which transmit diseases such as Zika virus, dengue fever, yellow fever, encephalitis and malaria, the deadliest of them all. Waterlogging has many causes including in areas of inadequate drainage where:

◆ rainfall exceeds the rate that soils can absorb or the atmosphere can evaporate

◆ gentle relief restricts throughflow of infiltrating soil water

◆ relief basins or depressions encourage accumulation of water

◆ seepage from rivers, canals and reservoirs infiltrate soils

◆ soils include an impermeable clay layer or iron pan

◆ excessive irrigation water is used to flood fields.

Salinisation

Over long periods of time, soil minerals weather and release salts. Additional salts are deposited via dust and precipitation. In well-drained areas with sufficient precipitation or efficient irrigation, these salts are leached out of the soil by infiltration and percolation, so increasing salinity in the soil (salinisation, see *AQA Geography A Level & AS Physical Geography*, 2.13) will not be a problem. However, where poor drainage leads to waterlogging, the water table will rise, bringing dissolved salts towards the surface. Furthermore, when the water evaporates, a crust of concentrated salt is left on the surface compounding the degradation (Figure **4** and *AQA Geography A Level & AS Physical Geography*, 2.6). Only salt-tolerant crops, such as cotton, can withstand salinisation. (Sugarcane, for example, is very salt sensitive.) In general, if the salts are alkaline and soil pH rises to above 11.0, plants become infertile. If the salts are acidic and soil pH falls below 4.0, plants cannot absorb nutrients. Either way, crops fail.

▲ **Figure 2** *Severe gulley erosion following excessive winter rainfall in Devon, February 2014*

▲ **Figure 3** *A farmer in Ayutthaya province of Thailand after heavy monsoon rain in 2011*

▼ **Figure 4** *Surface salt crust as a consequence of poorly managed irrigation in Kazakhstan*

Structural deterioration

Most farmers are aware of how soil texture (the proportions of sand, silt and clay) influences its characteristics and workability. They talk of the mixture in loams, such as a clay loam or silt loam depending on which particle is dominant. In most soils, however, the structure (how the individual particles are grouped together) is just as important as the texture – two soils with the same *texture* can behave very differently depending on their *structure*. A clay soil, for example, with a crumbly structure of rounded aggregates can be easy for air, water and roots to move through, but that same texture can be almost impenetrable if its structure has been destroyed by compaction by heavy machinery, or salinisation.

Soil management

Soil management is all about conservation. In its simplest form, this means preserving and protecting vegetation cover – essential on steep slopes by means of forest, orchard, vineyard or pasture. For example, afforestation and reforestation provide the best long-term solution to soil erosion because once the trees have grown, their foliage shades the soil from the sun and intercepts rainfall, and their roots help to bind the soil together and reduce surface runoff. However, agriculture is the purposeful tending of crops and animals – not trees – and soil conservation also includes protecting soils from waterlogging, salinisation, reduced fertility and, at worst, exhaustion. Consequently, soil management must include water management and both are integral to good agricultural land management.

There are many examples of soil management, in both modern and traditional farming systems. These include:

- terracing (Figure **5**)
- contour ploughing (Figure **6**)
- crop rotation and cover crops
- strip cropping (Figure **7**)
- direct drilling (no-till farming)
- selective afforestation with shelter belts
- controlled grazing
- improved drainage to prevent waterlogging and salinisation
- careful management of irrigation to reduce the risk of waterlogging and salinisation

> **Did you know?**
> Worldwide, as much as 10 per cent of all irrigated land may suffer from waterlogging, and up to 20 per cent from salinisation. The cost of irrigation-induced salinity is equivalent to an estimated US$11 billion per year.

▶ **Figure 5** *Terraced rice cultivation in the Philippines; note the retaining banks (bunds)*

Soil management measures vary according to both the physical environment (such as relief, climate and soil types) and human factors (such as those determined by levels of economic development). Socio-economic, behavioural, cultural, scientific and political/governmental influences may all have a bearing in determining what conservation can be achieved, and how and where.

Conservation need not be complicated and both theoretical expertise and practical experience can be applied in simple ways to make a big difference. For example, towards the end of the twentieth century, extensive livestock rearing in hilly regions of Costa Rica was unmanaged, leading to significant soil erosion. Now, based upon careful selection of areas most suitable, livestock rearing is sustainable (Figure **8**). Improved pasture is grazed in rotation along with stall feeding of fodder crops fertilised with their manure. Severely degraded land has regenerated naturally, agricultural production is more diversified, biodiversity has increased, cattle quality has improved and there have been 'spectacular' increases in the production of meat and milk.

🔺 **Figure 6** *Contour ploughing in Overberg, South Africa*

🔺 **Figure 7** *Strip cropping in Queensland, Australia*

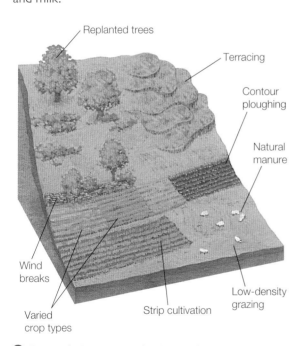

Replanted trees
Terracing
Contour ploughing
Natural manure
Wind breaks
Varied crop types
Strip cultivation
Low-density grazing

🔺 **Figure 9** *Prevention of soil degradation*

🔺 **Figure 8** *Brahman cattle reared for beef, veal, dairy products and leather in Costa Rica; Arenal volcano in the background*

ACTIVITIES

1 a Explain the distinction between soil texture and soil structure?
  b Why is it important for farmers to understand this?
2 Look at Figure **9**.
  a Explain the causes of soil degradation.
  b Outline methods adopted to prevent it.

STRETCH YOURSELF

It is notable how often crop rotations, cover crops, strip cropping, mulshing and direct drilling underlie organic agricultural systems. Research organic farming before commenting on the extent to which 'organic farming is conservation farming at its best'.

In this section you will learn about strategies to ensure food security

What is food security?

The World Food Programme (WFP) and the FAO consider people as food secure when they 'have availability and adequate access at all times to sufficient, safe, nutritious food to maintain a healthy and active life'. In determining this, food security analysts look at the following three elements:

◆ *Food availability*: Food must be available in sufficient quantities and on a consistent basis. This considers stock and production and also the capacity to bring in food from elsewhere, through trade or aid.

◆ *Food access*: People must be able to regularly acquire adequate quantities of food, through purchase, home production, barter, gifts, borrowing or food aid.

◆ *Food utilisation*: Consumed food must have a positive nutritional impact on people. This covers cooking, storage and hygiene practices, individual's health, water and sanitation, and feeding and sharing practices within the household.

Food Security Risk Index

Look at Figure **1**. The Food Security Risk Index is based upon the above criteria and identifies those regions whose food security is at risk due to either internal or external factors. The map infers areas suffering from hunger and food shortages. For example, as would be expected, most of the areas identified as being of extreme risk are in sub-Saharan Africa, together with Afghanistan and Haiti.

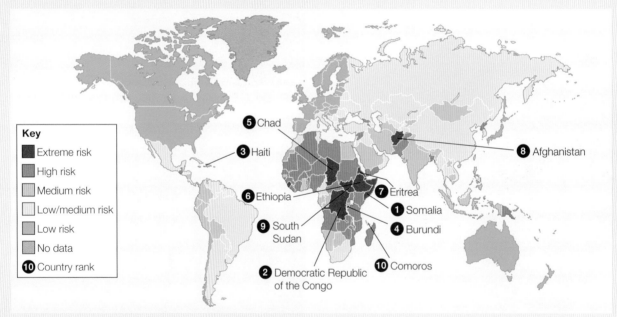

Key
- ■ Extreme risk
- ■ High risk
- ■ Medium risk
- □ Low/medium risk
- ■ Low risk
- ■ No data
- ❿ Country rank

❺ Chad
❸ Haiti
❽ Afghanistan
❻ Ethiopia
❼ Eritrea
❶ Somalia
❾ South Sudan
❹ Burundi
❿ Comoros
❷ Democratic Republic of the Congo

🔺 **Figure 1** *Food Security Risk Index (2013)*

We have already established that in 2014–16 one in nine of the world's people were suffering from chronic undernourishment (see 4.2). Progress towards fully achieving the Millennium Development Goals (MDGs) food security targets by 2015 was hampered by:

◆ challenging global economic conditions – not least in 2008, a year of financial and food crises

◆ extreme weather events associated with climate change, natural disasters, political instability and civil strife.

Inevitably perhaps, hunger rates in countries enduring protracted crises remain more than three times higher than elsewhere. These countries have the most food-insecure people on the planet. Yet, the hunger targets set by the MDGs have almost been achieved, despite the world population growing by 1.9 billion since 1990.

Food insecurity in Somalia

Look at Figure **2**. In August 2014, nearly 3 million people in Somalia were in severe crisis. Conflict between insurgents and government forces was restricting trade and humanitarian access in some areas. Even in Somaliland and Puntland, two of the region's most stable areas, a third of people were not able to afford food, medicine and school fees if they did not receive money transfers (remittances) from families and friends abroad. Poverty, food price inflation, an inadequate capacity to respond to extreme weather events (such as the 2011 drought) and poor food storage and transport infrastructure compounded the problems.

By August 2015, food insecurity was predicted to worsen. Poor rainfall had resulted in below-average agricultural production and continued displacement.

Strategies to ensure food security

The near-achievement of the MDGs hunger targets shows that the elimination of hunger in your lifetime is not unrealistic. A number of important lessons have been learnt from the MDGs experience and it is important that these are carried forward in future. For example:

◆ *Improved agricultural productivity*, especially by small and family farmers, leads to important gains in hunger and poverty reduction. Since the 1950s, major improvements to agricultural output have been mostly attributed to the Green Revolution. Improvements have led to higher yields per hectare through the use of high-yielding varieties of seed (HYVs), agrochemicals, and mechanisation. Progress in education and research has also had positive effects. Furthermore, marginal land has been brought into production using irrigation in drylands, drainage of swamp and bog lands and land clearance. While much of this change has been in LDEs and EMEs, progress in HDEs has been equally if not more dramatic. For example, in the EU between the 1960s and 1980s, the Common Agricultural Policy (CAP) encouraged farmers to increase food production by offering grants, subsidies and guaranteed prices. Now, however, the emphasis is more on creating a balance between efficient food production and environmental stewardship. Many countries are now also investing in research and evaluation of genetic modification of crops. In short, improved agricultural productivity is key, real and constantly evolving. Consequently, in North Africa, for example, severe food insecurity is close to being eradicated. Indeed there is now growing concern in the region, about a rising prevalence of overweight and obesity!

⬤ **Figure 2** *Somali refugees line-up to receive aid during the 2011 famine; this humanitarian disaster was triggered by the worst drought in more than half a century, compounded by decades of conflict, high inflation and increasing global food and fuel prices*

◆ *Economic growth* is always beneficial, not least because it expands the fiscal revenue base necessary to fund social transfers and other assistance programmes. But it needs to be inclusive to help reduce hunger.

◆ *Expansion of social protection* correlated strongly with progress in hunger reduction and in ensuring that all members of society have the healthy diets to pursue productive lives. Whether through cash transfers to vulnerable households, food vouchers, health insurance or school meal programmes, the FAO estimate that 150 million people worldwide are prevented from falling into extreme poverty thanks to social protection. There is still a long way to go: more than two-thirds of the world's poor still do not have access to regular and predictable forms of social support.

First Green Revolution...

During the 1960s, in countries such as the UK, Italy, the Philippines and Mexico, hybridisation (cross-breeding) experiments were generating much publicity by developing new varieties of crops. In Mexico, for example, hybrid wheat and maize strains were developed to withstand heavy rain, strong wind and diseases. Maize yields consequently doubled, while wheat tripled! Animals were also cross-bred, in order to improve their tolerance to difficult environmental conditions, such as aridity. But it was the development of improved 'miracle' rice varieties in the Philippines, such as IR8 (improved rice version 8), which gained most attention. Six-fold increases in yield were reported from the first harvest of IR8, and before long 10 per cent of India's padi fields were planted with it, but such high-yielding varieties (HYVs) proved to be vulnerable to pests and new diseases. Consequently, they required ongoing redevelopment which continues to this day (Figure **3**).

The increased use of agrochemicals also proved to be of enormous significance. Chemical fertilisers alone doubled crop yields in tropical areas, and locust plagues, for example, were controlled by the pesticide, aldrin. Synthetic hormones were also developed to control plant sizes and growth rates, so allowing growing seasons to be adjusted.

The Green Revolution also saw the extension of water control and irrigation schemes. These ranged from the installation of relatively inexpensive wells and pumps, to the development of large-scale, multi-purpose river projects, such as Egypt's Aswan High Dam.

Other innovations included the improvement of crop storage, handling and processing, in order to reduce wastage. Appropriate mechanisation was often accompanied by land reform, such as the reorganisation of small, irregular-shaped fragmented plots into more regular fields. Soil conservation measures, such as contour ploughing and strip cropping were also introduced (see 4.6).

Despite the massive increase in yields, the Green Revolution introduced a number of problems. The economic and social costs associated with many of these initiatives have proved to be very high. For example, only the richer farmers could normally afford to buy and run tractors, which, in turn, increased unemployment and rural depopulation.

Poorer farmers took out loans to buy the new seeds and fertilisers needed by the improved varieties, but were often unable to repay them, leading to debt and forcing the sale of their land. The demand for agrochemicals created industrial jobs, but their production and use could be dangerous (Figure **4**). Furthermore, education was needed to ensure productive cultivation of HYVs, given their frequent special irrigation and agrochemical requirements.

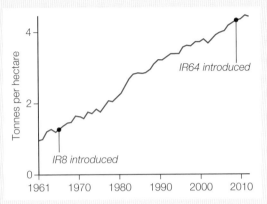

Figure 3 *Global rice yields; rice yields have more than tripled since 1961, keeping up with Asia's growing population; rice is the most important food crop in the world, providing more energy to humanity than any other food source*

Figure 4 *Crop spraying in Punjab has helped India double farm yields in 50 years but the agrochemicals have infiltrated into increasingly scarce water sources contaminating both soil and food. It has been alleged that 34 000 deaths from cancer in the Punjab between 2008 and 2013 may be related to increased use of pesticides.*

Did you know?

The top ten GM agricultural products are maize, soy, cottonseed, papaya, rice, rapeseed, potatoes, tomatoes, dairy produce and peas.

... then Gene Revolution

Recent biotechnology research and development in genetic modification (GM) has proved to be particularly controversial. This involves taking genetic DNA from one plant and introducing it to another in order to make it more resistant to drought, specified pests or diseases. Worldwide adoption has been rapid, but not uniformly so. By 2014, about 18 million farmers in 28 countries were growing GM crops on 181 million hectares (13 per cent of the world's arable land). Adoption in the USA has been particularly prevalent in soy, cotton and maize (Figure **5**).

Concerns about the possible, largely unknown, implications for human health and the environment are widespread. No GM crops are being grown commercially in the UK and nearly 40 countries have banned their cultivation. Yet imported GM commodities, especially soy, are being used mainly for animal feed and, to a lesser extent, in some food products.

Crop wild relatives – strengthening global food security for the future?

Crop wild relatives are wild species closely related to common food crops. These wild cousins of crops are vital to food security because they contain greater amounts of genetic diversity, making them more resilient in the face of climate change, waterlogging, salinisation, pests and diseases, and other new threats.

Exciting research, as part of the Adapting Agriculture to Climate Change project, is underway at the Royal Botanic Gardens at Kew in London. They have begun to collect seed from the wild plant relatives of 29 common crop plants whose genetic diversity can be used to breed new and useful traits into commercial crops so that they can better adapt to future climates and other threats. The crops studied include apple, banana, barley, carrot, lentil, oat, pea, potato, rice, sweet potato and bread wheat.

Kew aims to safeguard the wild relatives of food crops before they disappear. Furthermore, there may be other wild plant relatives of common crops not yet discovered, or that become extinct before we learn how they might be useful to us. By safeguarding their seed in Kew's Millennium Seed Bank, they are ensuring that these useful plant species are not lost to us forever (Figure **6**).

▲ **Figure 5** *In October 2015, Sustainable Pulse reported a 'growing swell of government level support worldwide for bans on GM crop cultivation for both health and environmental reasons'*

▲ **Figure 6** *Cleaning seed at Kew's Millennium Seed Bank; the seed bank aims to collect and store a quarter of the world's species by 2020*

STRETCH YOURSELF

' ... the elimination of hunger in your lifetime is not unrealistic.'

Research crop wild relatives further and with reference to both the Green Revolution and GM crops, explain why such initiatives are crucial to realising this objective.

ACTIVITIES

1 Study Figure **1**.
   a Describe the pattern of global food security.
   b Comment on the parallels between this and Figure **1**, 4.2.
   c Why are the majority of countries at 'extreme risk' in sub-Saharan Africa?

2 What do you think is the most significant reason to help explain continuing extreme food insecurity in Somalia? Justify your answer.

3 Study Figure **3**. What does the graph tell you about 'ongoing redevelopment' of rice HYVs?

4 The Green Revolution has successfully lessened the threat of food insecurity throughout the world, yet many describe it as 'socially divisive'. Why?

5 Why is biotechnology research and development in genetic modification (GM) so controversial?

In this section you will learn about global patterns of health, mortality and morbidity

'I am at the moment deaf in the ears, hoarse in the throat, red in the nose, green in the gills, damp in the eyes, twitchy in the joints and fractious in temper from a most intolerable and oppressive cold' (Charles Dickens 1812–70). We've all been there but much has improved in public health care since then but the emergence of new diseases and changes in lifestyle, have kept the study of disease (epidemiology) high on the political agenda of most countries.

What does 'health' mean?

Health is your physical, mental and social well-being and not just the presence of disease. This broad definition of 'how we feel' makes it difficult to compare the health of the global population. How do you compare work-related stress or mental illness in HDEs to physical exhaustion of manual work in LDEs? There are many different indicators of health, including life expectancy.

Mortality means death. The most common indicator of mortality is death rate (see 4.17). **Morbidity** refers to illness or poor health of a population. Indicators include:

◆ *Prevalence rate* – the total number of cases of a disease in a population at a given time divided by the total population.

◆ *Incidence rate* – the rate or time at which persons become ill. It is usually used as a measure of numbers of new cases of an illness.

Global patterns of health

The most deadly diseases have high prevalence and high incidence rates – in other words, much of the population is infected and the disease is spreading quickly. Without intervention, the disease is likely to become widespread in the country (**epidemic**), even affecting multiple countries or continents (**pandemic**).

🔽 **Figure 1** *Comparison of life expectancy and GDP per capita for 182 countries of greater than 100 000 population*

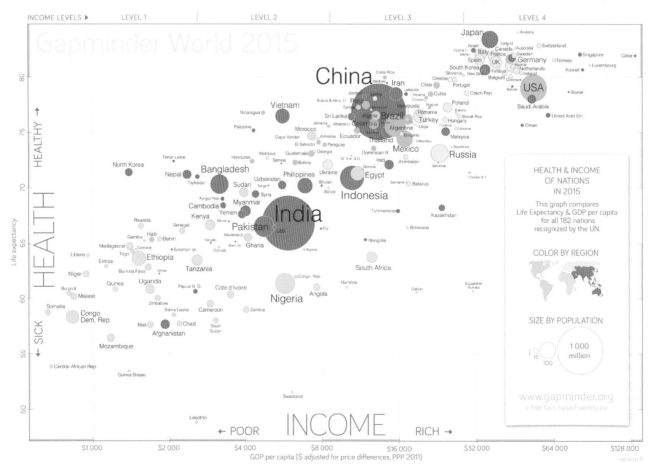

Health in world affairs: Ebola

The first case of Ebola was recognised in West Africa in 1976. When this highly infectious disease was reported in Guinea in March 2014, few in the developed world gave it much attention. The epidemic that followed, however, brought the disease to public consciousness and to world affairs.

The spread over time and space (or spatial diffusion) of the disease is significant (Figure 2). The **index case** of Ebola in 2014 was traced to a toddler in Guinea who had died in 2013. The disease then quickly spread to Liberia, Sierra Leone and elsewhere in West Africa. Several European countries, including the UK, have all treated patients who contracted the virus from the almost **endemic** region of West Africa. In the first 12 months of the epidemic, Ebola killed more than 10 000 people.

While the World Health Organisation (WHO) was criticised for an initial slow response time, volunteers from around the world risked their lives to help treat victims. Conversely, images of medical response teams in space-age-looking biohazard suits, combined with sensational news headlines and poor understanding, created mild hysteria. Amid fears of an outbreak in the UK, demand soared for biohazard suits. Emergency meetings of **COBRA** produced contingency plans to protect the UK, including increased screening of travellers from West Africa (Figure **3**).

It is in Africa that lives have been most changed. Ebola is passed on through close contact with the bodily fluids of those infected. As governments appealed for international help, quarantine and isolation were encouraged. Schools across West Africa were closed for six months, Christmas was all but cancelled and individuals became suspicious of their neighbours in case they were infected. Further outbreaks of Ebola are likely in the future, and it remains to be seen if the global community is better prepared.

In January 2016, the WHO declared Liberia, the last of the countries to be infected, 'Ebola-free'. Liberia suffered 4809 deaths from the disease – over 40 per cent of the global death toll.

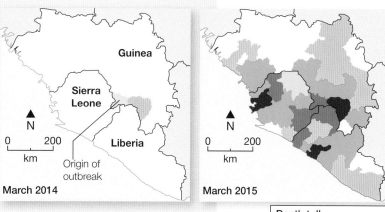

March 2014

March 2015

Death toll
- ☐ 1–10
- ☐ 11–50
- ☐ 51–100
- ☐ 101–250
- ☐ 251–500
- ■ >500

🔺 **Figure 2** *Spread of Ebola in West Africa (March 2014–March 2015)*

Ebola in West Africa
Information for the public

If you have returned from **Guinea, Liberia** or **Sierra Leone** or **cared for someone with Ebola** in the past **21 days**

and

You have a **fever** or **feel unwell**

Without touching anyone, **tell a member of staff** or **call 111**

Ebola facts:
- the risk of Ebola to the UK public is **very low**
- people with **early symptoms** (such as fever or sore throat) are **unlikely to spread Ebola**
- Ebola is **not spread through the air**
- however, someone with Ebola can be infectious if they are suffering from **diarrhoea, vomiting or bleeding**

For more information visit www.gov.uk/phe or www.nhs.uk/ebola

🔺 **Figure 3** *Ebola campaign poster, Public Health England*

ACTIVITIES

1 Look at Figure **1**.
  a Comment on relationships (correlations) shown on the scattergraph.
  b Suggest reasons for the correlations identified.

S 2 Go to the Gapminder website and try creating your own different scattergraphs using the data (click on 'Data' and 'Visualise'). Examine the relationships between different indicators of health and wealth. What happens to the way the data is presented if you switch from a *linear* to a *logarithmic* scale on the *x* or the *y*-axis,, or both? Which type of scale best presents any correlation or trend most effectively?

3 Using evidence from Figure **2**, describe the spatial diffusion of Ebola.

4 Comment on the usefulness of the information shown in Figure **3**. (Search the internet for 'Ebola poster'.)

5 Why was more not done to control the disease before it became an epidemic in spring 2014?

STRETCH YOURSELF

'In order to change we must be sick and tired of being sick and tired.' Discuss this statement with reference to health and world affairs.

In this section you will learn about:
- regional variations in health and morbidity
- factors that influence these variations

Regional variations in health and morbidity in the UK

When Glasgow hosted the Commonwealth Games in 2014, it was ironic that the athletes were competing in one of the least healthy cities in the country. Scotland's biggest city had the lowest life expectancy in the UK – just 75 per cent of boys and 85 per cent of girls born in Glasgow were expected to reach their 65th birthday. Look at Figure **1**. The **Office for National Statistics (ONS)** regularly produces a snapshot showing wide changes in life expectancy across the country.

Factors influencing health and morbidity

- Those living in poverty are likely to suffer poorer health than average. Thus, in areas that have traditionally suffered economically, such as the **deindustrialised** regions of Tyneside or Merseyside, life expectancy is lower.

- Occupation can influence health in other ways – for example, working at home as a young carer may restrict chances of gaining educational qualifications and also increase the risk of mental health problems.

- High alcohol consumption, lack of exercise, poor diet and smoking increase the risk of cancer and heart disease – the largest contributors to mortality rates in the UK. Partly as a result of aggressive policy-making and changing views, smoking rates in most parts of the UK are in decline. On the other hand, excessive alcohol consumption is now a feature of the middle aged and middle classes ('wine o'clock' has become part of everyday language) and obesity is a national epidemic. The causes of cancer are complex – lung cancer risk is linked to smoking, cervical cancer to unprotected sex and mesothelioma to breathing in of asbestos fibre.

- There is increased availability of health care services within urban areas but inner city hospitals and clinics are also those likely to be under most pressure. In short, a range of factors influence people's health.

	1992–4	2002–4	2012–14
Males			
United Kingdom	**73.7**	**76.2**	**79.1**
England	73.9	76.4	79.4
Wales	73.4	75.8	78.4
Scotland	71.7	73.8	77.1
Northern Ireland	73.0	75.8	78.3
Females			
United Kingdom	**79.0**	**80.7**	**82.8**
England	79.2	80.9	83.1
Wales	78.9	80.3	82.3
Scotland	77.3	79.1	81.6
Northern Ireland	78.7	80.6	82.3

▲ **Figure 1** Life expectancy in the UK at birth

ANOTHER VIEW

The north–south health divide

The existence of a north–south health divide within the UK is, at first glance, a stark reality – with increasing distance from London and the south-east of England, there is decreasing life expectancy. However, geographers should always consider contrasting scales (micro, meso and macro) across time and space. For example, Figure **1** does not provide detail on life expectancy variations at a street or city level, nor how the gap has closed between the highest and lowest life expectancies in different regions.

Consider a journey east on the London Underground. Between Lancaster Gate and Mile End on the Central line (a 20-minute journey), the average life expectancy of residents decreases by 12 years (see life.mappinglondon.co.uk). Or search for 'Fairness on the 83' to find out about health and other social disparities within Sheffield over the north to south number 83 bus route. Variations in morbidity across the UK do not necessarily match patterns of mortality across the country. While a broad north–south divide exists, different illnesses have contrasting distributions. For example, Cornwall has a high risk of malignant melanoma (skin cancer, see Figure **2**) but one of the lowest rates for heart disease in the UK. Other diseases, such as lung cancer are strongly associated with centres of population, particularly urban areas. The quality of health care services across the UK also varies.

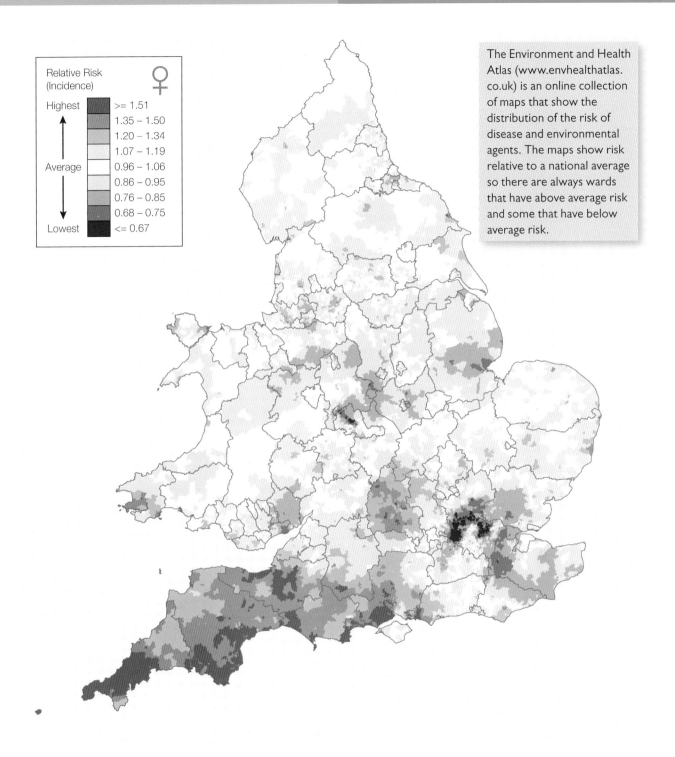

Relative Risk
(Incidence)

♀

Highest
>= 1.51
1.35 – 1.50
1.20 – 1.34
1.07 – 1.19
Average
0.96 – 1.06
0.86 – 0.95
0.76 – 0.85
0.68 – 0.75
Lowest
<= 0.67

The Environment and Health Atlas (www.envhealthatlas.co.uk) is an online collection of maps that show the distribution of the risk of disease and environmental agents. The maps show risk relative to a national average so there are always wards that have above average risk and some that have below average risk.

▲ *Figure 2* *Female skin cancer (malignant melanoma) incidence, (Environment and Health Atlas 2010)*

ACTIVITIES

1   Study Figure **1**. Describe changes in life expectancy across the UK over a 20-year period. (Hint: describe the broad pattern first using named examples and then identify exceptions.)

2   To what extent is our health controlled by our own choices and also by where we live?

STRETCH YOURSELF

Using the Environment and Health Atlas (www.envhealthatlas.co.uk), produce an illustrated report on morbidity variations in England and Wales. Include at least three annotated maps to support your answer. Click on 'Go to the atlas' to access the data about health conditions and environmental agents. For example, a map of average sunshine hours across England and Wales might be used to help interrogate the pattern of skin cancer (see Figure **2**), although the age profile of different regional populations might also be of relevance.

In this section you will learn about influences of age, gender, wealth and environment on lifestyle, nutrition and on access to healthcare in the UK and sub-Saharan Africa

Factors of mortality in HDEs such as the UK

The greatest cause of mortality in HDEs is lifestyle-related illness, such as diabetes, heart disease and cancer. Even though many of our chronic illnesses are preventable, as a society we continue to fail to listen to advice, adopt **sedentary** lifestyles and eat inappropriately. Several interrelated factors influence health in HDEs:

◆ *Wealth* – healthy foods such as fish, fruit and vegetables tend to be more expensive than those that are less healthy, such as those high in carbohydrates and fat. A study in Norfolk between 1990 and 2008 found that the number of takeaway outlets rose by 45 per cent. The biggest expansion in proportional terms was in the county's poorer areas (58 per cent) compared to the least deprived areas (30 per cent).

◆ *Age* – different age groups perceive their health needs differently. For example, older groups are more likely to have plans in place for future health needs (e.g. health insurance), be more informed of preventative strategies (e.g. vaccinations) and routinely have health checks.

◆ *Gender* – women are more likely to make 'healthy decisions', both in adopting preventative measures such as regular GP visits (e.g. cholesterol checks and mammograms) and lifestyles choices such as exercise or healthy diet.

◆ *Environment* – the relationship between climate and health is long established. For example, the UK is more prone to pneumonia, influenza and even the common cold in the cooler winters. In contrast, some cancers are closely linked to overexposure to solar radiation. The availability and quality of water supplies is also significant. For example, links have been made between the occurrence of Alzheimer's disease and high concentrations of aluminum in water related to industrial activities. Urban environments with higher densities of population present unique challenges to health (see 4.11) – they are at greater risk of depression and schizophrenia ('urban stress') even though infrastructure, socio-economic conditions, nutrition and health care services are better than in rural areas.

Factors of mortality in LDEs such as in sub-Saharan Africa

Sub-Saharan Africa is the region with the highest prevalence of hunger. One person in four is undernourished and malnutrition is a contributory factor in 4 million sub-Saharan deaths annually – around one-third of the global total. The contrasts with the UK could not be greater.

◆ Poverty results in a lack of access to food, health care, clean drinking water and sanitation, and is the dominant cause of hunger.

◆ Health choices are further constrained by misunderstanding and prejudice. Women, often the main food preparers, have limited access to education and control over household income. Poor understanding of dietary needs (as well as the often high cost of protein) is likely to lead to imbalanced meals, high in carbohydrates but lacking in other nutrients.

◆ During pregnancy and child rearing, women are also unlikely to have adequate rest leading to increased risks of poor health for both mother and child.

◆ Environmental problems, including drought, desertification, insect infestation, soil erosion and wetland degradation, add to the difficulties of survival on marginal lands. These, amongst other factors, combine to further reduce levels of productivity and so add to vicious cycles of misery (Figure **1**).

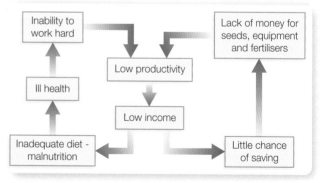

◆ *Figure 1* Cycles of misery

Bet of the century

In 2007, Alec Holden from Epsom, Surrey, celebrated his 100th birthday and won £25000 as a result, along with his telegram from the Queen (Figure **2**). He had placed a bet ten years previously that he would live to see his 100th birthday. Mr Holden explained that the secret to his longevity was a healthy diet, active brain, positive outlook on life and 'remembering to keep breathing'! As a nation, the population of the UK is living longer, which is ironic as our lifestyles are generally less healthy. Since 2007, bookmakers have raised the age of so-called 'age wagers' to 110!

⊙ **Figure 2** *Alec Holden with his betting slip*

The epidemiological transition

As Thompson's Demographic Transition Model recognises, the population dynamics of a country change over time (see 4.18). Birth rates and death rates change according to the the day-to-day, social and economic reality of living somewhere, as well as a population's prospects – the likelihood of survival into adulthood. How do those survival prospects change over time? The theory of the **epidemiological transition** outlines the way in which populations shift from being defined by high rates of infant mortality (and low life expectancy), as a result of infectious diseases and famine, to a state in which average life expectancy is much higher (50+ years) with degenerative, man-made diseases being more likely causes of death – primarily affecting the elderly. The theory was proposed by Abdel Omram in 1971.

Improved sanitation, as well as nutrition, was the secret of HDEs' epidemiological transition in the nineteenth century. However, in the twentieth and twenty-first centuries, LDEs with lengthening life expectancies are witnessing change as a result of healthcare and anti-disease programmes developed and financed internationally (see 4.14). So, it is argued, the transition isn't 'homegrown' but the shift is real. The number of older people in most sub-Saharan countries is growing at a higher rate than in most HDEs – to over 67 million by 2030 – and of these, the over-80 age group is the fastest-growing age group.

In HDEs, whether chronic degenerative diseases have increased over the twentieth and twenty-first centuries or whether we are simply getting better at diagnosing diseases like cancer is still a matter of debate.

Did you know?
Hunger kills more people every year than AIDs, malaria and tuberculosis combined.

⊙ **Figure 3** *Life expectancy in most African countries is increasing, so there are larger numbers of healthy octogenarians*

ACTIVITIES

1 Suggest reasons why in HDEs like the UK, we might be 'living longer even if … our lifestyles are less healthy'.

2 Explain why poverty is self-sustaining in some countries in sub-Saharan Africa.

3 Outline the theory of the epidemiological transition, with reference to named countries. Use data to support your comments on what the biggest killers are in your chosen countries, both now and, for example, 20 years ago. An EME would be a good starting point (in 2016, *The Lancet* reported that China now has the largest number of obese people in the world). Compare one or more of these countries to an LDE such as Afghanistan, Ethiopia or Mozambique where poverty and limited life expectancy persists. Look at www.gapminder.org for data about how the health of national populations has changed over time (see Figure **1**, 4.8).

In this section you will learn about:
- the links between location and health
- the influence of economic development and environmental factors on well-being

Longevity and Blue Zones

Where in the world do people live the longest? Logic suggests that wealthy urban centres would have the largest percentages of healthy older adults. In fact, the answer is the small Japanese island of Okinawa. Other examples of *Blue Zone* communities (an area where people live longest) include Nicoya in Costa Rica and Sardinia. Whilst access to public health services is important, geography or place also has a role to play – several Blue Zones are island communities and/or geographically isolated.

Did you know?
Three of the top five countries with the highest life expectancy have populations of less than one million (Monaco, Macau and San Marino).

The city of Glasgow

Scotland has the lowest life expectancy in Western Europe and the city of Glasgow the lowest in Scotland. Glasgow shares many economic and environmental characteristics of other cities in the UK – for example, Liverpool and Manchester – yet its residents are about 30 per cent more likely to die young. Furthermore, 60 per cent of those deaths are triggered by just four things – drugs, alcohol, suicide and violence.

The causes of Glasgow's health problems are complex and often inter-related. Epidemiologists have struggled to understand why traditional health approaches have largely failed in the city – they even refer to the city's health issues as simply the *'Glasgow Effect'*.

Some Glaswegians eat, smoke and drink in excess but no more than in other European urban centres – indeed, Scotland was the first part of the UK to ban smoking in public places and introduce minimum pricing of alcohol.

One theory is that deindustrialisation is the underlying cause of the Glasgow Effect. The collapse of traditional industries such as shipbuilding left parts of Glasgow with high unemployment rates and abandoned industrial sites.

Look at Figure **1**. Whole communities, such as the peripheral social housing estates of Riddrie and Cranhill, lost a sense of identity, pride and 'togetherness' and, it might be argued, a collective responsibility for health. Within such degraded urban environments, drugs and alcohol have filled the gap left by deindustrialisation. Certainly, health campaigns have largely failed to get across their message and Glaswegians' well-being and mental health in particular lag alarmingly behind the rest of Europe.

ANOTHER VIEW

What are urban environmental health problems?
There are four scales of urban environmental health problems.
- Within the *house*, such as indoor air pollution from open fires or damp and mould.
- *Neighbourhood* health hazards, such as inadequate or polluted local water supplies and sanitation, discarded drug needles and threats of physical violence.
- *City-wide* problems, such as air pollution from traffic congestion, ineffective waste management, river pollution and the urban heat island effect may lead to health concerns across the entire urban population.
- *Extra-urban impacts* that take place immediately beyond the city boundary or hinterland will also impact on the well-being of the urban population. For example, loss of 'green spaces' as a city expands.

▲ **Figure 1** *Cranhill housing estate; many of Glasgow's problems are no worse than those of other UK cities yet mortality rates remain higher*

Sri Lanka

Look at Figure **2**. Sri Lanka shares a similar life expectancy of 71 years with Glasgow yet few other indicators are comparable. Sri Lanka has many economic characteristics of an LDE, yet health provision indicates a country that is developing economically and moving through Stage 3 of the demographic transition model (see 4.18).

The government has made a strong commitment to health and has achieved the health-related Millennium Development Goal targets. Public healthcare is accessible to all and is almost entirely free of charge, similar to the NHS. The vision of the Sri Lankan health system is of 'a healthier nation that contributes to its economic, social, mental and spiritual development'. There has been investment in public health units and hospitals (Figure **3**) as well as on specialist equipment and training of doctors and nurses. Widespread vaccination and health education programmes have almost eliminated diseases such as polio, malaria and leprosy. Reducing this so-called 'disease-burden', combined with female empowerment, has helped to reduce the birth rate and promote economic development.

Public health intervention has been less successful in reducing the prevalence of non-communicable diseases such as cancers, diabetes and respiratory diseases that account for around 70 per cent of deaths in the country. Smoking continues to increase, despite increased taxes on cigarettes and banning smoking in public spaces, and is a significant factor in these deaths. Sri Lanka also needs to address the health consequences of an ageing population, to develop mental health services and to rebuild the health systems in the north and east of the country following a long civil war. These health needs will place increased pressure on government spending.

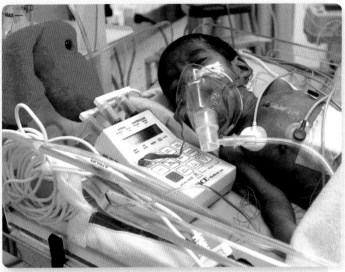

⬤ **Figure 2** *Sri Lankan medical facilities rate alongside those of many EMEs and some HDEs*

Indicator (2014)	Sri Lanka	Glasgow (Scotland)
Life expectancy	76 years	71
% Population under 15	25%	17%
% Population over 60	12%	18%
World Bank income group	Lower-middle-income	High-income
Total expenditure on health as % of GDP	3.4%	7.3%
Human Development index rank out of 186 countries	73	14

⬤ **Figure 3** *Contrasting indicators of health and economic development for Sri Lanka and Glasgow (2014)*

ACTIVITIES

1 Are small island or isolated communities more or less at risk of infectious disease? Suggest reasons for your answer.

2 Is it fair to directly compare the health needs of Sri Lanka and Glasgow? Justify your answer.

3 Are urban environments in HDEs more or less likely to lead to health problems than in LDEs? Explain your answer with reference to named countries. Hint: in your introductory paragraph, outline two or three specific urban environmental health problems that you intend to discuss later in greater detail.

STRETCH YOURSELF

To what extent is economic development a reliable indicator of health and well-being? Select an aspect of health to investigate in order to address this rather broad question, and/or look at a specific group within the population, such as women. Reducing maternal mortality was a Millennium Development Goal (MDG 5). The WHO's website provides a range of data on the progress made to reduce the death rates of mothers between 2000 and 2015 and factors that affected the pattern of decline worldwide, including economic development. The organisation will continue to report on this issue given Sustainable Development Goal 3 also relates to reducing maternal mortalilty rates going forward.

In this section you will learn about links between disease and the physical environment

Links between disease and the physical environment

The links between disease and the physical environment have not always been known. In 1854, John Snow proved that cholera was transmitted by drinking contaminated water. By mapping cases of cholera as a dot distribution map, Snow showed that cases of cholera in Soho, London, were linked to the Broad Street water pump (Figure **1**). When he had the handle of the pump removed, the cases of cholera immediately fell.

Today, our understanding of the relationship between our interaction with the environment and disease is much improved. Nevertheless, our exploitation of resources and use of new technologies has created new environmental hazards for health, such as storage of radioactive or toxic waste and shale gas extraction. Looking ahead in the **Anthropocene** (the epoch which is influenced by human activity), human influence on climate will, no doubt, further change the range of threats to health (see 4.27).

In general, in HDEs environmental hazards have a bigger impact on non-infectious diseases, such as heart disease, whereas in LDEs the impact is higher on infectious diseases, such as malaria (Figure **2**).

The impact of the environment on disease

Environmental hazards may be grouped as follows:

◆ Water, sanitation and hygiene – unclean drinking water and stagnant water that attracts the carriers (**vectors**) of disease such as mosquitoes (see 4.13) or rats.

◆ Chemical exposure – such as exposure to lead and asbestos, which is closely linked to heart disease and cancer.

◆ Radiation – including radiation from radon and ultraviolet sources (such as the sun), which leads to cancers.

◆ Air quality – including indoor (e.g. smoke from cooking, a cause of child pneumonia) and outdoor pollution (e.g. smog resulting from industrial and vehicle emissions cause a whole range of medical conditions (Figure **3** and 4.15).

◆ Chemical traces in foods – food may travel thousands of food miles before being consumed. This means that poisons may be harder to track back to the source.

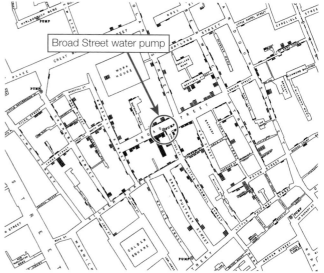

Broad Street water pump

△ **Figure 1** *Snow's dot distribution map of the cholera outbreak (1854) proved the existence of the link between contaminated water and disease. It showed a clear distance decay relationship – the number of deaths reduced with increasing distance from the Broad Street water pump.*

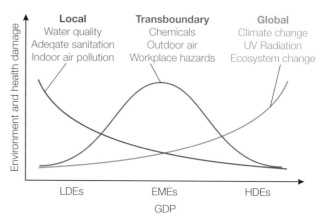

Local	**Transboundary**	**Global**
Water quality	Chemicals	Climate change
Adeqate sanitation	Outdoor air	UV Radiation
Indoor air pollution	Workplace hazards	Ecosystem change

Environment and health damage

LDEs · EMEs · HDEs

GDP

△ **Figure 2** *Public health and environment – global differences; environmental hazards are not the same everywhere (WHO, 2011)*

Did you know?
Pollution from vehicles, industries and energy production kills 800 000 people per year.

The lack of investment to address environmental health hazards

Despite environmental hazards causing around 25 per cent of all diseases worldwide, less than 5 per cent of global health spending is on prevention (Figure **4**). Where money is spent on prevention, the economic, social and environmental returns are significant. For example, local sustainable responses to climate change, such as clean cookstoves (Figure **5**), reduce the risk of burns and air pollution in the home. In 2011, the WHO noted that 'indoor smoke from solid fuels [was] responsible for the deaths of approximately 1.6 million people annually, including nearly one million children under the age of five, largely as a result of respiratory diseases' (WHO Public Health and Environment Global Strategy).

▲ *Figure 3* *Runners brave the smog during the 2015 Shanghai marathon. China's cities have seen extremely high levels of air pollution in recent years. In 2015, red alerts were issued to ten cities after smog blanketed China's coal-reliant north-east. Residents were advised to stay indoors.*

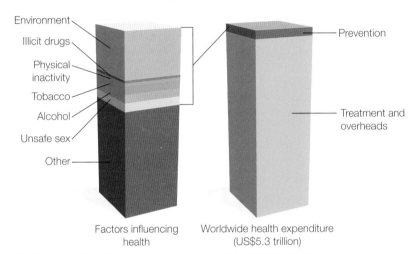

Environment
Illicit drugs
Physical inactivity
Tobacco
Alcohol
Unsafe sex
Other

Prevention
Treatment and overheads

Factors influencing health

Worldwide health expenditure (US$5.3 trillion)

▲ *Figure 4* *Health determinants vs spending (WHO, 2011)*

ACTIVITIES

1 Study Figure **1**. To what extent does Snow's dot map of cholera deaths show a distance decay relationship with the contaminated water pump?

2 Do you think that our use of the environment has made us more or less at risk of health issues?

3 When the causes of so many diseases are known, why do health and other authorities spend relatively little on investing in preventative measures, and instead invest in treating the symptoms?

4 Do all EMEs need go through a stage of economic growth that compromises the health of its citizens? Discuss this with a partner.

STRETCH YOURSELF

What risk might fracking pose to health in the UK? This is a contentious issue that is much debated amongst pro- and anti-fracking groups. Find out what both sides say about the risks and links between operations to exploit underground shale gas and disease. What can be learnt from the development of fracking in the USA? (See 5.4 and 5.16.)

▲ *Figure 5* *Use of appropriate technologies for stoves is environmentally sustainable as less fuelwood needed; these stoves are also socially acceptable as the flames and smoke are contained, leading to fewer burns and breathing complications*

In this section you will learn about:
◆ the worldwide prevalence and incidence of malaria
◆ links between the disease and the physical and socio-economic environment

The disease

The symptoms of malaria are fever, headaches and tiredness. Sufferers may also experience diarrhoea and vomiting. The parasite infects red blood cells causing anaemia and jaundice (a yellowing of the skin). Left untreated, malaria may cause kidney failure, seizures, coma and death. Symptoms of malaria only appear a week or even two weeks after you have been bitten; without thorough treatment, the disease can return up to 50 years later. You only need to be bitten once!

What is it like to have malaria?

'I awoke to what felt like lightning going through my legs, and then spreading through my body and in my head. Probably the worst headache, body aches, and chills you could possibly imagine. It felt like I was being stung repeatedly by an electric shock gun and could barely control my movements. The pain was so intense...'

How do you catch it?

There are over 100 species of *Anopheles* mosquito, which injects the parasite *Plasmodium* (which causes malaria) into its victim when it bites (Figure **1**). However, only about a third of these are considered malaria vectors of major importance: commonly transmitting the disease in areas of the world where malaria is found today. Not every mosquito bite results in malaria being passed on. The rate of new cases in an area depends on several factors linked to the natural environment and the human population (Figure **2**).

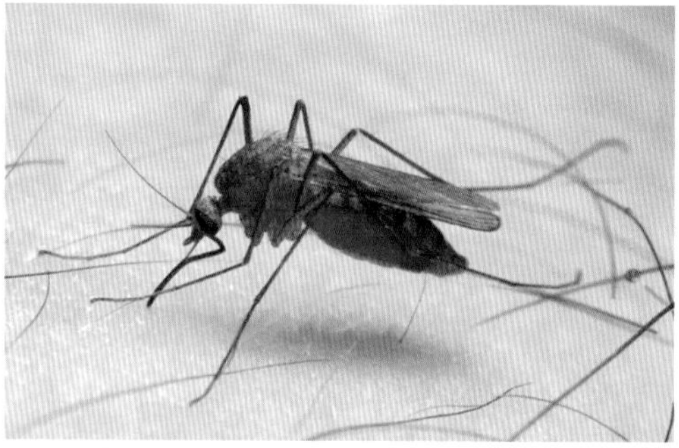

Malaria can also be caught via a blood transfusion, organ transplant or shared use of contaminated needles. It may also be transmitted from a pregnant mother to her unborn child. In fact, pregnant women have reduced immunity making themselves particularly vulnerable to the disease. These 'at risk' groups are targeted by international agencies and NGOs working on the eradication of the disease (see 4.14).

The severity of malarial infections is affected by the specie of parasite. The parasite most often found in sub-Saharan Africa, *Plasmodium falciparum* causes severe, potentially fatal disease and some are now resistant to antimalarial drugs.

◆ **Figure 1** *The Plasmodium parasite that causes malaria is transmitted in the saliva of an Anopheles mosquito when it bites*

◆ **Figure 2** *Factors affecting incidence of malaria*

Species of mosquito	Lifespan	The longer the lifespan of the insect the greater time for the parasite to complete its lifecycle too, inside the mosquito.
	Preference for biting humans vs other animals	Some species of mosquito prefer to bite humans!
		According to the WHO, the combination of these two factors form the main reason why 90% of the world malaria cases are in Africa.
Human immunity	Children	Partial immunity is developed over years of exposure to the disease which reduces the risk of severe illness. Therefore young children are most at risk as they lack the immunity that their parents benefit from.
	Immigrants	Epidemics may occur among immigrant populations; migrant workers or refugees who travel from regions where they lack regular or seasonal exposure to the disease.
The environment	Climate	Mosquitoes breed in water. Puddles, ponds and pools are abundant all year in a tropical wet climate. In tropical regions characterised by dry as well as wet seasons, transmission of malaria is most intense during and just after the rains – outbreaks can be predicted accurately by mapping rainfall.
	Rural vs urban	The geographic distribution of malaria within countries is complex. For example, there is a risk of infection in rural areas of countries in south-east Asia, where urban areas have been declared disease-free.

The global impact

Is malaria a global disease?

As of 2015, the WHO estimated that:

◆ 3.2 billion people are at risk of contracting malaria; more than 40 per cent of the world's population

◆ 214 million new cases occur every year

◆ Malaria is present in 97 different countries.

Malaria is prevalent in Africa, south-east Asia, Latin America and, to a lesser extent, the Middle East (Figure **3**). Of all the people living with malaria, 80 per cent live in just 15 countries, 13 of which are in Africa (Figure **4a**). So is malaria really a 'global' disease? International NGOs striving for the disease's eradication certainly see it as a concern for the whole of humanity (see 4.14).

Is malaria a global killer?

In 2015, the WHO estimated that there were about 438 000 deaths from malaria – mostly African children. Seventy-eight per cent of all deaths from malaria are accounted for in just 15 countries (Figure **4b**), of which just one is outside Africa. Climate change already seems to be moving the distribution to higher latitudes and there is a risk that this will continue towards the poles. Look at Figure **5**. This map of malaria death is worth a thousand words on the subject!

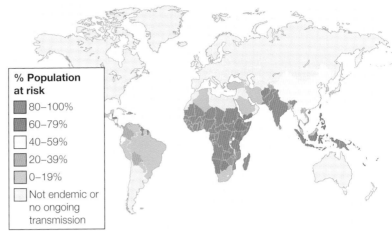

◆ **Figure 3** *Percentage of population at risk of malaria (2013)*

◆ **Figure 4** *Estimated proportion and cumulative proportion of the global number of a) malaria cases and b) malaria deaths in 2015*

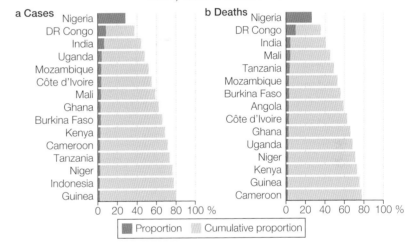

ACTIVITIES

1 Study Figure **2**. Find out more about one of the factors affecting the incidence of malaria and share your findings with the class.

2 Study Figures **3**, **4** and **5**.
   a Describe the distribution of malaria cases worldwide
   b Describe the distribution of deaths caused by the disease.
   c Choose either your answer to **2a** or to **2b** and try to explain the pattern you observe. Hint: referring back to Figure **2** may help.

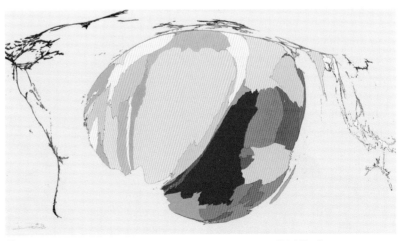

◆ **Figure 5** *Worldwide distribution of malaria deaths (2010). Countries are sized in proportion to the number of people who died from malaria in one year.*

In this section you will learn about:

◆ management and mitigation strategies employed to reduce the impact of malaria and their success

◆ the role of NGOs and international agencies in the fight against malaria

Malaria in Uganda

In 2015, there were 3.6 million confirmed cases of malaria in Uganda (Figure **1**) and almost 6000 deaths. Unlike in other parts of Africa, where malaria outbreaks tend to occur after the rainy season, transmission in Uganda happens all year round because of high temperatures and rainfall throughout the year (Figure **2**). There are also large areas of water, such as Lakes Kyoga, Albert and Victoria, which act as breeding grounds for mosquitoes, or reservoirs of infection. It is also in these areas where the population density is at its highest (Figure **3**).

In the case of malaria, little can be done to change the environmental conditions in which mosquitoes breed. Improved drainage and irrigation, particularly within rural areas, has had some success. The use of insecticides has only been partially successful because mosquitoes become resistant and alternative insecticides have to be developed. Insecticide-treated mosquito nets are another protective strategy that are widely used but they are not always used properly.

If mosquitoes can never be entirely eliminated, then the best approach to malaria control is a combination of early diagnosis and prompt treatment (Figure **4**). Antimalarial medicines are available but they are not only expensive but are also becoming less effective, so they need to be used with other drugs. In 2009, the first malaria drug factory was opened in Kabale by Afro Alpine Pharma. This is significant as, not only does it increase production and lower drug costs, but the plants used to produce the drugs are grown locally by more than 5000 farmers.

Total population	37 800 000
GNI per capita	$670
Percentage living on less than $1.25 a day	37.8
Life expectancy	54.9
Households with at least one insecticide-treated bed net	70–90%
Under 5s receiving anti-malarial drugs	65%

⬆ **Figure 1** *Health and development indicators for Uganda (2014)*

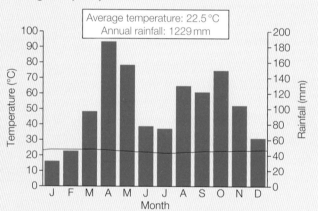

Average temperature: 22.5 °C
Annual rainfall: 1229 mm

⬆ **Figure 2** *Climate graph for Apac District near Lake Kyoga, Uganda*

▶ **Figure 3** *Population density map of Uganda*

◀ **Figure 4** *A laboratory technician in Amuria, Uganda, examines a patient's blood for signs of malaria*

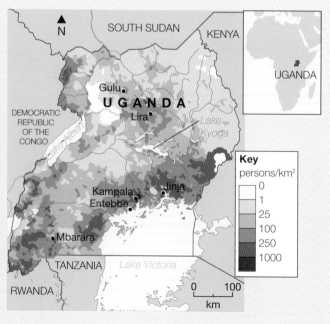

Key
persons/km²
0
1
25
100
250
1000

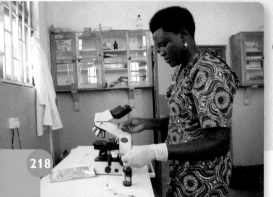

Management and mitigation of malaria

A Millennium Development Goal met

At a global scale, malaria's impact has been significantly reduced over the last 15 years (Figure **5**). The Millennium Development Goal to 'have halted by 2015 and begun to reverse the incidence of malaria' has been met convincingly (Figure **6**).

Malaria's impact in sub-Saharan Africa has also been reduced and the disease is no longer the leading cause of death among children in the region (Figure **7**).

⊙ **Figure 5** *Projected changes in malaria incidence rates (WHO 2000–2015)*

Cases reduced to zero since 2000

On track for >75% decrease in incidence 2000–2015

On track for 50%–75% decrease in case incidence 2000–2015

Less than 50% change in incidence projected, 2000–2015

Increase in incidence 2000–2015

Insufficient data

Non-endemic or no ongoing malaria transmission

Did you know?

On average, a person living near Lake Kyoga, Uganda, will receive more than 1500 infectious mosquito bites each year!

Indicators	2000	2015	% change
Incidence rate associated with malaria (per 1000 at risk) and Death rate associated with malaria (per 100 000) at risk)	146	91	−37%
	47	19	−60%
Proportion of children under 5 sleeping under insecticide-treated mosquito nets	2%	68%	>100%
Proportion of children under 5 with fever who are treated with appropriate antimalarial drugs	0%	13%	>100%

⊙ **Figure 6** *Malaria MDG indicators (2000 and 2015)*

⊙ **Figure 7** *Leading causes of death among children aged five and under in sub-Saharan Africa (2000–2015)*

—— Measles   —— Prematurity   —— Diarrhoeal diseases

—— Malaria   —— Birth asphyxia and birth trauma   —— Acute respiratory infections

Strategies that worked

'It is estimated that 1.2 billion fewer malaria cases and 6.2 billion fewer malaria deaths occurred globally between 2001 and 2015.' (World Malaria Report, WHO 2015)

How have such strides in disease reduction been achieved? Who funded them and what sort of strategies were employed to reduce malaria's death toll?

A number of different strategies have been employed to reduce the impact of malaria in affected regions:

◆ *control of the vector*: use of insecticides, for example, inside dwellings, on walls where mosquitoes rest

◆ *use of physical barriers to infection*: insecticide-treated mosquito nets

◆ *use of chemical barriers to infection*: the proportion of women receiving doses of preventative treatment in pregnancy has increased and limited use of seasonal anti-malaria drugs in children

◆ *investing in the swift diagnosis*: the sooner the disease is diagnosed, the better chance there is of recovery

◆ *drug treatment of the disease*: in particular, highly effective, artemisinin-based combination therapies (ACTs) (Figure **8**).

Despite recent progress in reducing the impact of this disease worldwide, it is notable that rates of decline of malaria incidence and mortality are slower in the countries that bear the highest burden from the disease. Although the overall number of cases of malaria worldwide has declined by a third over the last 15 years, in countries outside the WHO's list of 15 countries most affected (Figure **4**, 4.13) cases of malaria have, in fact, halved.

Perhaps, unsurprisingly, countries where there remains a high risk of suffering from malaria are those with a weaker health system (with high ratios of medical staff to population) and lower incomes (Figure **9**).

▲ **Figure 8** *Tu Youyou of China received a Nobel Prize in the sciences in 2015, for discovering artemisinin, a drug now part of standard antimalarial prescriptions. Her review of traditional Chinese recipes identified the plant* Artemisia annua *or sweet wormwood as a powerful medicine that can be used to reduce parasites in a patient's blood.*

▼ **Figure 9** *Gross national income per capita versus estimated number of malaria cases, by WHO region (2015)*

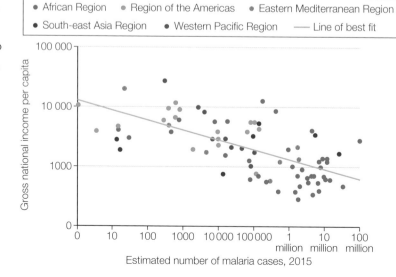

Estimated number of malaria cases, 2015

Did you know?

Nearly 500 million insecticide-treated mosquito nets were delivered to countries in sub-Saharan Africa between 2013 and 2015.

The role of international agencies and NGOs in combating malaria

There are a number of organisations working to eradicate malaria in Uganda and across the globe. These include the US President's Malaria Initiative, the international Roll Back Malaria Partnership, the World Health Organisation, the UK's Malaria No More charity (Figure **10**) and the Bill and Melinda Gates Foundation (Figure **11**).

Global financing of malaria control programmes increased from US$960 million to US$2.5 billion between 2005 and 2014 – a 260 per cent increase! Of the total funds invested in the international campaign against the disease in 2014, international investments (rather than those made by domestic governments in affected areas) made up 78 per cent. In 2016, the UK government and Microsoft founder, Bill Gates, pledged £3 billion (US$4.2 billion) over the next five years to fund research and to support efforts to eliminate malaria. The Gates Foundation had already donated significant funds to the cause since 2007, when the super-rich philanthropists Bill and Melinda Gates were introduced to the impacts of malaria and the problems of fighting the disease.

Eradicating malaria

'Any goal short of eradicating malaria is accepting malaria; it's making peace with malaria; it's rich countries saying: "We don't need to eradicate malaria around the world as long as we've eliminated malaria in our own countries." That's just unacceptable.' (Melinda Gates of the Gates Foundation)

▲ *Figure 10* *Andy Murray wears the Malaria No More logo. 'This disease claims a young life every two minutes, yet it is preventable and it costs less than a pack of tennis balls to treat and help save a life. As a new dad this really hits home,' he said of his involvement in the campaign.*

▲ *Figure 11* *The Bill and Melinda Gates Foundation headquarters in Seattle, Washington*

STRETCH YOURSELF

Find out about how a named non-governmental organisation is tackling malaria at an international scale. In your answer, give examples of how the natural (e.g. temperature, rainfall) and human environments (e.g. rice fields, homes) are being changed and the range of barriers used to disrupt transmission of the disease.

ACTIVITIES

1 Study Figure **2**. Calculate the temperature range and average monthly rainfall. Use these calculations and the data on the graph to help to explain why malaria is so prevalent in Uganda.

2 Using online resources explore the Malaria Atlas Project (www.map.ox.ac.uk), which sheds light on the impact of malaria control in Africa OR create your own maps using the Global Malaria Mapper site www.worldmalariareport.org. What do the maps you've found or created show about:
   a who is at risk?
   b how the disease is being controlled?

3 Study the quote from Melinda Gates above. What is your view? Select one of the following topics to discuss:
   • Is malaria Africa's problem or something we all need to solve?
   • Do you think eradication is possible? If not why not?

In this section you will learn about:
- ◆ the global prevalence and distribution of asthma
- ◆ links between asthma and physical and socio-economic environments

Defining the disease

The WHO defines asthma as 'a chronic disease characterized by recurrent attacks of breathlessness and wheezing, which vary in severity and frequency from person to person'.

An 'asthma attack' is associated with a narrowing of the bronchial tubes that allow air to pass into and out of the lungs when a person breathes. This narrowing of the airways reduces airflow leading to chest tightness and coughing. It can be a killer. However, in comparison to other long-term diseases, asthma has a relatively low fatality rate. Although it can't be cured, for most sufferers the symptoms can be controlled (see 4.16) – often by self-medication, via an inhaler.

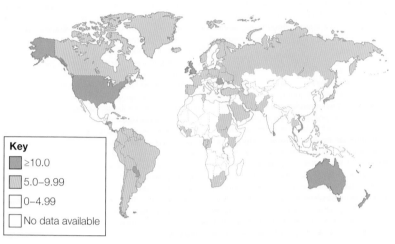

Key
- ≥10.0
- 5.0–9.99
- 0–4.99
- No data available

⚫ **Figure 1** *Percentage of children aged 13–14 suffering from asthma (2009)*

The global prevalence and distribution of asthma

An estimated 300 million people have asthma worldwide. International organisations like the Global Asthma Network suggest further research is urgently needed into the current scale of the impact of this disease, especially given the recent increase in the number of sufferers.

'The historical view of asthma being a disease of HDEs no longer holds: most people affected are in low- and middle-income countries, and its prevalence is estimated to be increasing fastest in these countries.' (Global Asthma Report 2014)

The burden of asthma is most prevalent in the elderly, aged 75 to 79 and also in the 10 to 14 age group – 14 per cent of the world's children experience asthma symptoms. Figure **1** illustrates the distribution amongst 13–14 year olds.

Mortality rates

Less than 1 per cent of all deaths worldwide are linked to asthma. However, the mortality rate linked to asthma varies greatly between countries. Figure **2** shows mortality rates for selected countries. 'Most asthma-related deaths occur in low- and lower-middle income countries,' noted the WHO in 2013.

⚫ **Figure 2** *Asthma mortality rates for all ages (2001–2010)*

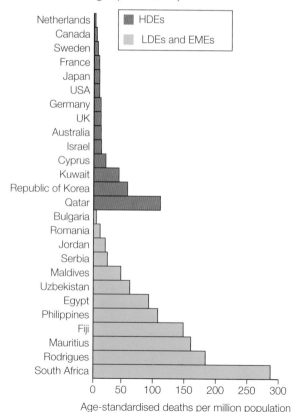

HDEs
LDEs and EMEs

Netherlands
Canada
Sweden
France
Japan
USA
Germany
UK
Australia
Israel
Cyprus
Kuwait
Republic of Korea
Qatar
Bulgaria
Romania
Jordan
Serbia
Maldives
Uzbekistan
Egypt
Philippines
Fiji
Mauritius
Rodrigues
South Africa

0   50   100   150   200   250   300

Age-standardised deaths per million population

Asthma's triggers

The causes of asthma are not completely understood but what is clear is that both environmental factors and also family history influence the overall risk of developing the disease (see 4.16).

A family history of asthma is a high risk factor for developing the disease, particularly before the age of 12 years – after which environmental factors are more likely to be the cause. Different socio-economic lifestyles also influence your chances of suffering from asthma (Figure **3**).

Environmental triggers

Exposure to different **allergens** that trigger (or exacerbate) asthma symptoms may occur at home or in the workplace. A wide range of workers have been identified as being at high risk of developing occupational asthma – they range from bakers to carpenters, hairdressers to employees of the automobile industry. The time of the year is also significant as exposure to some key allergens varies throughout the year.

Asthma and a 'Western' lifestyle

A further increase in the prevalence of asthma is anticipated, particularly in eastern European countries where acquisition of a Western lifestyle has already been shown to increase the prevalence of childhood asthma to as much as 20 per cent.

There is a positive correlation between the increased affluence of a society and the incidence of asthma. Exposure to parasites and other pathogens (viruses and bacteria) has decreased in our 'Western', 'cleaner' homes with a subsequent greater incidence of asthma. The idea that to fully develop our immune system we need exposure to a wide range of pathogens is known as the 'hygiene hypothesis'.

▲ **Figure 3** *There are many different triggers for asthma; some are environmental, others link to lifestyle*

EXTENSION

Breaking the mould

A study of rental accommodation in Cornwall has found an unintended consequence of **retrofitting** older social housing with insulation and double glazing: damp and mould. Researchers who surveyed the properties found higher levels of humidity in homes that had been insulated. From questionnaire data, they also noted that families suffering from **fuel poverty** were less likely to properly ventilate their property (open windows or switch on extractor fans), in order to conserve heat and save money.

'We need to make sure that increasing household energy efficiency is conducted alongside measures to raise awareness of the need to adequately heat and ventilate the home in order to help avoid exposure to damp and mould.' (Richard Sharpe, a researcher based at the University of Exeter).

ACTIVITIES

1 'Globally, asthma caused 489 000 deaths in 2013.'
   a Using Figure **2** describe and explain the global pattern of asthma mortality rates
   b Discuss the level of certainty we can have with regard to such statistics.

2 Using online resources, find out about one of the environmental triggers of asthma. Compare your findings with a partner who has investigated a different environmental trigger.

3 Study Figure **3**. Make a list of possible reasons why asthma is increasing most rapidly in lower- and middle-income countries.

4 A 2013 study of 8–12 year olds in 20 countries by the International Study of Asthma and Allergies in Childhood (ISAAC) found a link between damp housing conditions and respiratory symptoms, in both affluent and non-affluent countries. How might an awareness of these findings influence policy-makers at a regional and national level?

In this section you will learn about:
♦ the impact of asthma on health and well-being
♦ management and mitigation strategies

Impact on health and well-being

The following account of living with asthma illustrates the impact of this disease on one family. While the prognosis for most asthma sufferers is good (at least in HDEs), the day-to-day reality of living with asthma should not be underestimated. Globally, asthma is the fourteenth most important disorder in terms of the extent and duration of disability inflicted on sufferers.

Dealing with asthma

'I've been dealing with asthma my whole life. I was diagnosed at four and had a few bad asthma attacks. Once I nearly died. Since the age of nine, though, I haven't had an asthma attack – I do everything I can to stay on top of it. Later, my grandad was diagnosed with late-onset asthma in his 60s. So it wasn't a huge surprise when both my sons were diagnosed with asthma too.

I used to worry a bit about the boys taking steroids every day, but I know for a fact I wouldn't be here if I hadn't had them when I was four – they saved my life. There's a panic attitude around about steroids, but when you really think about it the benefits far outweigh the risks.

Cold and damp weather is the worst trigger for all of us. In November, both boys tend to get a night-time cough. During a cold spell, we always make sure we wear neck 'tubes' around our mouths and noses – they're a bit like scarves; apparently wearing them warms up the air before we breathe it in and helps to prevent asthma symptoms.

Catching a cold can cause their symptoms to flare up. Dust mites are another trigger. It's hard to avoid both of these so using our preventers means we are less likely to react when we come across them. And we know to have our relievers handy if we have a cold or we're visiting a dusty house.

The boys aren't embarrassed to use their inhalers in public. We've always made it a very matter-of-fact part of life. I try not to let asthma stop us doing anything.'
(Adapted from Asthma UK)

🔺 **Figure 1** *Living with a chronic disease*

Measuring the burden of asthma

Asthma takes effect much earlier in life than other chronic diseases, and is also something that they have to live with for the rest of their lives. Consequently, it imposes a high lifetime burden not only on those that are affected but also on carers, family and the community. Asthma has a global distribution with a relatively higher burden of disease in Australia and New Zealand, some countries in Africa, the Middle East and South America, and north-western Europe (Figure **2**).

The **disability-adjusted life year (DALY)** is a measure of overall disease burden, expressed as the number of years lost due to ill-health, disability or early death. It combines morbidity and mortality in a single measure that can be used to compare the overall health and life expectancy of different countries. Figure **2** presents the DALYs associated with asthma. The higher the DALY, the higher the number of healthy years lost and the poorer quality of life overall; asthma sufferers may die young or have their education, career (and earning potential) seriously curtailed by ill health.

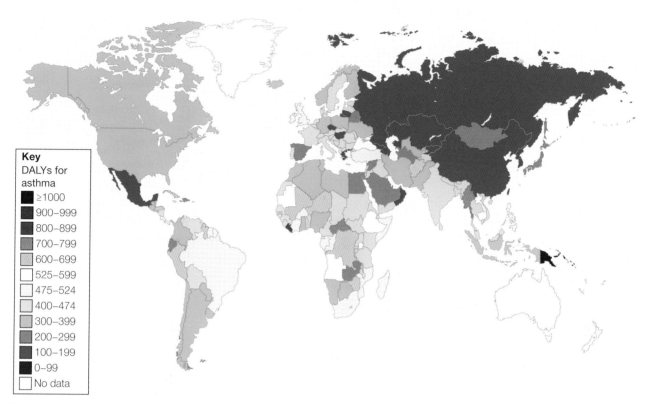

Key
DALYs for asthma

- ≥1000
- 900–999
- 800–899
- 700–799
- 600–699
- 525–599
- 475–524
- 400–474
- 300–399
- 200–299
- 100–199
- 0–99
- No data

⬆ **Figure 2** *World map of disability adjusted life years (DALYs) attributed to asthma per 100 000 population by country (2010)*

Asthma in the USA

Of the 1 in 14 Americans that suffer with asthma (7.4 per cent of adults and 8.6 per cent of children), 45 per cent reported having an asthma attack in the last year. Since the 1980s, asthma has been increasing in the US, in all age, sex and racial groups.

In 2010, 18 per cent of stays in hospital for children were asthma related, compared to just 13 per cent for adults. Almost 30 per cent of African American in-patient hospital stays were asthma-related compared to just under 9 per cent for white patients. This demonstrates the differing impact of asthma by race and community.

Asthma was the cause of 3651 deaths in 2014 – the equivalent of ten people each day. Again, the impact of asthma on mortality is differentially greater amongst African Americans – the death rate linked to asthma was almost three times that for their white counterparts.

Of all the ethnic groups, the rate of asthma and asthma attacks is highest among Puerto Ricans. Ethnic differences in asthma prevalence, morbidity and mortality are highly correlated with poverty, urban air quality, indoor allergens and inadequate medical care.

The overall cost to the US economy of asthma is huge – estimated at $56 billion (2012). The impact is felt in terms of medical costs (medication, hospitalisations) but also in terms of the indirect cost of work absenteeism. Asthma is one of the main reasons for absence from work in the US.

⬇ **Figure 3** *A US study of 17 000 families estimated that 10.1 million school days are lost across the country in a single year, which will have an effect on America's future workers*

Management and mitigation strategies

'Asthma is a common condition, affecting all levels of society. Olympic athletes, famous leaders and celebrities, and ordinary people live successful and active lives with asthma.' (Global Initiative for Asthma)

Once prescribed the appropriate medication and given proper support, asthmatics can manage their own health in three ways:

◆ avoiding trigger factors

◆ taking preventative medication (often every day) such as inhaled corticosteroids, that work continuously to reduce the narrowing of patients' airways

◆ using a different set of prescription drugs to reduce the effects of the disease. Salbutamol is an example of such 'rescue relief'.

However, if incorrectly diagnosed or if trigger factors are poorly understood, severe asthma attacks can occur which is a common reason for hospitalisation.

Poorly controlled asthma is expensive to society. For example, almost a fifth of overnight stays in hospital by children in the USA are asthma-related.

Role of international agencies and NGOs

There are a number of international agencies and NGOs involved in promoting health and combating asthma at a global scale. These include the Global Initiative for Asthma, the Global Asthma Network, the World Allergy Organisation and, of course, the WHO. Indeed, these organisations draw upon the WHO database to inform their work.

Many other groups of scientists, asthma experts and patient groups work at regional and national levels, such as the European Respiratory Society and the charity, Asthma UK.

Organisations with a global reach such as the Global Initiative for Asthma (GINA) aim to alleviate global suffering via a range of activities (see Figure **4**).

Research and the role of NGOs

International organisations promoting global health initiatives, including the WHO and international NGOs, interrogate and disseminate the work of scientists including the findings of the 20-year International Study of Asthma and Allergies in Childhood (ISAAC). Research into environmental triggers as well as the genetic causes of asthma (and how the two interact), remains a major focus, especially in the context of the increasing prevalence of asthma in both LDEs and HDE's (Figure **5**).

Did you know?

Inhalers contain the asthma drug salbutamol that was originally discovered by GlaxoSmithKline in the UK, and marketed as Ventolin from 1969 onwards. Today, it is available as a **generic medication** at a much lower cost. Salbutamol is on the WHO's *List of Essential Medicines* recommended for a basic health system.

▲ **Figure 4** *World Asthma Day*

Figure 5 *The role of international agencies, NGOs and GINA*

The role of international agencies and NGOS is to...	e.g. The Global Initiative for Asthma (GINA)...
raise the profile of the disease and the plight of the many people across the globe affected by it	organises World Asthma Day... an annual event to improve asthma awareness and care around the world.
educate medical staff in diagnosis, short- and long-term care	presents key recommendations for diagnosis and management of asthma... building capabilities of primary health care providers
educate policy-makers, to inform investment in treatment	provides strategies to adapt recommendations to varying health needs, services and resources
	reports on barriers to implementation with regard to its own recommendations, both amongst health care providers and patients
educate sufferers and their families, to inform preventative action to reduce the symptoms of asthma as well as promote self-management	publishes *Patient Guide: You can control your asthma*
	hosts asthma education videos featuring international experts and scientists
	provides a useful links guide to patient and advocacy groups
promote further research	identifies areas for further investigation of particular significance to the global community

The effects of fast food

The International Study of Asthma and Allergies in Childhood (ISAAC) has studied data from more than 319 000 13–14 year olds from centres in 51 countries, and more than 181 000 6–7 year olds from centres in 31 countries.

The results of this study linked the eating of three or more servings of fast food per week to the severity that asthma, eczema and rhinitis affects children in the developed world. In other words, a fast food diet may contribute to the increase in numbers and severity of these conditions. This could have huge implications for public health globally, given the popularity of fast foods and the expansion into new markets (Figure **6**).

The study found that three or more weekly servings of fast food were linked to a 39 per cent increased risk of severe asthma among teenagers and a 27 per cent increased risk among younger children. Conversely, eating three or more portions of fruit per week reduced the severity of symptoms by as much as 14 per cent.

Figure 6 *Western-style fast food restaurants are on the increase in China. In 2016, Macdonalds announced plans for 1250 new outlets throughout the country.*

ACTIVITIES

1 Study Figure **2**
  a Explain why the DALY is of particular interest to those studying the impact of asthma in different countries.
  b Compare and contrast Figure **2** with Figure **1** in 4.15.
2 Reread Figure **1**. How do the family members manage their asthma?

3 What might be the purpose of a World Asthma Day?
4 Using online resources, find out about the current objectives of GINA or another international organisation, the Global Asthma Network. How are its broad aims translated into action for sufferers? By whom?

In this section you will learn about birth and death rates, infant mortality rates, fertility and net replacement rates

'And the world's total population today is...'

You may have an app on your phone or tablet that enables you to check the world's population at any given time. It is interesting to watch the total tick up and then return to it later to see how much it has increased. As geographers we recognise that the total isn't an absolute truth, but an estimate (generated by organisations such as the UN and the US Census Bureau) based on a projection of past data and assumed trends. Statisticians use the best information available at the time. Not all governments are able to provide up-to-date numbers – there may not have been a recent census or an accurate nationwide survey, in which case estimates have to be made.

Of course, the total number of births isn't the only contributory factor that affects population change. To keep track of population numbers (and plan for the associated changes in demand for services), governments must also record the number of people who die in any given year.

The number of migrants into and out of a country are two other key statistics that governments try to track, as the implications of an influx of job seekers or conversely a 'brain drain' from a population can be huge. Note that these figures are not included in a calculation of **natural population change**.

What are the vital rates of natural population change?

There are a number of key terms, called the *vital rates*, which you need to understand in order to discuss natural population change – birth rate, death rate, infant mortality rate, replacement rate.

Figure 1 *A precious moment but how many more? Fertility rate is defined as the average number of children a woman is expected to give birth to, across her child-bearing years*

Did you know?

At current rates, 255 babies are born every minute, globally, or more than 4.3 new lives a second.

Birth rate

The *birth rate* is the most common index of fertility. It is expressed as a rate per thousand per year to enable comparisons to be made between countries with varying sizes of population. Birth rates are highest in the poorer countries of the world, where infant mortality rates are high and children are valuable assets in supporting their parents.

$$\textbf{Crude}\ \text{birth rate} = \frac{\text{total number of live births in 1 year}}{\text{total mid-year population}} \times 1000$$

Death rate

The *death rate* is an important measure as it is the decline in mortality, rather than any increase in fertility, that is largely responsible for population growth (Figure **2**). The twentieth century saw a global reduction in death rates as a result of improved diet, medical services and immunisation campaigns against infectious diseases. Average life expectancy was increased and infant mortality reduced.

$$\textbf{Crude}\ \text{death rate} = \frac{\text{total number of deaths in 1 year}}{\text{total mid-year population}} \times 1000$$

Infant mortality rate

The *infant mortality rate* is the number of deaths of infants under one year old expressed per thousand live births per year. These rates vary enormously throughout the world but tend to be highest in LDEs where child mortality rates are also high.

Replacement rate

The *replacement rate* shows the extent to which a population is replacing itself. This may be crudely measured as the difference between births and deaths. More sophisticated measures take into account age structure and/or gender (e.g. total fertility rate and net reproduction rate):

◆ *Fertility rate* is the average number of children a woman is expected to give birth to within her lifetime. Sometimes stated as the total fertility rate (TFR), this assumes that she survives from birth through to the end of her reproductive life. Rates can exceed 5 in LDEs, but in HDE's rates of less than 2 are normal.

◆ *Replacement level* is the number of children needed per woman in order to maintain a population size. This averages 2.11 which allows for deaths early in life. Replacement level calculations assume that migration is zero. Surprisingly, almost half of the world's countries are currently estimated to be below replacement level, which suggests that world population could start to decline by 2050.

◆ *Net reproduction rate* is a measure of the average number of daughters produced by a woman in her reproductive lifetime. A stable population has a rate of 1, with growing or declining populations more or less than 1 respectively. A growing population will have more females alive in the next generation than at present. So a net reproduction rate of 2 is a population that doubles each generation and 0.5 shows a population that will be half the size.

Figure 2 *High death rates are associated with ageing populations in Europe today; they are not high in absolute terms, compared to the level of death rate that characterised natural population change in previous centuries*

Country	Birth rate (per 1000)	Death rate (per 1000)	Natural increase (%)	Infant mortality rate (per 1000)	Fertility rate	Net reproduction rate
Mali	45.53	13.22		104.34	6.16	2.5
Nigeria	38.03	13.16		74.09	5.25	2.1
India	19.89	7.35		43.19	2.51	1.1
South Korea	8.26	6.63		3.93	1.25	0.6
Japan	8.07	9.35		2.13	1.40	0.6

Figure 3 *Rates of natural population change (2014)*

Natural increase = Birth rate – Death rate

ACTIVITIES

1 Outline the meanings of fertility, birth, death and infant mortality rates.

2 Study Figure **3**. A high rate of natural increase would be more than 2.5 per cent; a low rate would be less than 1.0 per cent.
   a Copy and complete the table.
   b Describe and comment on the vital rates of natural population change.

3 Which is more useful – replacement levels or net reproduction rates? Give reasons for your answer.

4 How is the strength of an economy likely to impact on natural population change?

Models of natural population change

In this section you will learn about models of natural population change and their application in countries at different stages of development and in contrasting physical environments

The physical, environmental and human world is highly complex. In an attempt to find order in this complexity, geographers produce models. As a simplified representation of reality, a model in geography can be:

◆ a theory or law

◆ an hypothesis or structured idea

◆ a relationship, equation or graph.

Good models fall between the extremes of being too simple and generalised to be of real value and being almost as complex as reality. Thompson's *Demographic Transition Model* illustrates the value and flexibility of a very good model.

Thompson's Demographic Transition Model (DTM)

Look at Figure **2**. This is the most useful and influential model of natural population change.

Thompson acknowledged the problems caused by unsustainable population growth by stating the obvious – fewer people demanding resources means more resources are available for individuals and countries to develop economically. He classified all countries, according to population and wealth, into one of three groups:

◆ Group C countries – the world's poorest with high birth, death and infant mortality rates (Figure **1**)

◆ Group B countries – rapidly growing with falling death rates, but creating wealth through industrialisation

◆ Group A countries – the world's wealthiest with low birth and death rates.

The DTM then plots a country's progression from Group C to B and finally to Group A as it develops economically through industrialisation. The process originally followed four stages, although a fifth has recently been added.

Figure 1 *Famine relief in Kenya. Both Malthus (see 4.26) and Thompson believed that famines like that experienced in Kenya were due to overpopulation. But in 2006, the United Nations' FAO (Food and Agriculture Organisation) reported that recent food shortages were more likely to have been caused by civil war and climate change than by population pressure.*

Think about

Adaptation of the DTM

Think about the age of the original DTM and its origins in the study of Western European and North American populations. Clearly birth, death and therefore population growth rates tend to decline with economic and technological progress – this helps to explain why contemporary rapid population growth is concentrated in LDEs. Progression from simple farming economies to complex, modern, urban–industrial ones generally fits Thompson's model. However, countries in Africa, Asia and Latin America have widely differing environments and also differing racial, cultural and historical backgrounds to Western Europe and North America. So, to assume that they will all go through similar transitions would seem unrealistic. Especially since LDE and EME base populations are higher and also do not have any major outlets for the massive overseas migration which relieves population pressure –this happened during Western Europe's phase of rapid growth.

Foreign aid and investment in agriculture, education and family planning programmes will also shorten the time-scale in each stage and this assumes that entirely unique stage characteristics do not evolve regardless! Just as contemporary arguments for a fifth stage reinforce these criticisms, so do suggestions that a sixth stage is needed to account for those countries now experiencing marked net immigration.

Stage 1: High birth and death rates as population is checked periodically by disease, war and famines. There is no birth control, life expectancy is short, and population growth slow and intermittent. Only remote ethnic groups within the most inaccessible regions of Amazonia and south-east Asia are in this stage today.

Stage 2: Birth rates stay high and may increase marginally, but death rate declines progressively. The high birth rate reflects lack of birth control, and factors such as women marrying earlier and children having an 'economic value' because they can work. The falling death rate is related to economic growth and improvements in personal hygiene, sanitation, medical care and diet. As the gap between birth rate and death rate widens there is a population explosion. The poorest LDEs today, such as Mali, are in this stage.

Stage 3: Birth rate starts to fall with the availability of birth control and women marrying later. This is possible because economic development and education starts to weaken cultural traditions, more women are **emancipated**, smaller families are desired and child labour is often replaced by education. The lower death rate reflects the control of major diseases and improved standards of health and sanitation. Population growth continues but at a progressively slower rate. Industrialising and urbanising countries today, such as India, are in this stage.

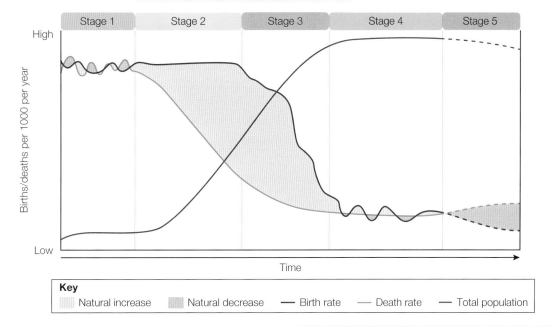

Stage 4: Birth and death rates fluctuate around a low level. Fluctuations in birth rate are primarily associated with periods of economic recession (causing falls) and optimism (rises). Consequently, although there is stability overall, there are occasional periods of actual population decline. When Thompson published his model in 1929, the world's richest countries were in this stage. Many are now changing further, hence the addition of a fifth stage.

Stage 5: Birth rate falls for mainly economic reasons and/or death rate rises again as a result of an ageing population, **diseases of affluence** and increases in environmental and social problems in urban–industrial societies. Everything from fatty, sugary and salt-rich foods to insufficient exercise and high-stress lifestyles increasingly challenge the assumption of longevity for all. The world's most economically developed countries today, such as Japan, are in this stage.

ACTIVITIES

1 Practise sketching and annotating a five-stage DTM. Make sure that you include birth and death rates and add labels showing the reasons for the key characteristics of each stage.

2 Add a sixth stage to your DTM. Suggest countries that might now be in this stage. Justify your choices.

3 What is the best unit of time to measure population change (e.g. seconds, minutes, hours, days, weeks and so on)? Explain your answer fully.

▲ **Figure 2** *Stages in the demographic transition model*

STRETCH YOURSELF

Discuss the applicability and validity of the demographic transition model.

In this section you will learn about population structure

What is population structure?

If you could take a selfie with everyone else in your country, you'd need quite a camera, or 'selfie stick'! Such a snap-shot would be of huge significance to **demographers**. In that moment, you would have captured a demographic history of births, deaths and migration changes during the lifetime of the oldest person. Your photo would show the country's population structure – this has great social, economic and political significance. This one image would show the type and range of welfare services needed for the population – from maternity care, to preschools through to retirement provision. It would show the nature of the labour force, and likely trends of supply and demand – not least evidence of race, culture, wealth and level of development. That amazing selfie would also show where the country was going – and so how to plan for its future.

How do we show population structure?

Look at Figure **1**. Population structure is best shown diagrammatically using age–sex (population) pyramids. These mirror bar graphs show long-term changes in fertility, mortality and so growth rates – their shape gradually evolving through time.

Interpreting age–sex pyramids

The shape of an age–sex pyramid shows us characteristics of the population distribution.

♦ A rising growth rate results in a youthful population shown by a flared base (see Figure **1a**).

♦ A falling or even negative growth rate results in an ageing population with a much more top heavy pyramid. This will also highlight female longevity (Figure **1c**).

♦ Lesser influences are also shown, such as slight bulges for net immigration and indentations for net emigration. Proportions in each age group can be shown using raw population data or as a percentage of the total.

⟩ *Figure 1* *Age–sex structure for countries at different levels of development: a) Mali, b) India, c) Japan (2014)*

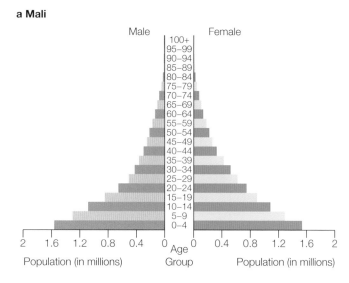

a Mali

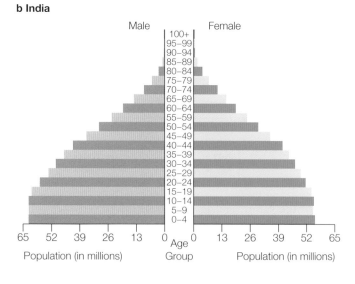

b India

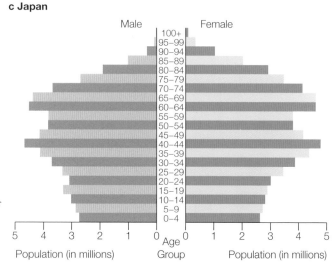

c Japan

Age–sex pyramids also reflect the social and economic character of individual countries – their state of development, the nature of their society and even their future prospects. They even show specific aspects of the country's demographic history –such as wars and natural disasters. For example, major conflicts lead to a reduction in the number of young people and also birth rates followed by post-war 'baby boom' bulges that rise through the pyramid in the decades that follow. The post-Second World War 'boomers' reached maturity in the 1960s and are now retiring or retired (Figure **1c**).

Dependency ratio

In practical terms, children and the aged may be considered unproductive and so dependent on the wealth-producing, economically active adults. So, the smaller the adult group relative to the other two, the more difficult it is for a country to be economically viable. This can be shown using a dependency ratio, which compares the working population to the dependent population:

$$\text{Dependency ratio} = \frac{\text{dependent population}}{\text{working population}} \times 100$$

In LDEs, it is more common for children to work at a young age. In HDEs, the reverse is true. It is not just children that are non-productive but also many young adults, because tertiary education is more widespread. Furthermore, in LDEs, the aged have always been more productive because of the usual absence of pensions. Conversely, in HDEs they became less productive as retirement ages fell, but will become more productive in the future as pension ages rise. As a result, dependency ratios require careful interpretation.

Describing graphs

When describing graphs (including age–sex pyramids), trends, examples and anomalies (TEA) should be identified:

◆ Trends cover the general direction – use adjectives like rising, falling, steady, accelerating, flared and so on.

◆ Examples illustrate the general trend and should include the highest and lowest values.

◆ Anomalies should cover any exceptions to the general trend – examples that stand out as different, outliers in scattergraphs, radical changes in gradient in line or bar graphs, and so on.

Did you know?

In Japan more nappies for adults are sold than for babies! An ageing population structure or what?

ACTIVITIES

1 Study Figure **1**.
  a Compare and contrast the population structures.
  b Which stage of the demographic transition model would each age–sex pyramid represent? Justify your answers.

2 Comment on the relative benefits of using raw population data or percentages in age–sex pyramids.

3 Study the table below:

	Mali	India	Japan
0–14 years (%)	47.6	28.5	13.2
15–64 years (%)	49.4	65.7	61.0
65 years and over (%)	3.0	5.8	25.8
Population change (%)	+3.00	+1.25	−0.13

  a Calculate the dependency ratio for each country.
  b With reference also to the rate of population change, comment on the results.
  c Why might 15 not be the best age to measure when people start to become productive adults in (i) Mali and (ii) Japan?
  d Not all people become dependent and stop being productive at 65. Suggest some exceptions to this figure.

In this section you will learn about how socio-economic factors and cultural controls have affected natural population change in China and Bangladesh, and some of the consequences of these changes

Managing population change

Having a large family in rural communities in LDEs often carries a high status. This is, in part, because of the economic value that a large family brings and also the security that children supply in old age. If children are a `gift from God' then many religious people believe that artificial techniques should not be used to influence family size. Cultural norms influence the choices of individuals every day, but what of the state that must protect and provide for its population?

Governments in different countries also approach the same problem differently. 'Local' economic, social, cultural, religious and political issues all play a part in how nation states manage population change. A population may be ageing, and in decline, so it would be wrong to assume that a government's role is always to restrict population growth. However, the following examples both relate to birth control.

Addressing population problems – enforcement or persuasion?

The approaches taken over the last 30–40 years to reduce population growth in China and Bangladesh have been very different. The consequences of these approaches have similarities, but also key differences, in particular regarding the scale of the gender imbalance now seen within these national populations.

China's one-child policy

China contains almost one-fifth of the world's people and has gone through an astonishing economic boom in recent decades (Figure 1). Yet in the 1970s, China faced excessive population growth and the fear of mass starvation by the end of the twentieth century.

Figure 1 *Birth rates in China (2000): (a) is a map of China showing birth rates, (b) is an equal population projection (a gridded cartogram transformation where each grid cell is resized according to the total number of people in an area), with a choropleth overlay of birth rate by region*

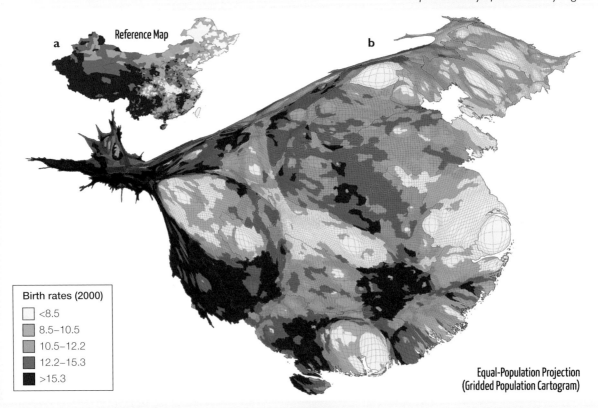

Reference Map

a

b

Birth rates (2000)

□ <8.5
▨ 8.5–10.5
▨ 10.5–12.2
▨ 12.2–15.3
■ >15.3

Equal-Population Projection
(Gridded Population Cartogram)

As a response, the one-child policy was introduced in 1979 (Figure **2**). This policy worked through various incentives and restrictions until its 'relaxation' in 2013, and eventual replacement with a two-child policy (effective from the start of 2016). The measures were undoubtedly successful, particularly in urban areas (see Figure **1**) – China now has low birth, death and natural increase rates. In addition, both maternal and infant mortality rates were reduced significantly and average life expectancy increased – the population is under control. As a consequence, there is a lower unemployment rate yet, at the same time, production output has not been affected – during this period the economy has grown with China becoming the major manufacturing hub that it is today.

◆ **Figure 2** *Effective propaganda? China's one-child policy.*

Quality of life has also improved with the demand on the social infrastructure and resources – healthcare, education, energy, food supply, land, water – being less than it would have been had the birth rate been allowed to grow unchecked. The often widely publicised problem of atmospheric pollution in China might also have been much more serious had the policy not been introduced.

However, internationally (and more recently, internally) the policy has also had its critics. Chinese culture and tradition had always been based around the large family, with male offspring being particularly important. Reports of selective terminations based on the sex of the foetus, female infanticide, infant abandonment and even child trading made uncomfortable headlines. As did the allegations of corruption allowing exceptions, accounts of forced late terminations and sterilisations.

China's population balance has certainly changed. Its population structure in younger age groups is male-weighted.

◆ China's 'missing women' (see page 237) was estimated to be 49.9 million in 2003 (Stephan Kasen and Claudia Wink, 2003).

◆ In 1982, China's sex ratio at birth (the number of boys born for every 100 girls) was 108, but by 2012 it had increased to 118.

This threatened a 'marriage squeeze' with insufficient brides and, over the last 20 years, there has been an increase in sex trafficking in China with abducted women being traded as brides, in tandem with a boom in prostitution.

◆ **Figure 3** *Smile please! China's only children will have to shoulder the burden of an ageing population soon.*

Furthermore, the gender imbalance may not be the only social consequence of more than 30 years of the one-child policy. Sociologists queried the future contribution of generations of only children who may have received excessive amounts of attention from parents and grandparents during their upbringing – dubbing them 'little emperors'. Ironically, in a country with an ageing structure, specifically a 'four-two-one problem' of one adult child having to provide support for his or her two parents and four grandparents, China's little emperors may have to grow up fast (Figure **3**).

Primary health care in Bangladesh

Look at the congestion in Figure **4**. Bangladesh is one of the most densely populated countries in the world. Almost 170 million people occupy an area the size of England and Wales, with one in three living in extreme poverty. This is because 80 per cent of Bangladesh comprises low-lying, fertile floodplain and delta, perfect for intensive rice cultivation but prone to flooding. Three-quarters of the population live in these hazardous rural areas. Heavier monsoons and sea-level rise associated with climate change add longer-term threats.

Legislation raising the minimum age of marriage, to 18 for women and 21 for men, has undoubtedly reduced the fertility rate in Bangladesh. However, it is a primary health care approach to improving access to contraception, maternal and child health care that is of greater significance.

Trials in the late 1970s showed the value of a doorstep service with trained female health workers. They helped mothers to pick a method of contraception best suited to them, to treat side-effects and also to provide basic maternal and child healthcare (Figure **5**). Children became healthier, fewer women died of pregnancy-related causes and child mortality fell. With fewer children to support, these families grew wealthier. Parents accumulated more farmland, built more valuable homes and gained access to running water. Their children stayed in school longer and women enjoyed higher incomes. Since these trials, the government has set about training tens of thousands of female **primary health care** workers.

Average birth rates have fallen from six children for each woman to slightly more than two, and Bangladesh has become one of the first LDEs to meet the UN Millennium Development Goal of reducing child mortality by two-thirds. Bangladesh's population was expected to double by 2050, but is now likely to reach just over 200 million before stabilising soon after.

▼ **Figure 4** *A rickshaw traffic jam in Dhaka*

Like China, Bangladesh's population suffers from a gender imbalance, with recent estimates of 2.7 million 'missing women'. However, research into the decline in infant mortality rate over the last 20 years has found that the rate of infant mortality amongst girls has fallen faster than the rate of male infant mortality (the rate amongst females was previously higher). Thus, Bangladesh has been inching closer to a situation in which boys and girls under five have an equal chance of survival. In particular, this trend (the faster decline in the rate of infant mortality amongst girls) was observed in areas with more easily accessible primary health care services.

Figure 5 *A frontline health worker administers polio vaccine to a child at a temporary vaccination centre, Dhaka, Bangladesh*

The 'missing women' why don't as many girls survive?

In countries such as Bangladesh and China, the norms and values of dominant social institutions often result in boys and girls being treated differently because of their sex. For example, girls often miss out on inheritance rights. Despite the undoubted impact of the use (or misuse) of prenatal scanning technology and selective terminations, research has shown that this tendency to favour boys over girls actually extends into the home. Children are offered different amounts of food according to gender, even varying levels of care during times of ill health: 'Parents do not engage in conscious discrimination between sons and daughters, but sex discrimination is embodied in cultural beliefs.' (Ingrid Waldron, 1987)

One result of such discrimination is the low male-to-female mortality ratio (or 'excess female mortality') amongst children aged under five. Around the world, comparative neglect of female children is, in general, worse in rural areas and more severe for later-born

children, i.e. those born into families that may already have too many mouths to feed.

The economist Amartya Sen described the phenomenon of excess female mortality in terms of the number of 'missing women' worldwide. In the late 1980s, he pioneered methods of calculating the additional numbers of females of all ages who would have been alive at that time, by comparing the actual sex ratio of populations with the expected sex ratio. He concluded that the number of missing women was larger than the combined death toll of both world wars (men and women). Later research refined Sen's methods but supported his infamous claim that, globally, 'more than 100 million women are missing': 'We find that the number of "missing women" has increased in absolute terms to over 100 million… and a deterioration [of the situation of women] in China is related largely to its strict family planning policies.' (Stephan Kasen and Claudia Wink, 2003)

ACTIVITIES

1 Both China and Bangladesh adopted a policy to reduce population growth. Explain the basic principle of government that lies behind such a policy.

2 Study Figures **1a** and **1b**. Assess the extent to which China's one-child policy may be seen as a whole-nation policy or one aimed at certain areas of the country. Hint: you may also wish to refer to an atlas or online resources to locate China's major cities.

3 'The greatest decline in the female infant mortality rate occurred in areas of Bangladesh where there was greater access to primary health care services.' Assess the reasons why this was the case.

In this section you will learn about the causes, processes and outcomes of migration change in relation to regions of origin and destination

What is migration change?

Migration is population movement. Temporary or permanent, voluntary or forced, it can be at all scales from local and regional (in-migration and out-migration) to international (immigration and emigration). On any one day in any one place, births and deaths change the population balance. People also move – some arrive, some leave. The total population at any one time is the balance between natural change and migration change . It can be expressed as an equation (Figure **1**).

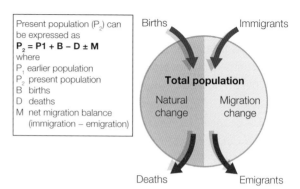

Present population (P_2) can be expressed as
$$P_2 = P1 + B - D \pm M$$
where
P_1 earlier population
P_2 present population
B births
D deaths
M net migration balance (immigration – emigration)

▲ **Figure 1** *Components of population change*

Types of migration

Temporary migration

Temporary migration includes the following:

♦ Diurnal (daily) movements to and from home, familiar to so many who commute from dormitory settlements to work or school.

♦ Seasonal movements associated with agriculture. For example, the annual cycle of **transhumance**, still practised in the Himalayas, and to a lesser extent in the Alps.

♦ International migrations, for example, British and American oil employees working monthly, or longer, shifts in the Middle East.

Step migration

Such are the difficulties associated with migration decisions that *step migration* is not uncommon. This is the process of migration in a series of shorter movements from the place of origin to the final destination – such as moving from a farm, to a village, to a town and finally to a city.

Forced migration

Forced migration follows natural disasters, persecution and wars but can also be as a result of gradual deterioration of economic opportunity, for example, as a result of desertification. It can be at local or international scales and the numbers involved can be staggering. For example, following tribal-based genocide in Rwanda, Africa, in 1994, approximately two million refugees fled to Zaire and half a million to Tanzania! Such **refugees** may be called **asylum seekers** until recognised by the country of destination where they make a claim.

Permanent migration

Permanent migration involves a permanent change of residence. Such migration demonstrates distance decay. This means that the shorter the move – such as a change of house within the same community – the less upheaval and so greater numbers of migrants are involved. As distances increase, the number of migrants decreases. For example, larger numbers move from rural to urban areas (urbanisation) within LDEs than move abroad (emigrate).

Most challenging are the issues faced by those emigrating overseas. Migrants face the demands of adjusting to new cultures, climates and languages as well as legal and financial disincentives.

Voluntary migration

Voluntary migration means that current circumstances and the hope that a better standard of living is possible elsewhere are considered in making a decision. These migrants may form a migration stream if there are many from a particular country, region or city heading to a certain destination. For example, immigration from the Caribbean to Britain in the decades immediately following the Second World War.

Push/pull factors

Motivation is the key difference between voluntary and forced migrations. Both push (negative) and pull (positive) factors are likely to be involved:

◆ *Push factors* repel: they range from physical factors such as soil exhaustion and natural disasters to economic and social issues such as poverty, marital/family breakdown and job opportunities.

◆ *Pull factors* attract: everything from seeking better job opportunities, a better standard of living and education to a better environment – whether it be the attraction of the bright lights of the city or a more peaceful existence. For example, the 'American Dream' attracting Mexicans across the border (Figure **3**).

Migration outcomes

🔻 *Figure 2* *Positive and negative outcomes of migration*

	Positive	Negative
Origin	Overpopulation pressures, such as on water and soil resources, may be eased. Less demand for services such as education and health care. Remittances (money sent back 'home') supports relatives.	Skilled labour shortages – especially professionals such as doctors. Gender imbalances – far more men than women migrate. Ageing population structures – those of working age are more likely to migrate reducing productivity and threatening the future of rural communities.
Destination	Labour pool increased and new trades and skills introduced – such as doctors, scientific researchers, university students. Migrant workers are mobile, pay taxes and spend money – which creates jobs and wealth. Cultural and racial variety promotes diversity and encourages integration and understanding.	A large influx of migrants can add to housing shortages and welfare systems that may already be under pressure. Cultural differences can lead to racial tensions – segregation, crime and violence. Education and health care services can become strained.

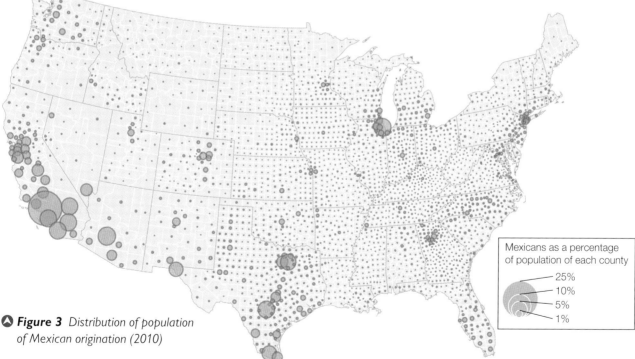

Mexicans as a percentage of population of each county
— 25%
— 10%
— 5%
— 1%

🔺 *Figure 3* *Distribution of population of Mexican origination (2010)*

ACTIVITIES

1 Outline contemporary examples of temporary, permanent, voluntary, step and forced migrations.

2 Look at Figure **3**. Describe the distribution of the Mexican population of the USA in 2010. Suggest reasons for the distribution you describe.

3 'Migration is a significant force for good.' Discuss.

In this section you will learn about causes, patterns and impacts of international migration from North Africa to Western Europe

A risk worth taking?

In October 2013, 302 coffins containing the bodies of men, women, children and babies were lined up on the Italian island of Lampedusa. All had drowned in, what was then, the largest single loss of life of migrants crossing the Mediterranean. The shocking images caused leaders around the world to vow to never let it happen again. On 23 April 2015, a boat capsized in the Mediterranean – an estimated 900 migrants lost their lives. The spring of 2015 claimed the lives of over 2000 migrants, mostly from North Africa, but with a significant number from West and sub-Saharan Africa (Figure **1**). By the end of the year, a total of 3700 people had died or were missing as a result of the many perilous journeys attempted to reach the EU; an illegal industry of such scale that it is likely to have netted smugglers 'at least a billion dollars' noted the International Organisation for Migration.

⬥ **Figure 1** *An overloaded African migrant boat*

'Colonises half the world; complains about immigrants'

It has been argued that the UK's 2016 Brexit vote was partly a backlash against the principle of freedom of movement for all EU citizens. In particular, the vote was seen as a backlash against the numbers of foreign workers from within the EU who have arrived in the UK in recent years. Indeed, the text 'Colonises half the world; complains about immigrants' was plastered across the Union Jack on social media to tweak the conscience of those that voted 'leave' based on the issue of immigration.

In fact, between them Germany, Hungary, Sweden and Austria received around two-thirds of the EU's asylum applications in 2015. As a proportion of its population, Germany received almost ten times the asylum applications made in the UK in the same period (Eurostat, 2015).

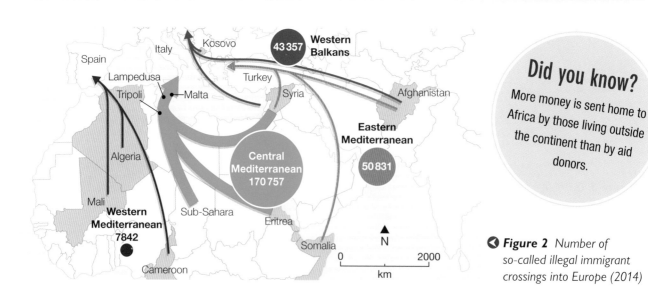

Did you know?

More money is sent home to Africa by those living outside the continent than by aid donors.

◀ **Figure 2** *Number of so-called illegal immigrant crossings into Europe (2014)*

Causes of migration

There is no single reason why African migrants make the hazardous trip across the Mediterranean into Europe. Conflict remains a strong push factor in several regions – for example, refugees fleeing civil war in Syria or Eritreans escaping forced conscription into the military. Elsewhere, immature governments, such as Libya, struggle to look after the needs of all of their citizens and are unable to control their borders. Africa has the fastest population growth rate in the world and while there are economic gains, the economic pull of Europe is strong.

Look at Figure **2**. Migrants make use of a wide network of people traffickers who charge several thousand pounds for transportation from North Africa to Italy. Migrants may be stuck in small coastal settlements for several months or even years until the money is raised – an example of step migration (see 4.21).

Impacts on northern and western Africa

Unlike other migration flows, many migrants leave their homes without telling their families. Many migrants are, therefore, vulnerable and act without the support and advice of traditional family safety nets. Human traffickers exploit this vulnerability, forcing them into debt or the sex trade – communities may never hear from them again. Many of the bodies found at sea have no means of identification.

Successful migrants do send remittances home. As most of this money is sent via informal channels, such as with friends on holiday (so-called family aid), it is difficult to be precise about the economic impact.

Impacts on Italy

The cost to Italy of the rescue and reception of hundreds of thousands of foreigners is roughly €200 million a year. By year end 2014, 170 100 **economic migrants** and asylum seekers had arrived in Italy by sea – about four times that of 2013 (Figure **3**). In addition, 3200 migrants died at sea and 85 000 had to be saved by the Italian navy's *Mare Nostrum* operation and more than 35 000 by the Italian coastguard. While more-informed migrants (e.g. from Syria) moved further north, many stayed in Italy, which has an unenviable reputation for racial intolerance (e.g. racial abuse aimed at footballers). Anti-immigration and far right political parties in Italy, including *Forza Nuova* and the *Northern League*, supported the cancellation of the humanitarian *Mare Nostrum* operation in late 2014. Public protests and crimes of racial hatred have also increased.

Did you know?

The acronym SEEP (Social, Economic, Environmental and Political) is a useful reminder of the four main headings to consider in a response to questions on causes and/or impacts.

Source country	Number of migrants
Syria	42 323
Eritrea	34 329
Mali	9938
Nigeria	9000
Gambia	8707
Palestine	6082
Somalia	5756

Figure 3 *Immigration to Italy (2014)*

ACTIVITIES

1 In what circumstances should 'we' (or would you) offer a migrant a home?

2 Copy and complete the following summary table on possible causes and impacts of migration into Europe:

	Causes	Impacts on destination
Social		
Economic		
Environmental		
Political		

3 By the end of 2014, the total number of forcibly displaced people worldwide was at its highest since the Second World War. Using Figure **2**, comment on the distribution of source countries of migrants to Western Europe.

STRETCH YOURSELF

In 2016, the focus of Europe's 'migrant crisis' switched from Italy to Greece. Find out about the routes, origins and experiences of immigrants who made the journey to Greece, many via the island of Lesbos. What 'deal' was reached between the EU and Turkey to reduce the numbers of people journeying across the Mediterranean Sea and how effective was it?

In this section you will learn about critical perspectives on the social, economic, political and environmental implications of migration

Australia: the perfect country...

Australia is an extraordinary country. It is one of the richest and most urbanised countries in the world (Figure **1**) – 89 per cent of the population live in urban areas – yet overall it has one of the lowest population densities of any country. Australia has enviable agricultural potential and enormous mineral resources. It is also respected globally for healthy lifestyles, freedom, opportunity, sporting excellence and for its high standard of living. Surely Australia is the ideal country?

... or is it?

Is this actually the case? Australia's vast area covers a wide variety of environments and climates. It faces many environmental challenges, including soil degradation, desertification, habitat destruction and urbanisation. It also has its hazards – cyclones, droughts, thunderstorms, floods and bush fires (see *AQA Geography A Level & AS Physical Geography*, 5.16). As a result, population is mainly concentrated outside the tropics and the hostile semi-desert and desert interior. In fact, in more temperate latitudes, this country's growing cities are demonstrating signs of **overpopulation** (as opposed to **underpopulation** or **optimum population**). For example, decades of diminishing rainfall and dramatic projections about the future impact of climate change on Perth have led the city's authorities to investigate every possible strategy to supply its increasing population with water, including plugging leaks, groundwater replenishment (with treated wastewater) and raising customer awareness of the need for water efficiency. Precipitation in the south-west corner of Australia is forecast to drop by up to 40 per cent by the end of the century.

So what does Australia's age–sex structure show us? Look at Figure **2**. It shows every indication of an ageing HDE with high life expectancy (82 years), low infant mortality and birth rates below replacement level. However, its growth rate (1.09 per cent in 2014) is higher than most HDEs and much influenced by net immigration.

Social, economic and political implications of net immigration

Until the 1970s, the Australian government encouraged young, mainly European families to migrate to Australia – some UK families were given assisted passages for just £10! The ethnicity, culture, religion and of course languages spoken in Australia today were shaped by these movements and Britons are still the largest group of foreign-born immigrants in Australia (Figure **3**).

Since the 1970s, a skill-based immigration policy has been in place to increase Australia's pool of skilled and professional people. Prospective immigrants must pass a points-based skills test, based on employment record (to match skills shortages), educational qualifications, age (under 30s are encouraged) and the ability to speak some English.

Figure 1 A view of Perth, the capital of Western Australia; one third of Perth's residents were born abroad with most coming from the UK, New Zealand and South Africa

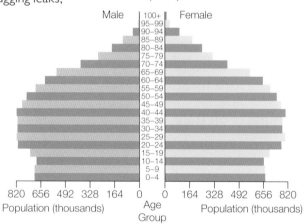

Figure 2 Age–sex structure for Australia (2014)

Male   100+   Female
95–99
90–94
85–89
80–84
75–79
70–74
65–69
60–64
55–59
50–54
45–49
40–44
35–39
30–34
25–29
20–24
15–19
10–14
5–9
0–4

820 656 492 328 164 0 0 164 328 492 656 820
Population (thousands)   Age Group   Population (thousands)

Figure 3 Country of birth of immigrants, shown as a percentage of Australia's population (top four countries only)

	2001	2011
UK, Channel Islands and Isle of Man	5.8	5.3
New Zealand	2.0	2.5
China (not including SARs and Taiwan)	0.8	1.8
India	0.5	1.5

Clearly, with its ageing population and increasing dependency ratio, the need for skilled workers in Australia could not be overlooked. However, for professions like dentistry, veterinary medicine and optometry, there appears now to be an oversupply of graduates looking for work, with reports of reducing wage rates. Some commentators have suggested that young professionals born in Australia are struggling to find work as a direct result of immigration, arguing for a further tightening of government policies, even after recent curbs on visa numbers.

Deterrence and detention: the Pacific Solution?

Australia has seen rising numbers of applications for asylum over recent years and as with the EU (see 4.22), until 2015 a significant proportion of those applicants arrived by boat (Figure **4**).

The approach of the Australian Government to so-called illegal maritime immigration has hardened over time. Mandatory detention centres for asylum seekers (see 4.21) were introduced in 1992. Offshore detention centres were subsequently established on Manus Island, Papua New Guinea, and Nauru (an island nation north-east of Australia) in the Government's notorious 2001 'Pacific Solution'. By 2002, the number of unauthorised arrivals in Australia by boat had reduced dramatically (see Figure **5**).

On returning to office in 2013, Prime Minster Kevin Rudd announced that 'asylum seekers who come here by boat without a visa will never be settled in Australia'. Figure **6** shows the number of visas granted to people from the top three countries of citizenship of illegal maritime arrivals, prior to the 2013 announcement. Such hardline policies have always been justified in terms of saving lives that would otherwise be lost at sea but, with reports of suicide, rape and medical neglect at offshore detention centres, the human cost of such policies has been criticised by the UN. In 2016, the Government of Papua New Guinea ruled that the Manus Island detention centre was illegal.

Fit and healthy?

Before being granted a visa to stay in Australia, applicants must be free from diseases or conditions that may be threats to public health, or likely to result in significant health care and community service costs.

Tuberculosis is a particular concern. Anyone applying for a permanent visa has to be tested, with chest x-rays for applicants aged over 11 years.

Two hundred days without an illegal boat arrival

'The Coalition Government's strong border policies had returned integrity to Australia's borders and immigration programme ... Operation Sovereign Borders has cut off the people smugglers' ventures; since this Government started turning back the boats there has been only one illegal boat arrival' said Peter Dutton, Minister for Immigration and Border Protection.

⬆ **Figure 4** *Press release, 12 February 2015*

⬇ **Figure 5** *Number of people arriving in Australia by unauthorised boat (1989–2014)*

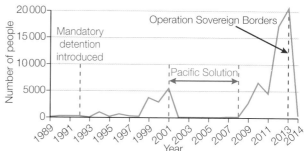

	2008–9	2009–10	2010–11	2011–12	2012–13
Afghanistan	176	1440	1336	1972	2352
Iran	4	67	333	1269	1020
Pakistan	0	6	14	94	469

⬆ **Figure 6** *Number of visas granted to immigrants arriving by unauthorised boats, by country of origin (2008–13)*

ACTIVITIES

1 In a table, summarise the social (including health), economic, political and environmental implications of net immigration into Australia.

2 Look again at Figure **5**. What impact have government policies on 'illegal maritime immigration' had?

3 Investigate the politics behind Australia's policy on arrivals by boat, and immigration more generally, by finding out what NGOs on either side of the debate say (e.g. Sustainable Population Australia vs. Amnesty International, or the Refugee Council of Australia).

In this section you will learn about environmental constraints on population growth

As a society, could we ever destroy ourselves – literally run out of food to eat and other resources on which to survive? This question might seem hypothetical, but there is a suggestion that it has happened at least once before to a society.

Population growth on Easter Island

Look at Figure **1**. Easter Island, *Rapa Nui*, a UNESCO World Heritage Site, is a remote 163 km² island in the Pacific Ocean – the next inhabited island is more than 1600 km away. Around 1000 years ago, a group of Polynesians settled on Easter Island and, over the next 600 years, a society grew and thrived in this apparently resource-plentiful land – its population peaked at several thousand. As many as 16 million trees, some towering to over 30 m, were removed, possibly as part of slash-and-burn farming. Quarried stone was used to build houses, farm buildings and the famous statues – almost 900 of them. However, after over half a millennium of good times, it all stopped. James Cook visited the island in 1774 and found only around 700 islanders living marginal lives. What happened?

▲ *Figure 1* *The UNESCO World Heritage site of Easter Island*

Where did it all go wrong?

Pessimists suggest that Easter Island is a reminder that there are finite limits on our environment. If we exist beyond our means, at some point, we will all go down together (see Malthus, 4.26).

Is there another view?

However, some **anthropologists** suggest that Easter Island is an unlikely story of success. There is a suggestion that rats, stowaways in the canoes of the first settlers, colonised the island. It was these foreign invaders that ate the trees and perhaps the islanders adapted their diet over time to include both rat (little other meat was available) and also vegetables grown from newly created rock gardens (Figure **2**). The argument is that, although the island was stripped of resources and the ecosystem was under severe pressure, the islanders adapted to survive (see the theories of Boserup and Simon, 4.26).

If there is one lesson to be learnt from Easter Island, it is that we need to learn from our past. If, as a global society, we continue to use the environment in an unsustainable way, then we will have to learn to live differently, live with less or ultimately face the consequences.

▼ *Figure 2* *Rock gardens on Easter Island provided wind protection, reduced soil erosion and reduced evapotranspiration*

What do we need to survive?

On a basic, physical level, new-born babies need resources (food, water, energy) and need more as they grow. They also produce waste and pollution. Of the planet's ecosystems, many are becoming rapidly depleted. To survive, an understanding of our *ecological footprint* (see 3.22) and the constraints that the environment places on our activities and population growth is crucial (see 4.25).

Have we learnt from Easter Island?

In the 1970s, the human population overtook the Earth's rate of replenishment of natural resources – since then we have been living unsustainably. Today, environmental constraints on population growth may be summarised as follows:

◆ *Food productivity* – around 20 per cent more food per person is produced today than 40 years ago, but there are still almost 1 billion people who go hungry. Globally, we grow enough food for all but many people don't have sufficient land to grow enough food, or the income to purchase it.

◆ *Water consumption* – only 2.5 per cent of the Earth's water is freshwater (see *AQA Geography A Level & AS Physical Geography*, 1.2) and this is reducing because of climate change and also greater pressure from population growth. Currently, around 1.1 billion people do not have access to freshwater, mainly in LDEs. Any major failure of irrigation and water distribution schemes in the future may lead to prolonged drought within regions unaccustomed to such environmental hardships, such as south-eastern and south-western Australia, the southern Mediterranean and California (Figure **3**).

◆ *Climate change* – greenhouse gases are being created faster than they are being absorbed by shrinking forests and oceans. Around 10 per cent of the world's population live less than 10 m above sea level. Any rise in sea level will put these homes, crops and livelihoods at risk. Knock-on effects on the productivity of natural and manmade ecosystems will be severe.

◆ *Natural hazards* – droughts, floods, wildfires and tropical storms are becoming more frequent, geographically concentrated and on an unprecedented scale. In May 2015, a record-breaking heatwave in India killed over 2500 people. The frequency of heatwaves in south Asia has increased since the 1950s and, despite warnings from UN climate scientists, the threat of heatwaves has been given little attention by authorities.

ACTIVITIES

1 In 1992, Maurice Strong, Secretary General of the 1992 Earth Summit said, 'Either we reduce our numbers voluntarily, or nature will do it for us brutally.' What do you think Maurice Strong meant?

2 Suggest a matching real-world example for:
   • optimum population • underpopulation • overpopulation.
   (Hint: start with 4.23 and the glossary.)

   All three depend on the balance between people and resources at a given level of technology. So, for example, irrigation or land reclamation from the sea may take a region from overpopulation to something closer to the optimum.

STRETCH YOURSELF

In areas with chronic water shortage, tourist arrivals from HDEs have exacerbated an already difficult situation. Find out about how water shortages on the islands of Bali and Flores, in Indonesia, created by tourism (in Flores, tourism is just taking off), threaten to reduce living standards of people already living at the margins of survival. Why do you think tourists are, in particular, implicated in water misuse?

⬤ **Figure 3** *The Colorado River is so heavily used for agriculture, industry and domestic needs, that this once mighty river rarely reaches its delta and the Gulf of California*

In this section you will learn about concepts of carrying capacity, ecological footprint and the demographic dividend

The balance between population and resources

Look back at Figure **2** in 4.1. Despite varying rates of population change particularly between HDEs and LDEs, global totals are projected to reach 9.2 billion by 2050. More people require more land, food, water, energy and other resources. Is this sustainable?

Each of us consumes resources and generates waste – currently over 30 times more per person in so-called *high-impact* HDEs compared to *low-impact* LDEs. But economic development in, and emigration from, LDEs is increasing the proportion of high-impact people – those with lifestyles that have a large effect on resources and on the environment. People in LDEs rightly aspire to HDE living standards but whether this is environmentally sustainable is questionable.

What is carrying capacity?

The world is facing some harsh ecological facts given our excessive demands upon the environment. The increasing imbalance between world population and material resources is demonstrated by hunger and poverty in LDEs, and pollution and environmental destruction in HDEs. Overcrowding in urban areas, for example, is associated with increasing vandalism and crime. On an international scale, it may be reflected in wars, as countries compete for limited resources.

As early as 1798, Thomas Malthus (see 4.26) predicted that the environment has an ultimate *carrying capacity* – a ceiling beyond which extra numbers cannot be adequately fed, housed or employed. How population growth could adjust to this is represented by the exponential growth model (Figure **1**).

What is our ecological footprint?

Our *ecological footprint* is a measure of the human demands we place on the ecosystems that support us and is expressed in terms of the amount of biologically productive land needed to produce the resources we consume and to absorb the waste we generate.

According to a report by the Worldwide Fund for Nature, we currently require one and a half Earths to

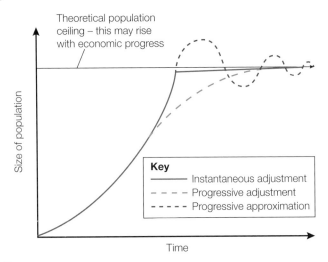

▲ **Figure 1** *The exponential growth model. Instantaneous adjustment of population to the limits of the environment seems unlikely but, given widespread birth control, progressive adjustment is much more likely. The so-called J-curve of progressive approximation, in which global population is shaped by periodic checks of famine, disease and war, is proposed by neo-Malthusian groups such as the Club of Rome*

The concept of carrying capacity

When thinking about the maximum number of people that a given environment can support over a sustained period of time, without environmental degradation, we need to take into account the disparity between the per capita consumption levels of rich and poor. The carrying capacity for people living at a subsistence level is much greater than that for people living a 'Western lifestyle'. This information should be fed into models or predictions about our future growth and prosperity. Or maybe if we all aspire to a lifestyle of *high impact* consumption, perhaps models of the impact of our future global population should just use data taken from current Western consumption and extrapolate up. What do you think?

sustain our demands. If we maintain our current lifestyle and consumption patterns, this will rise to more than two Earths by 2030!

Calculations of our ecological footprints highlight the need for us to make significant changes if we want to create a sustainable future.

What is the demographic dividend?

Given that about one quarter of the world's population is between 10 and 24 years old, and birth rates in many parts of the world are falling, there is the potential to reap a *demographic dividend*. Falling birth rates result in a smaller proportion of young, dependent ages and relatively more people in the economically active age groups. As long as capital investment and new technology harness the potential of the workforce, this may result in faster economic growth and fewer burdens on families.

Look at Figures **2** and **3**. This comparison of South Korea with Nigeria demonstrates the demographic dividend well. As South Korea's birth rate fell in the mid-1960s, school enrolments declined and funds previously allocated for primary education were used to improve the higher-level education. The population bulge is now of working age and the economy is booming. Nigeria, in contrast, shows larger numbers of young dependents.

However, the demographic dividend does not last forever – there is a limited window of opportunity. In time, the age distribution changes again, as the large economically-active adult population moves into the older, less-productive age brackets, with a smaller number of economically-active adults who were born during the fertility decline. Consequently, the dependency ratio rises again – this time involving the need to care for the elderly, rather than the need to take care of the young.

Taking a wider geographical perspective, east Asia's demographic dividend is now peaking with the prospect that it will fade steadily as populations age (Figure **4**). Their window of opportunity is beginning to close. Assuming declining fertility rates over the next several decades, sub-Saharan Africa, on the other hand, is just starting to enter its window. If these African governments, such as Nigeria, take actions that follow, even to some extent, those of east Asia, the demographic dividends may become real rather than potential. There is hope.

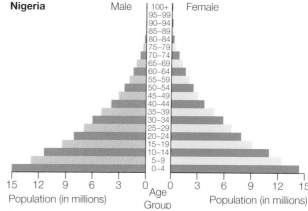

▲ *Figure 2* Age–sex structure in Nigeria (2014)

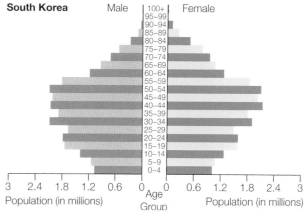

▲ *Figure 3* Age–sex structure in South Korea (2014)

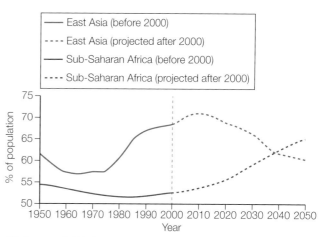

▲ *Figure 4* Working age population (ages 15–59) as a percentage of total population in east Asia and sub-Saharan Africa

ACTIVITIES

Outline the meanings of carrying capacity, ecological footprint and the demographic dividend.

STRETCH YOURSELF

In 2015, ahead of the Paris Climate Change Conference, India's Government suggested that developed countries should do more to curb climate change to bring about 'climate justice', thereby raising the temperature of the international politics around growth and sustainability. Clearly, the governments of HDEs cannot block the efforts of LDEs to develop and so increase their resource consumption. They struggle to suggest to their own people that they lower their own resource demands and living standards. So what can they do?

In this section you will learn about studies of various predictions of future global population

Predictions of future global population

Just as climate change today dominates much geographical thinking, concerns about whether there would be sufficient global resources to support our growing numbers preoccupied attention in the twentieth century. It was not so much the size of the global population that alarmed demographers but the rate of growth in LDEs that were least able to cope. But population projections always proved uncertain. Predictions have to balance the impacts of national population policies against assumptions about future wars, economic systems, food and energy supplies, climate change, resource and biodiversity depletion and overall environmental sustainability.

The United Nations produces alternative scenarios based on demographic surveys which are projected into the future to produce low, medium and high estimates of population. The medium scenario is most often quoted as what *should* be possible in the future but the low and high scenarios allow for variations in the predictions for birth and death rates. All scenarios show the population growing until 2050. The 2010 estimate gave a global population ranging from 6 to 16 billion by 2100 (Figure **1**). The more likely scenario is that the planet will be home to between 9.6 and 12.3 billion people by 2100.

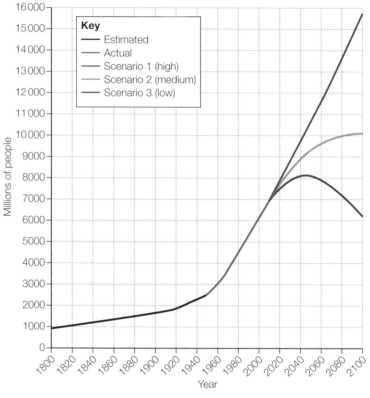

Key
— Estimated
— Actual
— Scenario 1 (high)
— Scenario 2 (medium)
— Scenario 3 (low)

Millions of people / Year

Figure 1 *UN world population estimates 1800–2100. Scenario 1 shows the population continuing to increase rapidly. Scenario 2 shows a gradual slowing in the growth rate. Scenario 3 assumes that drastic action will have been taken to reduce population growth.*

Pessimistic predictions

The English demographer Thomas Malthus (1766–1834) gloomily forecast that, unless population growth was slowed by, for example, later marriages, the exponential rise would outstrip food supply and lead to disastrous 'checks' by famine, war and disease (Figure **2**).

Some global groups still refer to these forecasts in arguing for solutions to overpopulation. For example, the Club of Rome is an international think tank of industrialists, diplomats and scientists who used computer modelling in the late 1960s to predict 'unavoidable' limits to growth within a hundred years, if population growth and increasing demand for resources continued unchecked. Given conservation, recycling, resource substitution and other technological changes since then, their predictions have proven to be, at best, alarmist.

Paul Ehrlich (an American biologist who first used the phrase 'population explosion' in 1968) is another so-called neo-Malthusian who, in *The Population Bomb*, introduced millions to the suggestion of famines, civil wars and environmental catastrophes as indicating a finite future for a planet in crisis.

Figure 2 *Thomas Malthus; his 1798* Essay on the Principle of Population *has had far greater influence than he could possibly have imagined*

Optimistic predictions

Others, such as Danish economist Ester Boserup (Figure **3**), argued that agricultural innovations, such as the Green Revolution which increased yields, proved that population growth stimulated the changes necessary to support the extra numbers. In her book *The Conditions of Agricultural Growth: The Economics of Agrarian Change under Population Pressure* (1965), Ester Boserup proposed a different dynamic in terms of the relationship between population growth and the environment, arguing that 'with increased population density, agricultural production can also be increased by innovation and a greater input of labour.' Using evidence researched in LDEs such as Bangladesh, she demonstrated that 'necessity is the mother of invention'. Julian Simon (1981) continued the debate suggesting that at times of scarcity the greatest gains could be made from entrepreneurs substituting new resources and innovating. Bjørn Lomborg also backed the importance of innovation in *The Skeptical Environmentalist* (2001).

▲ **Figure 3** *Ester Boserup (1910–99)*

How much higher?

Unless there is some catastrophic Malthusian check – such as a calamitous pandemic – the world's population is likely to be around 10 billion by mid-century. This increase is driven by high fertility in sub-Saharan Africa, where population is forecast to more than double, and by further rises in Asia's population. However, the overall growth rate is declining steadily and the richest, most economically developed regions will remain largely unchanged at around 1.3 billion. Look again at Figure **2**, 4.1. This overall pattern should remain valid, with change only in regional detail.

Beyond mid-century, predictions become less certain – much depends on fertility rates in the next few decades. The promotion of family planning, particularly in Africa, will be crucial in achieving population stabilisation. Reproductive health is, in consequence, a key future UN priority.

ACTIVITIES

1 Study Figure **1**:
 a Outline the main factors behind each scenario.
 b What would be the effect on each scenario if the outcome of these factors was not as predicted? Explain your answer.

2 Why are fertility rates in the next few decades so crucial to determining world population growth trends in the latter half of the twenty-first century?

3 Using online resources, further investigate the positive and negative perspectives on the sustainability of future global population discussed in this section. (Hint: start with an online encyclopedia, such as Wikipedia.)

 You might start with Ester Boserup vs. Paul R. Ehrlich. Bjørn Lomborg has his own website, where you'll find papers and shorter press releases.

 Try to find out about their key ideas, how influential their ideas were (or still are) and the criticisms levelled against their work.

EXTENSION

A circular economy

A circular economy is arguably the most exciting, forward-thinking illustration of the so-called Boserupian response to population pressure – innovation promoting radical change.

The traditional economy works on the principle of producing, using and disposing. A circular economy takes recycling to a whole new level by keeping resources in use for as long as possible. Through innovative design and changing perspectives on ownership, the maximum value from resources (whilst in use) can be extracted. At the end of their serviceable life, products such as car engines, carpets and washing machines are then recovered and regenerated into new, improved replacements. This would reduce the use of finite resources and also the environmental impacts of production and consumption.

Unfortunately, explanations of a circular economy are notoriously complicated. But a good starting point to understand what might be the long-term future for us all, is to investigate websites for WRAP (www.wrap.org.uk) or the Ellen MacArthur Foundation (www.ellenmacarthurfoundation.org).

In this section you will learn about health impacts of global environmental change – ozone depletion and climate change

Changing attitudes to health

Attitudes towards health have changed over time as scientific research, education and media coverage have increased our understanding (Figure **1**). Nevertheless, just because we are aware of health risks does not necessarily result in changes in our behaviour to these risks. For example, obesity is now, arguably, a global pandemic (see 4.2).

Health impacts of global environmental change, including ozone depletion and climate change, present significant and new challenges. These include changing not just our lifestyles and behaviours but rethinking established opinions and attitudes.

Ozone depletion

Depletion of the stratospheric ozone layer by chlorofluorocarbons (CFCs) has led to increased solar UV radiation at the surface of the Earth. The effects of reduced ozone (O_3) include increased risk of skin cancer and eye diseases, such as cataracts, reduction of crop yields and decrease in oceanic plankton, disrupting marine ecosystems (see *AQA Geography A Level & AS Physical Geography*, 6.10).

Figure 1 *Attitudes towards health have changed considerably over the years*

Skin cancer

One in every three cancers diagnosed worldwide is skin cancer (non-melanoma) and it is the most common form of cancer in the UK (see Figure **2**, 4.9). It is also one of the few cancers where the cause of most cases is known. Risk factors include:

♦ Exposure to the sun (UV radiation) – a history of sunburn, particularly in childhood, and unprotected exposure to the sun over a lifetime (Figure **3**)

♦ Outdoor working – increased exposure to the sun: farm workers, builders, gardeners and so on

♦ Being fair skinned – those with light-coloured hair and skin tend to be more likely to burn than to 'tan'

♦ Age and family history – older people and a family history of skin cancer are more at risk

♦ Other – radiotherapy, weakened immune system and other skin conditions.

Look at Figure **2**. The high albedo of snow (80 per cent of the sun's UV radiation is reflected) and thinner atmosphere of higher altitudes (where less UV radiation is absorbed) combine to place skiers and snow-boarders in a potentially high-risk health category for cancer. Any exposed skin, particularly exposed facial parts such as the nose and eyes are at risk – hence the importance of both high-factor sun cream and ski goggles.

Think about

Health risk factors

To prevent disease, it is necessary to identify and deal with their causes – the health risks that underlie them. The WHO defines a health risk factor as 'any attribute, characteristic or exposure of an individual that increases the likelihood of developing a disease or injury'. Examples include high blood pressure, tobacco use, physical inactivity, alcohol consumption, unsafe water and poor sanitation. Risks for burden of disease are measured in disability-adjusted life years (DALYs, see 4.16) – being underweight and unsafe sex are two of the leading global causes of DALYs.

Each risk factor usually involves an interrelated sequence of causes (socio-economic, environmental, community and individual behaviours). Intervention in this sequence, by responding to background causes such as education and income, may have an 'amplifying' effect as they have a knock-on effect to other linked factors.

Figure 2 *UV radiation exposure increases 4 to 5 per cent with every 300 m in altitude gain*

Cataracts

Cataracts are an inflammation of the eye and usually develop slowly. They can affect anyone but are common in older people and, if left untreated, may lead to permanent blindness. According to WHO estimates, some 12 to 15 million people worldwide become blind from cataracts annually, of which up to 20 per cent may be caused or enhanced by UV radiation.

While risk factors for cataracts are similar as for skin cancer, they have a different distribution. In contrast to the fairer skinned populations of Europe, cataract is the main burden of disease attributable to UV radiation in Africa, the Middle East and in some South American countries.

2020

2060

◮ *Figure 3 Numbers of extra skin cancer cases related to UV radiation; note that in 2060 those that are most likely to be affected (the over 60s) would have been children when the problem was at its most severe*

Tackling the ozone issue

In 1985, a hole in the ozone layer was discovered over Antarctica. The realisation of the potential health impacts of this rapidly expanding hole quickly led to the Montreal Protocol (1987) banning the CFCs responsible. It seemed that the global community understood how quickly major changes to the atmosphere can be caused by human interference and how long it takes for nature to recover from them.

Climate change

Climate change illustrates that the lessons learnt were short-lived. Greenhouse gases continue to be emitted into the atmosphere and, unlike CFCs, there is no legally binding global agreement forcing alternatives to burning carbon (see 5.18). However, the 2016 Paris Agreement was signed by 174 countries and aimed to reduce carbon emissions so that global warming restricts temperature increase to 2 °C of pre-industrial levels.

Although global warming may bring some localised benefits, such as fewer winter deaths in temperate climates and gains in food production, the overall health impacts are likely to be negative and disproportionately impact populations of LDEs (Figure **4**).

Did you know?

Between 2030 and 2050, climate change is expected to cause approximately 250 000 additional deaths per year, from malnutrition, malaria, diarrhoea and heat stress (WHO, 2016).

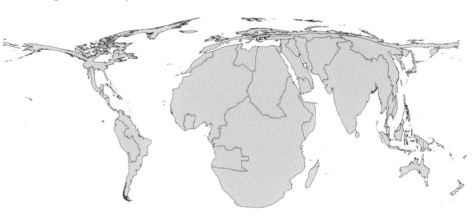

◮ *Figure 4 World map drawn proportionally to reflect mortality related to climate change*

Thermal stress

Global climate change is likely to result in an increase in the frequency and intensity of heatwaves, as well as warmer summers and milder winters. Furthermore, the impact of high summer temperatures on health may be heightened by an increase in humidity. Heatwaves have a greater health impact on the elderly and those with pre-existing conditions. Increases in mortality are mainly a result of cardiovascular, cerebrovascular and respiratory disease but morbidity figures also increase from, for example, heat stroke and heat exhaustion. Increased levels of harmful air pollution are also more common in cities and a heatwave exacerbates the health impacts, such as asthma. (See also the urban heat island effect in 3.21.)

Populations living in temperate climates, for example northern and western Europe, are potentially at greater health risk from heatwaves as they may be less aware of personal coping strategies and may take several days to acclimatise. In practice, populations in LDEs are at greatest risk of mortality because while 'normal' temperatures may be higher, fewer resources exist to allow adaptation (such as air conditioning or changes in working practices).

In contrast, some studies have concluded that climate change will actually result in an annual net reduction in mortality rates – decreases in winter mortality may be greater than increases in summer mortality (one study suggests a decrease of 20 000 cold-related deaths in UK by 2050). However, annual outbreaks of winter diseases such as influenza are not strongly linked with changes in monthly winter temperatures. Furthermore, in some locations it is low-frequency extreme winter weather events, such as snow blizzards in North America that cause marked spikes in mortality.

Agricultural productivity

Look again at section 4.3. Compare hill sheep farming in Snowdonia with arable farming in Herefordshire – clearly, agriculture is strongly influenced by weather and climate. Farming is finely adapted to local physical conditions – climate change represents a real threat to this balance and upsets established farming systems and practices (Figure **5**).

▼ **Figure 5** *Impacts of climate change on agricultural productivity*

	Impacts	Effects
Direct impacts	Changes in mean climate	Higher yields in mid and high latitudes; higher growing-season temperature results in a northwards extension of production of, for example, cereals, maize, sunflower and soy
		Lower yields in seasonally arid and tropical regions; as crops are already likely to be growing to their maximum capability, increased temperatures and uncertainties of precipitation patterns may result in crop failure
	Increase in extreme weather events	Extreme temperatures; only a few days of high temperatures, particularly if at the flowering stage, will dramatically reduce crop yields
		Drought; by 2050 drought-related yield reductions might increase by 50% for major crops
		Heavy rainfall and flooding; leading to lower quality and yield of crops. Secondary impacts include soil waterlogging, reduced plant growth and inability of heavy farm machinery to operate
		Tropical storms; cause economic and societal problems that impact negatively on agricultural systems, particularly in LDEs
Indirect impacts	Increase in pests and diseases	Increased CO_2 levels; pests such as aphids and weevil larvae will be more abundant
		Changing migration patterns; such as locusts in sub-Saharan Africa that respond to changes in rainfall patterns
	Increased water extraction for irrigation	Increased irrigation; precipitation reduction and/or extraction upstream may limit water availability for irrigation downstream
	Mean sea-level rise	Inundation of low-lying coastal agriculture; as a result of thermal expansion of water and additional water from melting ice, coastal agriculture may be flooded and affected by salinisation for years following
Non-climate impacts	CO_2 fertilisation	Increased atmospheric CO_2 concentrations; leads to increased photosynthesis and so yields for some crops may increase (although quality may be lower)
	Air pollution (ozone emissions)	Reduces rates of photosynthesis which, in turn, reduces crop yield

Nutritional standards

What we eat is influenced by many variables. If climate change forces food prices to rise, then consumers may choose less healthy food. Processed food with high sugar and fat contents is usually cheaper than 'fresh' alternatives because the cost of the food is a smaller component of the overall cost.

Climate change may force shifts in agricultural production (see Figure **5**). This would result in new crops that are bred to survive in different climatic conditions, as well as ceasing production of crops grown in current locations due to being no longer economically and/or environmentally viable. As food grown in different locations and by different methods has different vitamin, antioxidant and amino acid compositions, this would further constrain consumer 'choice'. Different methods of processing, storage, preparation and cooking also affect the nutritional content of food. For example, as fuels become more expensive and in shorter supply, higher fuel costs may reduce cooking options for lower income groups.

Climate change presents a challenge to public health organisations in both monitoring and responding subsequently to nutritional changes. For example, the Food Standards Agency (FSA) is responsible for food safety and food hygiene across the UK. It works with local authorities to enforce food safety regulations and its staff work in UK meat plants to check the standards are being met. FSA initiatives such as the Eatwell website and nutritional labelling of food help to protect public health and maintain nutritional standards.

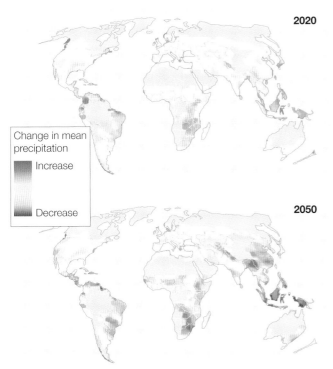

2020

2050

Change in mean precipitation

Increase

Decrease

⬤ **Figure 6** *Predicted changes in annual mean precipitation over global croplands*

Vector-borne diseases

The WHO estimates that one-sixth of the illness and disability suffered worldwide is owing to vector-borne diseases, with more than half of the world's population currently at risk.

Climate and weather conditions strongly influence the pattern of the vectors – temperature affects the survival and reproductive rates. Precipitation affects vectors that have aquatic breeding grounds (such as mosquitoes) and humidity levels influence levels of vectors such as ticks or sandflies.

Indirect effects of climate and weather include impacts on human systems that, in turn, may affect vector ecology and risk exposure. For example, drought may lead to increased use of irrigation systems that might stimulate vector-borne disease. Furthermore, forced population movement as a result of water shortage might place previously unaffected populations at risk. Seasonal differences and natural variability caused by, for example, El Niño, can also affect the prevalence of vector-borne diseases.

ACTIVITIES

1 **a** Explain how and why the global community responded both promptly and with success to reducing the health impacts of ozone depletion.

   b Suggest reasons why there has not been a similar level of response to the health impacts of climate change.

2 **a** Suggest reasons why measuring health risks of a population may be problematic.

   b Explain why public health campaigns (such as wearing sun protection) are not always effective.

3 Study Figure **3**.

   a Describe the distribution of additional skin cancer cases.

   b Suggest reasons for any pattern.

4 Where should policy-makers start when tackling health impacts of climate change? For example, education, public health campaigns, prevention, socio-economic conditions and so on. Justify your answer.

5 Study Figure **6**. Compare and contrast the annual mean precipitation over global croplands projections for 2020 and 2050.

STRETCH YOURSELF

To what extent are LDEs bearing the heaviest burden of the health impacts of global environmental change? Make reference to health impacts of ozone depletion and/or climate change.

In this section you will learn about:
◆ prospects for the global population and projected distribution
◆ critical appraisal of future population–environment relationships

Prospects for the global population

Look at Figure **1**, 4.26. It shows UN estimates for future population change in three scenarios. Scenario 1 suggested the rate of world population growth continuing to increase rapidly. This is as unthinkable as it is unsustainable. Global carrying capacity would be breached in a catastrophic Malthusian check. Fortunately, the world has not stood by and ignored this possibility. The influence of organisations such as the UN Commission on Population and Development and the UN Foundation are enabling nations to tackle population issues and have made Scenarios 2 and 3 much more realistic. Indeed, such is progress on the empowerment of women – their emancipation and improvements in reproductive health (see 4.29) that Scenario 2's gradual slowing in growth rate now seems established. The implications of this cannot be underestimated. World population growth results in more land, air, water and noise pollution. It also leads to soil erosion, desertification, acid rain and human-enhanced global warming.

Did you know?

Air pollution is said to cause 130 000 deaths every year in the USA alone.

Projected population distributions

The bulging youth populations of today will ensure that rapid population growth will remain a feature for the majority of Africa (Figure **1**). Conversely, with an already ageing population and low fertility rates, Europe will be one of the biggest net losers of population.

❤ **Figure 1** *Projected annual population growth rate (2010–50)*

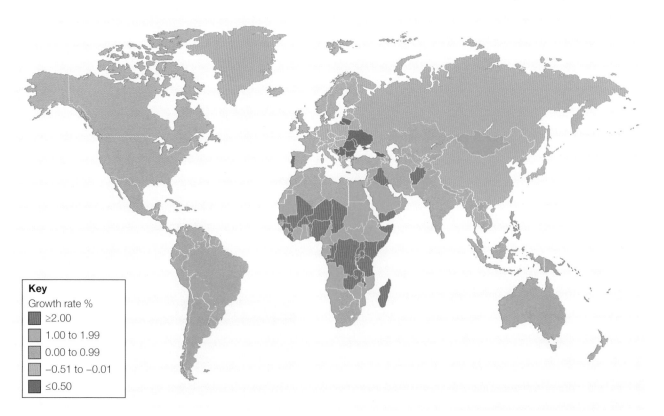

Key
Growth rate %
■ ≥2.00
■ 1.00 to 1.99
■ 0.00 to 0.99
■ −0.51 to −0.01
■ ≤0.50

Factors in future population–environment relationship

◆ *Climate change*: arguably the most pressing environmental issue of our time. Human-enhanced global warming is changing climates, weather patterns and sea levels. As world population continues to rise, more fossil fuels are burned at an accelerating rate. Continental and sea ice is melting, and permafrost thawing. Millions of people in countries like Bolivia, China, India and Peru depend upon replenishable meltwater from mountain glaciers for domestic and industrial water supplies, irrigation and hydro-electric power. Projected future sea-level rise is threatening low-lying regions such as Bangladesh, Egypt's Nile Delta, the Netherlands and Florida, USA.

◆ *Destruction of natural habitats:* within the next half century, another quarter of the remaining forests on Earth will be removed for uses such as roads, golf courses and urban areas.

◆ *Loss of biodiversity*: biodiversity has been described as 'the foundation of all life on Earth' because it is crucial to the functioning of ecosystems (see *AQA Geography A Level & AS Physical Geography*, 6.1). A significant proportion of the world's wild species and genetic diversity has already been lost, and climate change threatens more.

◆ *Unsafe water supplies*: over 1 billion people in LDEs lack access to safe drinking water. Some politicians predict that future world conflicts will be driven by the need for water security. Most of the world's freshwater in rivers and lakes is already being used for domestic and industrial consumption and irrigation.

Throughout the world, underground aquifers are being depleted more rapidly than they are being replenished.

◆ *Soil erosion*: we have already established that 40 per cent of soil used for agriculture around the world is either degraded or seriously degraded (see 4.6). Erosion by wind and water is between 10 and 40 times the rate of soil formation. Leaching, salinisation and acidification have further degraded many of the world's soils (see 4.6).

◆ *Loss of wild foods*: about 2 billion people, mostly in LDEs, depend on the oceans for protein. Wild fish stocks could be managed effectively but, in reality, overfishing has resulted in the decline or collapse of valuable fisheries worldwide.

◆ *Fossil fuels*: oil, natural gas and coal remain the world's most important energy resources. However, burning fossil fuels is both highly polluting and a significant contributor to global warming. These resources are also finite with reserves of oil and gas arguably best measured in decades rather than generations.

◆ *Toxic chemicals*: toxins from industry are both manufactured and released as effluent. Reduced sperm counts, birth defects and mental health problems have all been associated with exposure to toxic chemicals. Despite their tiny concentrations in the air, soil, oceans, lakes, rivers and groundwater, toxins such as refrigerator coolants, detergents and plastics affect us all. We breathe, swallow and even absorb them through our skin

ACTIVITIES

1 Study Figure **1**.
  a Describe the distribution of projected annual growth rate (2010–50)
  b What role is migration likely to contribute to changes in future population distribution?
2 How might 'unforeseen events' such as war, famine, disease, technological innovation and political upheaval affect future population distributions? Illustrate your answer by reference to specific events and regions that may be affected.
3 **a** Working in pairs, use the internet and other resources (such as textbooks and magazine articles) to conduct additional research on one of the environmental threats to population outlined in this section. (Be sure to coordinate with the rest of the class in order to ensure that you have a good range of problems.)
  b Prepare a short PowerPoint presentation of your findings for showing to the rest of the class.

STRETCH YOURSELF

No other animal species has ever made such immense or rapid changes to natural ecosystems as humanity, but no other animal species has the power to regulate its own numbers. Neither do other animal species have the knowledge or means not just to destroy the environment, but to manage and improve it to ensure a sustainable future for us all.

In pairs or small groups, examine the messages in the statement above. Why do environmentalists repeatedly insist that 'the future starts today'?

In this section you will learn about:
◆ character, scale and patterns of population change in Iran
◆ environmental and socio-economic factors influencing population change in Iran

Population change in Iran

How is success judged? Who is in the best place to make this assessment? Iran (Figure **1**) is often quoted as the 'Population Success story', having witnessed a remarkable fall in population growth rate over the last 30 years as a result of a highly successful family planning programme. Indeed, fearing a population decline, the Iranian government now believes that the drive to reduce birth rate has been too successful and population policy has changed again. Will this next phase of demographic transition be judged equally favourably?

The birth of a new state

Iran was part of the great Persian Empire (Figure **2**) but the last 40 years have been dominated by political change, armed conflict and almost continual dispute with the 'Western World', particularly the USA.

Following the 1979 revolution, the rule of the increasingly autocratic US-backed Shah was ended and the Islamic Republic of Iran was established – so began the first wave of the US-imposed trade sanctions. The following year, a large-scale invasion by Iraq of the western borders of Iran started a conflict that lasted eight years. Following claims of Iranian state sponsorship of terrorism, tensions increased with the US and the 'West' throughout the 1980s and fuelled anti-West sentiment. In 2006, after Iran refused to suspend its uranium enrichment programme, sanctions were increased and remained in place until a nuclear agreement was reached in 2016. Both the Iran–Iraq war and unprecedented sanctions have significantly impacted the economy and so everyday lives of Iranians.

▲ *Figure 1* *Location of Iran*

▲ *Figure 2* *Iran has a culture influenced by some of the world's great dynasties – including Alexander the Great and Genghis Khan. Persepolis dates from 521 BCE and is testament to the splendour of the Persian Empire.*

Population rises ...

Look at Figure **3**. The age–sex structure for the Islamic Republic of Iran is typical of a country in stage 3 of the demographic transition model (DTM; see 4.18). Following the revolution in 1979, and professing to be the 'government of the oppressed', Iran attempted to prioritise the most basic needs of the population – such as food, basic health care and universal access to primary education (Figure **4**). At this time Iran was fairly typical of a country in stage 2 of the DTM. Although the urban middle classes were already starting to practise birth control and to limit their family size, the rural poor were still mostly choosing to have large families to try to ensure their own economic security.

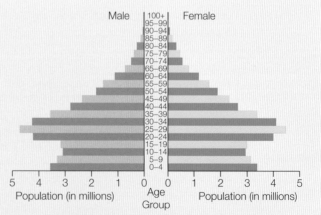

▲ *Figure 3* *Population age–sex structure of Iran (2016)*

The bulge of population in lower/middle-age groups (Figure **3**), represents the 'baby boomers' of the 1980s. During the early years of the revolution, there were no official population policies (indeed, existing family planning policy was suspended). Subsequent 'policy' included widespread encouragement for early marriage and campaigns outlining the benefits of larger families. The new leaders considered any population programme as a tool of the West intended to dominate and limit the numbers of new Muslims. Large families were also considered a core revolutionary value, as a means to increase numbers in the army in future years to defend against the armed hostility of Iraq. In addition to higher birth rates, the population also increased as a result of an influx of refugees from neighbouring countries affected by the Iran–Iraq conflict. Furthermore, despite the shock of conflict and sanctions, improved primary health care resulted in significant falls in infant and maternal mortality rates between 1981 and 1986.

Article 29: It shall be the universal right of all to enjoy social security covering retirement, unemployment..., and health, medical treatment and care services through insurance, etc. The Government shall be required, according to law, to provide the aforesaid services and financial protection for every individual citizen of the country...

Article 30: The Government shall be required to provide free education and training for the entire nation...

▲ *Figure 4* Articles 29 and 30, Constitution of Independent Republic of Iran

...growth rate falls...

Between 1976 and 1986, the population grew a staggering 40 per cent. Crippled economically by the costs of war and the effects of sanctions, the threat of a Malthusian crisis of hunger was very real. Government officials sought advice on both the socio-economic costs of large families on quality of life and also on the moral and ethical aspects of family planning. In 1989, population control measures were included within the first, and subsequent, National Five Year Socio-economic Development Plans. The Plans had a dramatic impact on fertility rates (Figure **5**). The success of the family planning programme was a result of a combination of factors:

- Involvement of religious leaders (Ayatollahs) – permission was granted for both the distribution of contraception and also male sterilisation.

- Model of service delivery – family planning services were integrated throughout the health care network, including at more remote rural health care level and coordinated centrally by the Ministry of Health and Education.

- Improved living conditions – the development of a modern infrastructure (clean drinking water and electricity, transport and communication networks), education, health care, modern agricultural and industrial practices all helped to change attitudes towards family planning.

- Role of women – initiatives such as Women Health Volunteers and expansion of the network of rural midwives, have helped to promote family planning and to encourage female empowerment.

Core components of the Iranian family planning programme

- Freedom to choose contraception method
- Education on population issues
- Respectful of religious and cultural values
- Spacing of pregnancies of between 3–4 years
- Preventing high-risk pregnancies of women under 18 and over 35 years
- Free voluntary sterilisation
- Baby-friendly hospitals

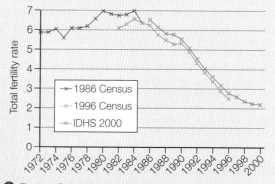

▲ *Figure 5* Iran's total fertility rate (1972–2000)

- Public education campaigns – families and young people, particularly girls, were targeted in a widespread campaign on the benefits of family planning (this included education in schools, mosques and even in military training).

... and rises again?

Look again at Figure **3**. Iran's 'baby boomers' from the 1980s provide a demographic dividend for today's growing economy – in other words, in a population that still has a youthful structure, they provide a much-needed and abundant (and therefore relatively cheap) workforce. Yet the republic now also has a high life expectancy (nearly 76 years) meaning that Iran needs to face the prospect that the 'baby boomers' will soon become ageing dependents. The proportion of Iran's population aged 60+ increased from 5.3 per cent in 1966 to 8.1 per cent in 2011 and is predicted to be 12.3 per cent by 2025. Characteristics of this ageing population include:

♦ Regional contrasts – differences in fertility and migration levels combine to create an uneven geography of elderly. In the absence of accurate data, it makes targeting resources to the elderly more difficult.

♦ Increasing proportion of aged people living alone – as more elderly are women, who are more likely to be economically inactive compared to men, there is a dependence on kinship support from other family members.

♦ Poverty – formal and informal support programmes, such as pensions and aid, are inadequate and a coordinated approach is needed by the government.

♦ High employment rate – old people are working after the retirement age of 65 (nearly 40 per cent of those aged 65+ are economically active).

♦ Low literacy rate – a result of past cultural differences prior to the Revolution, men are more likely to be able to read and write. In some rural areas, the literacy rate is as low as 5 per cent for women.

♦ Ill health – those aged 50+ make up the highest proportion of individuals on hospital waiting lists (non-communicable and chronic diseases are the biggest cause of health problems, but these require lengthy and costly treatments).

Fertility decline in Iran

While it is clear that fertility reduction did not happen at all under the Shah, it is also misleading to assume that it happened as a consequence of family planning policies alone under the new regime. Firstly, following crippling costs of war and sanctions, the new regime simply did not have the economic means to spread the message of the benefits of smaller family size. Secondly, social and economic reforms, particularly in rural areas where birth rates were highest, have improved living conditions. For example, as a result of a fall in infant mortality rates, parents now invest more money in their education (as there is less risk of them dying in infancy). Younger generations are better educated and female empowerment has enabled women to have access to healthcare services and to make informed choices. In short, the seeds of modernity have been sown and with this social attitudes have changed. Whilst the revolution and Islamic rule brought immediate political changes, ironically perhaps the most enduring legacy is the adoption of Western reproductive behaviour.

Did you know?
Seventy per cent of Iran's science and engineering students are women.

Did you know?
Iran has one of the only condom factories in the Middle East.

▶ *Figure 6* *Western influences are now stronger in Iran than in previous decades*

A new baby boom

To combat this 'ticking time bomb', Iran is now promoting a second 'baby boom'. In 2010, government payments were reintroduced for each new child and by 2012 the Health Ministry's 'population control' budget was abolished. Birth control is no longer state subsidised and religious leaders and state television messages again urge women to have larger families. Even the incentive of gold coins to new-borns has been discussed. However, this approach is not proving as straightforward the second time around. In an economy where inflation can be as high as 60 per cent, young people are not wholly convinced by the government's arguments of the benefits of larger families. Young couples continue to marry later and prefer to continue their studies at university rather than have children. In particular, women are enjoying the benefits of a more liberal and modern society and are gaining higher salaries in male-dominant industries such as oil, gas, construction, mining and technology. Career progression, economic security and the lure of a Western lifestyle are foremost in the minds of young people to the detriment of the desire to have a large family (Figure **6**). Currently, total fertility rate remains low at around two children.

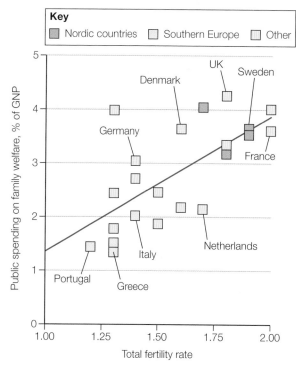

▲ **Figure 7** *Scattergraph of family welfare spending and total fertility rate*

ACTIVITIES

1 Copy and complete the timeline in Figure **8**. Limit the timeline to ten entries only. The Iran profile and timeline on the BBC News website will help.

2 Study Figure **3**.
   a Divide your page into two. In one half, sketch an outline copy of Figure **3**.
   b Annotate the population pyramid to show characteristic features of the population. (It might be helpful to consider each third of the pyramid in turn – top, middle and bottom.)
   c To what extent is Figure **3** typical of a country in Stage 3 of the DTM? Suggest reasons for your answer.
   d In the other half of your paper, sketch a second pyramid to suggest the population structure of Iran in 50 years time. Annotate the model as previously. (Hint: www.populationpyramid.net is a good starting point)

3 Study Figure **4**. Suggest how these changes in the constitution have affected population change.

4 Iran has no integrated national programme for the elderly. Suggest three priorities that any policy for the elderly should address. Justify your choices.

5 Study Figure **7**. What lessons might Iran learn from the relationship shown on the scattergraph?

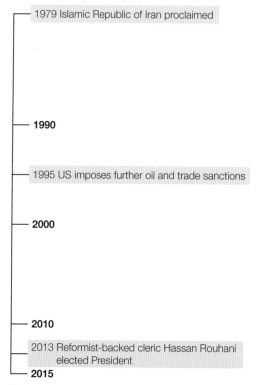

▲ **Figure 8** *Iran timeline showing significant changes in the politics, society and economy of Iran since 1979*

STRETCH YOURSELF

Compare and contrast the approaches of China and Iran in the management of population change.

In this section you will learn about the relationship between place and health in Abbey Ward, Lincoln

In 2003, readers of *The Idler* website voted Kingston upon Hull (Hull) as winner of the 'Crap Towns: The 50 Worst Places to Live in the UK'. Was this a fair perception?

In 2017, Hull is the UK City of Culture. The city council wishes not only to attract multi-million pound investment into the city but also to change the long-term perceptions towards the city. Cynics might argue that an award alone is papering over more systemic cracks in the economic and social fabric of the city. However, changing attitudes – encouraging civic pride, celebrating cultural heritage and bringing communities together – is of crucial importance in bringing about positive change to an area.

Lincoln – a city of contrasts

Lincoln is an historic city of almost 100 000 people and is situated 60 km south of Hull (Figure **1**) and not without its own urban problems. A 2016 report highlighted that Lincoln is home to some of the most deprived areas in England and that average earnings of residents are consistently lower than the national average (Figures **2** and **3**). Furthermore, when specific streets and neighbourhoods are examined in detail, acute levels of deprivation are identified.

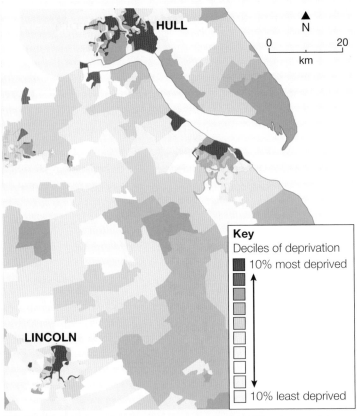

Figure 1 *Index of multiple deprivation for Hull/Lincoln area (also see 2.8)*

Figure 2 *Summary of changes in key indicators of Lincoln City Poverty Profile (2016)*

Theme	Indicator	3 year change	5 year change
Housing	Affordable homes delivered	Worse	Not available
	Mortgage repossessions	Better	Better
	Landlord possessions	Worse	Worse
Low income and inequality	Average earnings by residence (full-time and part-time)	Worse	Better
	Child poverty rate	Better	Better
	Children resident in out-of-work benefit households	Better	Better
	Fuel poverty rate	Better	Not available
	Indices of Multiple Deprivation – Number of areas in most 10% of deprived areas nationally	Not available	Worse
Homelessness	Statutorily homeless households	Worse	Worse
	Homelessness applications	Worse	Worse
	Homelessness preventions	Better	Better
	Households living in temporary accommodation	Better	Worse
Education	GCSE attainment rate	Better	Flat
	Key Stage 1 attainment rate	Better	Not available
	Key Stage 2 attainment rate	Worse	Not available
	Population with no qualifications	Worse	Worse

Worklessness, benefits and welfare reform			
Out-of-work benefit claiments (JSA)		Better	Better
Unemployment ration		Better	Better
Housing benefit recipients		Better	Better
Children living in families receiving tax credits		Better	Not available

Monks Road (Abbey Ward), Lincoln
Physical environment

The area of the city around Monks Road is a **zone in transition** (Figure **4**). Within one kilometre of the CBD, this busy arterial road has attracted mixed land use typical of an inner-city location. Late Victorian/Edwardian housing (1869–1919) lies in close proximity to areas of renewal and redevelopment (Figure **5**). Today, the area continues to change and adapt in response to both population dynamics and as a consequence of local and regional government investment.

Key findings of the Lincoln Poverty Profile

High proportion of single occupancy households

Shortage of affordable housing

Average earnings of residents decreased by £2102 between 2010 and 2015 and are more than £4500 lower than the national average

Increasing number of homeless applications

Nearly 25% of residents aged over 16 have no qualifications

◣ **Figure 3** *Lincoln Poverty Profile (2016)*

◢ **Figure 4** *Location of Monks Road, Abbey Ward, Lincoln*

◢ **Figure 5** *The Monks Road area is typical of an inner-city zone in transition*

Population and socio-economic character

Monks Road lies within Abbey Ward, an area identified as suffering from socio-economic and environmental problems, including ageing and deteriorating housing. Compared with all wards in England, this ward is in the 10 per cent most deprived for income; the 1 per cent most deprived for health and disability; and the 10 per cent most deprived for living environment.

There are higher than average concentrations of young adults (Figure **6**). This figure is partly made up of post-16 students and, since 2004, eastern European migrants (Figure **7**) – Monks Road is an example of a **multicultural** society. Most recent changes reflect the needs of the growing Polish community, for example, specialist food stores and the Polish Advice Bureau and the Polish Chaplain at the local church. Abbey Access Training offers training across a range of programmes, including literacy and numeracy for adults. These have been important in supporting the mixed local population, including migrants, to find employment.

Health

The government-funded schemes, the Abbey Renewal Area (1998–2008) and Community First (2011–15), have attempted to improve the health and well-being of local residents (Figure **8**). They have included refurbishment of the Abbey Sport Court, improvements to local homes (e.g. central heating) and support for community groups such as 'Chubby Cherubs' and 'City Health Walks'. A £3 million restoration project has overhauled the local arboretum and provided a range of recreational facilities, including a children's play area and a community access centre. A seven-day NHS walk-in centre opened in 2009.

Nearby, Lincoln City Football Club works in partnership with a number of organisations to empower local residents to make a difference and to be more active. The schemes include Active Lincoln (funded by Sport England) to encourage residents to take up sport and the Stand Up Speak Up: Be Active project (funded by the Health Lottery).

Age range	Abbey	Abbey %	Lincoln	Lincoln %
0–15	1744	15.2%	15351	16.40%
16–29	3761	33.0%	25833	27.60%
30–44	2574	22.5%	17846	19.10%
45–59	1810	15.8%	16311	17.40%
60–74	910	8.0%	11587	12.40%
75+	627	5.5%	6613	7.10%
Total	11426	100.0%	93541	100.00%

Figure 6 *Age structure, Abbey Ward (2011)*

Language	Number of speakers
Polish	580
Lithuanian	151
Latvian	105
Russian	70
Persian/Fardi	51
Kurdish	48
Portugese	47
Slovak	40
Tagalog/Filipino	38
Arabic	34
Total population	11426

Figure 7 *Top ten languages spoken in Abbey Ward (other than English) (2011)*

Figure 8 *The Abbey Renewal Area Neighbourhood Office was located in the heart of the community and offered help and advice on such topics as government grants, housing tenancy, and crime prevention and education courses*

Community and attitudes

Mobile populations, such as migrants and students, traditionally feel less ownership of an area that they may have lived in for a short period of time. Nevertheless, the area has a long history of community involvement. For example, Monks Road Working Men's Club recently celebrated its centenary year. The local residents' group, the Monks Road Neighbourhood Initiative was created in 1998 and produces a successful newsletter. It also holds regular meetings to keep local residents informed of key activities in the area. Action LN2 is a more recently formed community group. Links with the nearby primary school and the post-16 college are also important drivers in helping to foster a sense of community and distinctiveness (Figure **9**).

Action LN2

Formed in 2013, Action LN2 is a group of residents who volunteer in the Monks Road area. They aim to bring about positive change by involving local people in small-scale, practical projects that make a difference to people's everyday lives as well as celebrating the diversity, culture and heritage of the area. Smaller projects have included language and culture swap sessions to encourage social mixing, and also football sessions for children and families in the adjacent arboretum. The annual 'Our Big Gig in the Arboretum' helps to showcase the community and dispel some of the negative attitudes about the Monks Road area. Alice Carter, a volunteer with Action LN2, explains that the 'most rewarding part of our volunteering … is being able to step back and see the impact that we are having in the community. Change can be a long term process, but it has to start somewhere'.

▼ **Figure 9** *The school and college are important parts of the local community*

ACTIVITIES

1 Study Figure **1**. Describe the distribution of multiple deprivation.

2 Study Figures **2** and **3**.
   a What do you understand by the term poverty? (Hint: review the materials at www.lincolnagainstpoverty.co.uk)
   b Why is poverty a difficult term to define and to measure?

S

3 Using evidence from Figure **4** only:
   a Identify the location of the CBD. Suggest reasons for your answer.
   b Describe the site and location of Monks Road.

4 Look at Figure **7**. Work out the proportion of each language as a percentage of the total population of Abbey Ward.
   a Are the results what you would expect?
   b Do you think that a similar study in your home area would:
      i produce higher or lower percentages for each language
      ii include any languages that do not appear in Figure **7**?

5 a Comment on possible health issues of living in Monks Road. Support your answer with specific evidence, including census data (2011) and map evidence.
   b Suggest how the various government and community initiatives adopted in the area might benefit both the short-term and long-term health of the population? (Hint: make reference to both physical and mental health in your answer.)

STRETCH YOURSELF

To what extent do rural places offer a better quality of life than urban places? In your answer, consider the relationship between place, physical environment, socio-economic character and health. You may also wish to consider attitudes towards living in a rural area. The website for the Lincolnshire Research Laboratory (www.research-lincs.org.uk) provides an excellent starting point.

Now practise...

The following are sample practice questions for the Population and the environment chapter.

They have been written to reflect the assessment objectives of Component 2: Human geography Section C of your A Level.

These questions will test your ability to:

◆ demonstrate knowledge and understanding of places, environments, concepts, processes, interactions and change, at a variety of scales [AO1]

◆ apply knowledge and understanding in different contexts to interpret, analyse and evaluate geographical information and issues [AO2]

◆ use a variety of quantitative and qualitative and fieldwork skills [AO3]

1 How do agricultural systems vary in terms of their productivity? (6)

2 Study Figure **1**. How might a graph of leading causes of childhood deaths in Europe over this period differ? (9)

▼ **Figure 1** *Leading causes of death among children aged under five years in sub-Saharan Africa (2000–2015)*

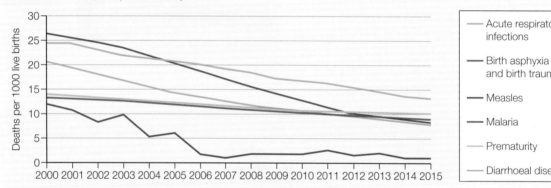

3 'Models of population growth often suggest there are environmental limits to growth but the last half century may be used as evidence to argue the contrary.'
Evaluate this statement with reference to a range of perspectives on population growth. (9)

4 How is air quality linked to health and well-being? (6)

5 Assess the impact of cultural controls on the natural population change of a named country or society. (9)

6 Evaluate the role of international agencies and NGOs in combating disease at the global scale. (9)

7 Study Figure **2**. Outline the factors that might account for the global pattern of food consumption. (6)

◐ **Figure 2** *Average daily calorie intake per capita (2011)*

Key:
- 3480–3769
- 3270–3479
- 3050–3269
- 2850–3049
- 2620–2849
- 2390–2619
- 2170–2389
- 1890–2169
- Less than 1890
- No data

8 Explain the prevalence and global distribution of a non-communicable disease, with reference to both physical and socio-economic factors. (9)

9 'Pollution forms part of a negative feedback loop that keeps the global population in check.' Evaluate this statement. (9)

10 For a country or region you have studied, what have been the most significant implications of migration? (6)

11 Study Figure **3**. Analyse the spatial pattern of change in the incidence of malaria cases from 2000 to 2015. (9)

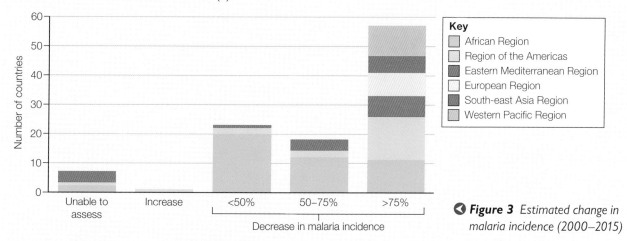

Key
- African Region
- Region of the Americas
- Eastern Mediterranean Region
- European Region
- South-east Asia Region
- Western Pacific Region

◁ **Figure 3** *Estimated change in malaria incidence (2000–2015)*

12 Give an example of a technological strategy used to improve food security and evaluate its success. (9)

13 To what extent is our global future tied to the health of our planet? (20)

5 Resource security

Young boys work in a copper mine in the Democratic Republic of Congo. Most of the copper they extract is exported to China — what factors may influence the destination of this valuable resource?

Your exam

'Resource security' is an optional topic. You must answer one question in Section C of Component 2: Human geography, from a choice of three: Contemporary urban environments or Population and the environment or Resource security. Component 2 makes up 40% of your A Level.

Your key skills in this chapter

In order to become a good geographer you need to develop key geographical skills. In this chapter, whenever you see the skills icon you will practise a range of quantitative and relevant qualitative skills, within the theme of 'resource security'. Examples of these skills are:

- Using atlases and other map sources 5.4, 5.6, 5.7, 5.15, 5.22
- Drawing and annotating maps 5.11
- Presenting data and interpreting graphs, 5.2, 5.5, 5.11, 5.12, 5.13, 5.16, 5.20
- Analysing quantitative data and geospatial data, including applying statistical skills 5.5, 5.6, 5.7, 5.12, 5.22
- Drawing and annotating diagrams or models of human systems and human processes, 5.8, 5.10

Fieldwork opportunities

While it may not be possible to investigate resource security at a national or international level in any depth on your own, it is possible to take a closer look at resource issues at a much smaller scale.

1 Investigating water supplied to you

Find out where the water you use at home or at school comes from: a river or an underground aquifer? Perhaps you live in an area experiencing water stress. If so, what plans are companies or authorities putting in place to cope with future growth in demand? Currently, what approach does your water company take to manage the different demands of domestic customers, industry and public service providers? Does the local level of abstraction have an impact on the river habitats of plants and animals downstream at particular times of year? (If so, the Environment Agency may be involved, as the Agency is the government department that grants abstraction licences and polices compliance with agreed levels.) Annotated field sketches or photographs taken at different times of year may inform your investigation of local water supply issues.

2 Calculate your water footprint

Use an online app to calculate how much water you consume on a daily, weekly or annual basis. How does your water footprint compare to the UK average and that of other people around the world? The calculator might estimate your consumption of 'virtual water' (water used to produce goods made overseas) as well as the volume of water you and your family use at home. Double-check the calculation against the rate of consumption stated on your family's water bill (if you have a water meter, your bill is likely to provide this information). How sustainable is your rate of water consumption?

3 Finding out about policy priorities of local voters

Government ministers today must walk a tightrope of securing the UK's energy supply ('keeping the lights on') while maintaining the low cost of energy for homeowners and British business, and being seen to invest in reducing carbon emissions. Design a questionnaire to investigate the priorities of voters in your area, with regard to energy: 'On a scale of 1 to 5 how important to you is investing in the development of renewable energy?' The results of your survey may be of interest to your local MP.

In this section you will learn about:
- resource classification
- natural resource development over time
- the resource frontier

Have you ever been described as resourceful? If so, be flattered as this means you're skilful, productive, capable, of value – in short, you are of use! You're a human resource. That is what natural resources are all about – all those parts of the environment that are of value and use to us, such as raw materials, energy sources, climate and soils.

Classifying natural resources

Such is the huge scope and variety of natural resources that we need to classify them, for example, as mineral, energy or even aesthetic. We talk of stock resource evaluation – namely measured reserves, indicated reserves, inferred resources and possible resources (Figure **1**). However, the standard distinction is simply between stock and flow.

Stock resources are non-renewable – they are finite and therefore exhaustible (Figure **2**). For example, uranium for nuclear power and fossil fuels such as coal, petroleum (oil) and natural gas. They are created at rates considerably slower than their use.

Flow resources are renewable – they are ongoing. They are either immediately available, such as tidal advance and retreat and geothermal power (Figure **3**), or created at comparable rates to their consumption, for example, trees for timber or fuelwood.

Think about

What is a useful resource?

Resources have to be perceived, understood and evaluated as useful. Technology is relevant here. For example, nuclear fusion (rather than fission which is exploited today) could yield vast amounts of energy but we cannot yet harness it. Religious and cultural taboos may restrict the use of certain animals as a source of food, or their excrement as fertiliser. Political considerations may prevent the exploitation of contentious methods of mineral extraction such as open-cast mining. Another consideration is that resources don't necessarily have to be practical. Resources can be **non-utilitarian** with only a social value – such as our expectations of clean air and water, or access to unspoiled recreational areas.

🔽 *Figure 1* Stock resource evaluation

Mineral resources do not have demonstrated economic viability, but have reasonable prospects for eventual economic extraction		
Measured and *indicated* mineral resources can be estimated with sufficient confidence to allow evaluation of the economic viability of the deposit	*Measured resources*: geological conditions including grade of deposit can be confirmed to allow detailed mine planning	
	Indicated resources: geological conditions including grade of deposit can be reasonably assumed to allow some planning	
Inferred and *possible* mineral resources are estimated using limited information	*Inferred resources*: economic viability cannot be evaluated in a meaningful way	
	Possible resources: it is reasonably expected that the majority of inferred resources could be upgraded to *indicated* mineral resources with continued exploration	
Mineral reserves are the economically mineable part of mineral resources. In effect, the ore which will be delivered to the processing plant		
Proven and probable reserves are the economically mineable part of *measured* and *indicated* resources for which at least a preliminary feasibility study demonstrates that economic extraction is justified	*Proven reserves*: economic extraction of the *measured* resource is justified	
	Probable reserves: economic extraction of the *measured* and/or *indicated* resource is justified	

🔼 *Figure 2* The McArthur River Mine in Northern Territory, Australia, is one of the world's largest zinc, lead and silver mines

🔽 *Figure 3* A geothermal power station in Iceland; geothermal power generates 25 per cent of Iceland's electricity

Natural resource development over time

Mineral exploration

Mineral and energy resources have been mined and exploited for centuries. Their distribution is uneven so finding, identifying, mapping and evaluating the size of available stocks have long been key roles of geologists. Nowadays, their work is aided considerably by aerial and satellite **remote sensing** using, for example, infrared photography and radar scanning.

Mineral exploitation

Many of the world's mineral supplies are found in the rocks of **cratons** such as the Baltic Shield (Europe), Laurentian Shield (North America) and the Siberian Shield (Asia). Valuable concentrations also exist in the ranges of fold mountains that occur along some of their margins. Other sources include the veins and cavities of basic igneous rocks, and within river and marine sediments – hence the global spread.

In general, the northern hemisphere is more important for mineral production than the southern hemisphere. This is mainly because there is less land in the southern hemisphere and the cratons are much smaller. The production of some minerals is strongly concentrated in only a few places. For example, South Africa produces 75 per cent of the world's gold and 66 per cent of uranium comes from North America. Mineral exploitation depends on three main factors.

The mineral content of the rock

Mineral exploitation is invariably accompanied by the extraction of waste rock. *Low-grade* deposits produce a comparatively high proportion of waste. *High-grade* **ores**, in contrast, will be worked in the most isolated locations and difficult environments. For example, iron ore with an iron content of around 60 per cent is found in northern Sweden. This high iron content encourages its exploitation despite extreme environmental challenges of the remote, Arctic location.

Geological conditions

If a mineral is found at shallow depths, open-cast extraction (which is cheaper than shaft mining) may take place. For example, Malaysia's dominance in the mining of tin is undoubtedly related to the ease of extraction from **alluvial plains**. Water jets and dredges are used – the heavier tin being separated from the sediments by gravity. In contrast, no working tin mines remain in Cornwall. It is no longer economic because of the difficulty in extracting from narrow veins in hard granite masses.

Accessibility in relation to the markets

Transport costs are generally unimportant in the production of minerals such as gold because values are high and bulk so small. It is because of this that South Africa can produce so much of the world's gold. Lower value minerals (in relation to bulk) are much more strongly influenced by transport costs – hence the importance of the development of bulk carrier ships that enable low value ores to be transported over great distances at relatively low cost.

The resource frontier

Frontiers as boundaries, between the known and unknown or the settled and unsettled, raise exciting geographical questions. For example, how often do you hear of Alaska, USA as 'the last frontier'? Resource frontiers are usually exploited once transport links can be established. Railway maps of both Africa and Australia show repeated examples of routes leading inland from a port. These were often built specifically to transport an isolated mineral ore to a purpose-built coastal loading facility. The iron ore mines of Kiruna and Gällivare in northern Sweden were, at first, linked to the harbours of Luleå in the Gulf of Bothnia. The railway was then extended to the ice-free Norwegian port of Narvik, to allow all-year export.

ACTIVITIES

Describe and explain the relationships between mineral reserves, transport costs and markets.

STRETCH YOURSELF

The exploitation of natural resources destroys non-utilitarian ones. Discuss.

In this section you will learn about:
- the concept of a resource peak and sustainable resource development
- Environmental Impact Assessments

Resource peak

In sport, all competitors peak, but an athlete's peak performances will only be obvious in retrospect – judged against the decline that followed. So it is with any stock resource – at some point in time, production will peak and then what is known as 'diminishing returns' will follow.

Look at Figure **1**. North Sea oil and gas production clearly peaked in 1999 and has been in decline since then. Is this simply evidence of stocks running out? As with any resource peak, it is not quite as straightforward as this.

Any graph of stock resource production will follow the bell-shaped pattern known as 'Hubbert's Curve'. The top of the bell marks the point of maximum output, which is roughly half-way through the reserves. This peak also indicates when stocks are cheapest, beyond which a buyers' market becomes a sellers' market. Furthermore, because the highest quality and most easily obtained product is extracted first, the decline in production may be hastened by more challenging extraction and production processes. This results in higher prices that reduce demand.

The challenge for resource management is to predict these peaks before they happen so that plans for the long-term future can be made. However, this is far from straightforward because:

- attempting to measure the size and nature of future resource demand is undoubtedly difficult
- resources are considered exploitable under the current economic and technological conditions. Reserves are constantly reassessed as societies and technologies develop.

Population growth and economic development increase the demand for stock resources and are easier to predict. Hence the need to conserve resources by recycling, reducing waste and developing alternatives (see 5.19, 5.21).

Sustainable resource development

The word 'sustainable' is often wrongly interpreted as meaning simply environmentally sustainable. Fossil fuel exploitation, for example, becomes 'unsustainable' in terms of depletion of reserves, pollution and climate change. However, stock resources can, and many would argue should, be exploited sustainably – socially, environmentally, economically and politically – in short, a responsible mining approach. For example, the Rössing Uranium Mine in Namibia has been in operation for four decades producing 2.3 per cent of the world's uranium for the nuclear power industry (Figure **2**). Its management of plentiful reserves, welfare commitment to nearly 900 workers, profitability and wealth creation for the Namibian economy makes it an excellent illustration of truly sustainable, and so responsible, stock resource development.

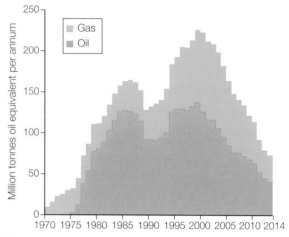

Figure 1 North Sea oil and natural gas production 1970–2014

Think about

Mineral peak

There will always come a point of maximum mineral production in any area, with decline in subsequent years as extraction and production becomes more challenging. This *mineral peak* is subject to changing demands which may, in turn, stimulate technological progress – for example, miners finding ways to extract deeper and lower grade ores with lower production costs. Demand for minerals can change rapidly. For example, asbestos demand grew through most of the twentieth century until public knowledge of the health hazards of its dust outlawed the use in construction and fireproofing in most countries. In contrast, petalite (first used as a raw material for glass-ceramic cooking ware) is a major source of lithium, which has increasing importance in the production of lightweight, efficient batteries in smart phones, tablets, digital cameras and electric vehicles.

The Rössing Uranium Mine, Namibia

The Rössing Uranium Mine occupies a 25 km² site in the Erongo region of the Namib Desert (Figures **2** and **3**). Rössing is operated by the Rio Tinto Group of companies – a TNC with mining operations in more than 40 countries. The uranium is mined from tough alaskite, then processed in a crushing plant to produce uranium oxide. The open pit currently measures 3 km in length, 1.5 km in width and 390 m in depth.

What makes Rössing sustainable?

Social sustainability

Social sustainability is met through its health, safety and welfare commitment to its workers. Attracting workers to such an arid, hostile and potentially dangerous location has prompted thoughtful solutions including:

◆ monthly checks for radioactive contamination, despite the ore being very low grade

◆ building the new town of Arandis with fully serviced housing, health clinic, hospital, shopping centre, town hall, primary school, pedestrianised areas, recreation centre and swimming pool.

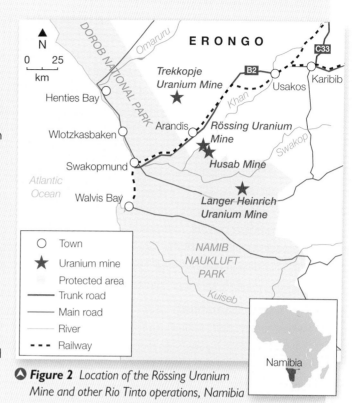

Figure 2 *Location of the Rössing Uranium Mine and other Rio Tinto operations, Namibia*

Figure 3 *The Rössing Uranium Mine, Namibia is one of the largest and longest-running open pit uranium mines in the world*

Environmental sustainability

Environmental sustainability is met through **stewardship**. Screening of the site is unnecessary because of the remote location and vastness of the surrounding hills. However, water is in short supply and management is crucial especially as dust needs to be controlled by continual spraying of the open pit, roads and tracks. All effluent is treated and solid waste pumped as a pulp into a nearby valley, which is dammed to retain the slurry and recycle the water. Leachates are controlled and local water sources are regularly tested for contamination.

Economic sustainability

Economic sustainability is ensured through Rio Tinto's financial backing even when uranium prices dip. Almost all (98 per cent) of the workforce are Namibians, who benefit from the stability of full-time contracts with training, career progression, insurance and pensions. Their standard of living is improved further by the Rössing Foundation, which was set up to strengthen links with the local communities that supply workers, goods and services. The Foundation has widened its remit to support charities, develop education facilities and even hosting an annual national marathon!

Political sustainability

Political sustainability is all about mutually assured benefits. Rössing is such a major operation that the revenue it generates through licences, **royalties** and taxation goes far beyond local or even regional significance. This, and other Rio Tinto operations in Namibia, generate a multiplier effect that the Namibian government values and nurtures.

Environmental Impact Assessment (EIA)

An Environmental Impact Assessment (EIA) is a process where impacts on the environment of a proposed development or project are gauged. Any unacceptable impacts can be reduced or avoided by taking relevant mitigation measures. An EIA, and the resulting Environmental Impact Statement (EIS), aims to ensure that decision-makers consider environmental impacts when deciding whether or not to go ahead with a project.

Look at Figure **5**. Nowadays, environmental law globally can be increasingly vague, complex and demanding. So much so that specialist contractors, such as the TNC Schlumberger (Figure **4**) can be hired to both prepare EISs and ensure their approval by the relevant planning authorities. Schlumberger is the world's largest oilfield services company, operating in 85 countries. But it is also one of many engineering and consulting companies diversifying into this important field. Each country will also have its own local contractors offering the same kind of service.

▶ **Figure 4** *The Schlumberger US executive offices, Houston, Texas*

The Environmental Policy of the European Union includes a mix of seven mandatory and discretionary procedures to assess environmental impacts	
Description of the project	This includes a site description, plans for construction, operations and decommissioning and all sources of environmental disturbance (such as air and noise pollution)
Alternatives that have been considered	An explanation of all alternatives, such as sources of fuel in a thermal or biomass power station (local, national or international)
Description of the environment	A list of all aspects of the environment that may be affected by the development, including flora, fauna, soil, water, cultural heritage
Description of the significant effects on the environment	This includes a definition of 'significant' in the context of the project. For example, in planning a wind farm, a significant effect might be noise pollution (the sound output of a wind turbine has been likened to a small jet engine). Another significant impact may be collisions with birds
Mitigation	Ways to avoid negative impacts should be developed (using information from row 4 of this table)
Non-technical summary	Preparation of a summary of the EIS, without jargon or complicated diagrams, to be understood by 'the informed lay-person' – for use in any public consultation during planning
Lack of know-how/technical difficulties	A final section advising on any areas of weakness in knowledge. This can then be used focus areas of future research

▲ **Figure 5** *EU directive for completion of an EIA*

Ⓢ

What is a comparative line graph?

A line graph is used to show continuous data, usually showing changes that take place over time. Data points are plotted using two axes and then connected by straight lines. The independent variable, for example time, is plotted on the horizontal axis and the dependent variable, such as energy production, on the vertical axis.

A comparative line graph shows more than one set of data on the same graph (see Figure **7** in section 5.3). In this instance the axes must be the same and follow the same scale.

Line graphs are useful when:

◆ continuous data is presented

◆ showing change over time (the gradient of the line presents a strong visual)

◆ important points, such as highest and lowest, need to be identified (and the exact value read).

ACTIVITIES

Ⓢ

1   Study Figure **1**.
   a   Describe the pattern of North Sea oil and gas production since 1970.
   b   What market forces could explain the production trends from 1987 to 1993?
   c   Why should it *not* be assumed that production will decline to zero after 2020?

2   Can stock mineral resource development ever be sustainable? (Use evidence from Rössing Uranium Mine in your answer.)

3   Using examples, outline what factors determine when a resource peaks.

4   Study Figure **5**. In which sections of the EU directive on the completion of EIAs would the use of local experts, such as the RSPB in the UK, be advisable? Give reasons for your answer.

STRETCH YOURSELF

What do you understand is the difference between the terms 'responsible' mining and 'sustainable' mining? You may find that www.iied.org/mining-big-progress-new-challenges-sector-searches-for-sustainability is of use.

In this section you will learn about:
- ◆ global patterns of production, consumption and trade/movements of energy
- ◆ global patterns of water availability and demand

When we flick a switch, we expect something to happen – the kettle to boil, the television to start or the room to fill with light. We are all familiar with the occasional power cut lasting an hour or two. In EMEs, such power outages are almost entirely a result of failures in the supply system rather than a shortage of power plants (Figure **1**). In contrast, in LDEs a lack of electricity-generating capacity is routine – for example, the cost to the African economy of load shedding is over 2 per cent of GDP. In EMEs, rapid industrialisation and increased living standards result in growing demand for energy. On 30 and 31 July 2012, high demand, inadequate supply coordination and transmission failures in India led to a massive power system collapse that affected 700 million!

Global patterns of production and consumption

The last 100 years, and in particular the last half-century, have seen dramatic increases in global energy production and consumption (Figure **2**). Stock resources have dominated but overall global production is shifting from coal and oil dominance towards natural gas and, since the 1950s, nuclear power. Large reserves of stock resources are often located in politically unstable parts of the world. So, resource-rich regions such as the Middle East are at the forefront of complex international relations (see 5.4).

The distribution of stock resources also varies globally because of local physical factors such as climate and landscape differences. Hydro-electric power (HEP) continues to be the largest contributor to global flow resources.

As a global society, we are still largely dependent on fossil fuels (5.12), which make up around 85 per cent of world energy consumption – this proportion is unlikely to change significantly in the near future. Nevertheless, the energy mix of individual countries does vary enormously with flow energy resources of greater significance in some regions, for example in Scandinavia. Look at Figure **3**. There remains a stark contrast in the global patterns of consumption with HDEs, particularly in North America, consuming far higher amounts than LDEs.

⊙ **Figure 1** *The North-east Blackout on 14 August 2003 affected 55 million people in the USA and Canada; it was caused by the failure of an electric generation plant in a Cleveland suburb*

Did you know?
Such is the frequency of power cuts in South Africa that newspapers regularly print 'load shedder recipes' for food you can prepare without electricity!

OIL	1985	1995	2005	2015
World	**2796.6**	**3286.4**	**3937.8**	**4361.9**
Saudi Arabia	172.1	437.2	521.3	568.5
USA	498.7	383.6	309.0	567.2
Russia	542.3	310.7	474.8	540.7
Canada	85.7	111.9	142.3	215.5
China	124.9	149.0	181.4	214.6

GAS	1985	1995	2005	2015
World	**1490.9**	**1905.7**	**2519.4**	**3199.5**
USA	427.9	480.9	467.6	705.3
Russia	376.3	479.3	522.1	516.0
Iran	9.2	30.4	92.1	173.2
Qatar	4.9	12.2	41.2	163.3
Canada	76.1	143.8	168.4	147.2

COAL	1985	1995	2005	2015
World	**4477.8**	**4640.9**	**6103.4**	**7861.1**
China	872.3	1360.7	2365.1	3747.0
USA	801.6	937.1	1026.5	812.8
India	157.5	289.0	429.0	677.5
Australia	166.6	248.1	378.8	484.5
Indonesia	2.0	41.8	152.7	392.0

◀ **Figure 2** *Main producers of stock energy resources 1985–2015 (million tonnes of oil equivalent)*

🔽 **Figure 3** *Energy consumption per capita (2014)*

Key
Tonnes of oil equivalent
■ >6.0
■ 4.6–6.0
■ 3.1–4.5
□ 1.6–3.0
□ 0–1.5

Pacific Ocean

Atlantic Ocean

Pacific Ocean

Indian Ocean

Global trade and movement of energy

The unequal distribution of energy sources means that there is a global trade in energy (Figure **4**). While flow resources (e.g. wind) cannot be physically moved, stock resources can be transported around the world. However, this is least economical for 'bulky' coal where friction of distance is an important factor. Overall, while energy prices can be volatile, it is increasingly costly to purchase and then to transport stock energy resources worldwide. This has led many countries to develop alternative domestic sources of energy and to review their energy mix (see 5.12).

Global patterns of water availability and demand

Look at Figure **5**. Around 1.2 billion people live in areas of physical **water scarcity** (see 5.5), while almost one-third of the population do not have access to safe drinking water. Population continues to increase as does demand for water from industries and agriculture. Unless there is significant investment in water conservation and extraction methods, increased water shortages are almost certain. Even in water abundant regions, water recycling measures are now largely commonplace.

🔻 **Figure 4** *Crude exports of oil 2017 and growth 2011–17 (predicted)*

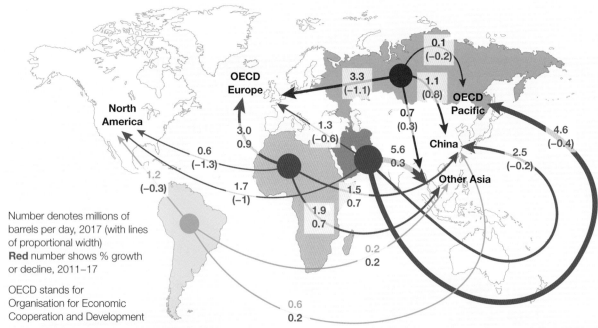

Number denotes millions of barrels per day, 2017 (with lines of proportional width)
Red number shows % growth or decline, 2011–17

OECD stands for Organisation for Economic Cooperation and Development

🔻 **Figure 5** *Areas of physical and economic water scarcity*

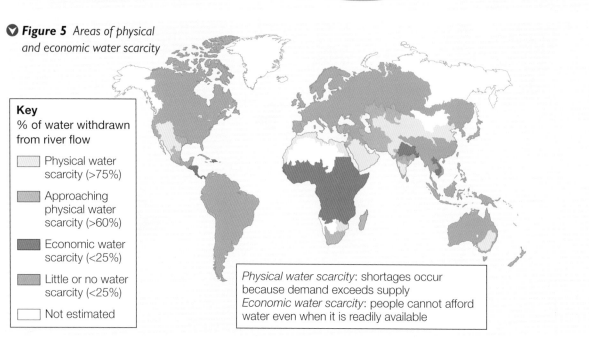

Key
% of water withdrawn from river flow

- Physical water scarcity (>75%)
- Approaching physical water scarcity (>60%)
- Economic water scarcity (<25%)
- Little or no water scarcity (<25%)
- Not estimated

Physical water scarcity: shortages occur because demand exceeds supply
Economic water scarcity: people cannot afford water even when it is readily available

What is a proportional pie chart?

A pie chart is a circle divided into segments to show the proportion of a total population (Figure **6**). Every 1 per cent contribution that a segment or category contributes corresponds to a slice with an angle of 3.6 degrees.

Pie charts are useful when:

◆ showing percentage or proportional data (they are visually effective)

◆ displaying data for six categories or fewer (too many and the naked eye struggles to interpret differences)

◆ presenting data in nominal or ordinal categories

◆ data categories have variation in size (similar-sized segments are difficult to interpret).

Proportional pie charts are used when comparing two or more sets of data when the categories are similar but there is a change in another variable, such as time or absolute total. The different-sized totals are shown by drawing the pie chart proportional to the totals they each represent. It is sometimes helpful if the same colours or shades are used to aid comparison between the charts even if this may mean that the second and subsequent pie charts are no longer presented in order of category size. Figures **6** and **7** use data from Figure **2**.

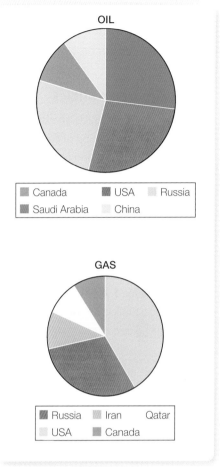

▲ **Figure 6** *Proportional pie charts showing five largest oil and gas producers (2015)*

ACTIVITIES

1 a What is load shedding? Suggest reasons why it occurs.
   b Do you think load shedding is more common in HDEs or LDEs? Explain your reasoning.

2 Study Figures **2** and **7**.
   a Construct a comparative line graph (see 5.2) to show changes in gas or coal production (1985–2015).
   b Describe the changes shown. Are there any surprises? Explain your answer.
   c Find a compound line graph on the internet. Construct a compound line graph to show changes in world energy production (1985–2015) for oil, coal and gas. Suggest reasons for any pattern(s).

3 Study Figures **2** and **6**
   a Draw a third pie chart to show the main producers of coal (2015). The chart should be proportional to that used for gas (2015). Assume the radius of the proportional pie chart for gas is 3 cm.
   b Compare your pie chart with Figure **6**. Suggest reasons for any similarities and/or differences.
   c Which nation(s) might be considered a world energy superpower? Explain your choice(s).

4 'Water, water everywhere and not a drop to drink.' Discuss the relevance of this statement with reference to global patterns of water availability and demand.

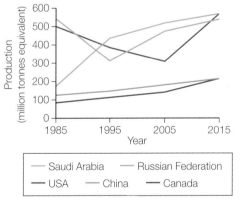

▲ **Figure 7** *Comparative line graph of oil production (1985–2015), using data from Figure **2***

STRETCH YOURSELF

To what extent are increases in world energy demand and consumption compatible with sustainable development? Consider the contrasting needs of countries in different stages of development in your answer. (Hint: read section 5.2 again.)

In this section you will learn about:
◆ the geopolitics of energy
◆ water resource availability and management

Energy geopolitics

At its peak in the 1920s, the British Empire was the largest formal grouping of countries in history. How did such a small island control nearly one quarter of the Earth's surface? While there is no one single reason, an understanding of energy **geopolitics** is central. (Geopolitics is the study of ways in which political decisions and processes affect the use of space and resources.)

For every successful international power throughout modern history, that success is based on energy resources. Look at Figure **1**. Coal, steam and the subsequent Industrial Revolution fuelled the British Empire and, in the last 100 years, oil helped establish the USA as a world superpower.

The dynamics of energy geopolitics means that power relationships can change quickly. In 2015, the self-styled Islamic State of Iraq and the Levant (ISIL) made between US$1 million and US$2 million each day from oil sales. Without its oil assets, its ability to wage conflict and to be self-sustaining would be severely compromised.

Since the programme of pit closures in the 1980s, the UK has imported an increasing proportion of its coal. Coal is largely used in the generation of electricity. The demand for coal increases when natural gas prices rise and it becomes less economic to burn gas in power stations (Figure **2**).

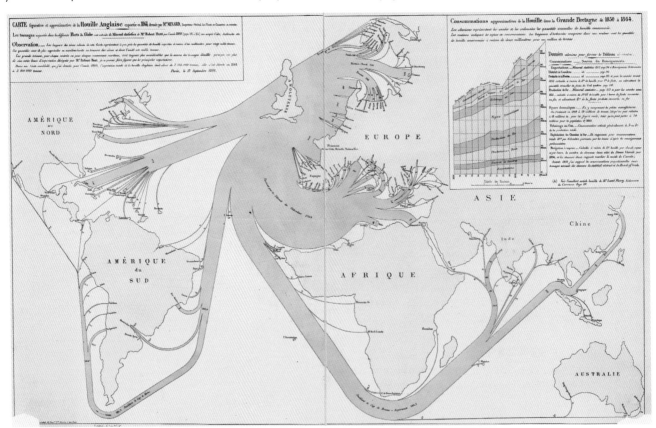

🔺 **Figure 1** *The British Empire relied on the strength of industrial production at home including the mining of coal. This historical flow line map produced by Charles Minard, an economic cartographer, shows British coal exports in 1864. Coal exports to Western Europe provided revenue for later expansion while existing British colonies such as Malta, Singapore and in the Caribbean were also large importers given their relative small population totals.*

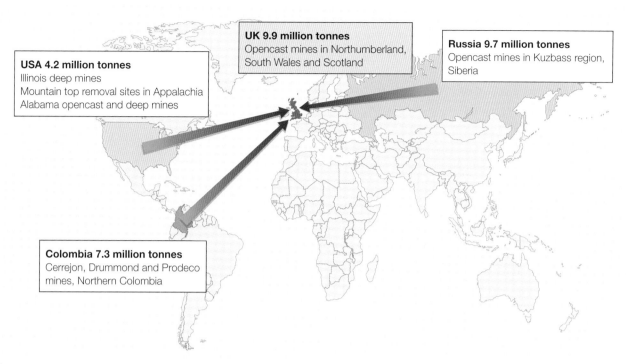

USA 4.2 million tonnes
Illinois deep mines
Mountain top removal sites in Appalachia
Alabama opencast and deep mines

UK 9.9 million tonnes
Opencast mines in Northumberland,
South Wales and Scotland

Russia 9.7 million tonnes
Opencast mines in Kuzbass region,
Siberia

Colombia 7.3 million tonnes
Cerrejon, Drummond and Prodeco
mines, Northern Colombia

Figure 2 *Where does the coal used in UK power stations come from (2014–15)?*

Alternative resources – shale gas

As oil reserves start to diminish, the discovery and exploitation of vast shale oil and gas reserves (**fracking**) are likely to help maintain the USA's position as the world's leading geopolitical power (Figures **3** and **4**). The USA has become a shale gas exporter – it exported its first shipment to Scotland in September 2016. Already, oil imports have been cut by around 50 per cent compared to 2007. However, many argue that as in the exploitation of other stock energy resources, notably coal, there has been insufficient thought as to the implications of the so-called 'shale revolution'. Aside from the significant environmental concerns, there are substantial geopolitical implications. For example, in contrast to the geography of oil reserves (Figures **2** and **4**, 5.3), exploitable oil and gas shale resources are far more widespread across the globe – including across Asia and Africa – and are located closer to centres of population. Particularly in LDEs, this raises the possibility of local centres of production, meeting local needs and even managed by local authorities rather than governed by global decision-making of energy TNCs.

|←——————— 20 cm ———————→|

Figure 3 Shale rock

Figure 4 *A fracking drilling rig in Colorado, USA*

Water geopolitics

For over five thousand years, water all over the world has been seen as a source of cooperation and peace. It has enabled the growth of settlements, agriculture and industries – the very existence of humanity has depended on a reliable water supply. Nowadays, however, climate change and population growth are creating an increased demand from dwindling water sources. More worrying is that, unless countries adopt legal and diplomatic policies towards water usage, such dramatic reductions in water levels could lead to so-called 'water wars'.

Look at Figure **5**. Lake Chad lies on the border between Chad and Cameroon – find it in an atlas. You may have to look hard. Fifty years ago, it had a surface area of 25 000 km². Today, it has less than 2000 km² and in your lifetime it may disappear altogether. Climatic changes and large and unsustainable irrigation projects built by the adjacent countries of Chad, Cameroon, Niger and Nigeria are major causes of the dramatic shrinkage. Overgrazing can result in loss of vegetation and lead to deforestation – these factors can combine to cause a drier climate. The changes in the lake have resulted in crop failures, livestock deaths, soil salinity, collapsed fisheries and other 'poverty related environmental activities' (see also 5.9).

Water scarcity

An assessment of water scarcity in an area requires more than a measurement of available/renewable water supplies. Water scarcity is a more sophisticated tool that examines the population-water equation. The answer may differ widely dependent on the geographical area studied – so, in countries that have low population densities water scarcity may 'appear' not to exist. Conversely, in settlements where demand significantly exceeds supply, the lack of usable water may be far more acute (e.g. much of sub-Saharan Africa).

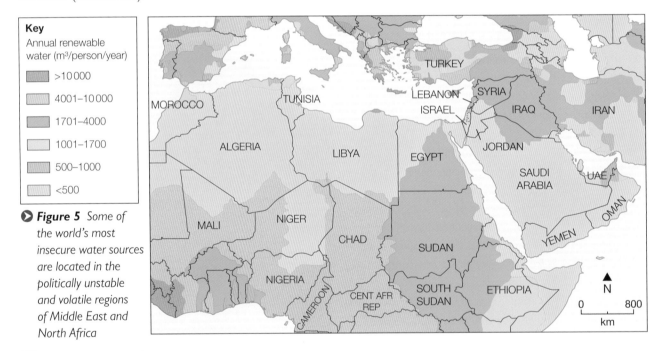

Key

Annual renewable water (m³/person/year)

- >10 000
- 4001–10 000
- 1701–4000
- 1001–1700
- 500–1000
- <500

▶ **Figure 5** *Some of the world's most insecure water sources are located in the politically unstable and volatile regions of Middle East and North Africa*

River management

Rivers are also at risk of severe shrinkage and/or pollution. Some of the world's greatest rivers, including the Ganges and the Nile, will see a reduction of at least 25–30 per cent in their flow over the next 30 years as a result of human activities and climate change. This will place increased strains on the collaborative management of transboundary river basins as actions of one state upstream will have consequences in another state downstream. For example, the Nile is loosely managed by ten states through a number of institutions at continental level (e.g. the African Union) and sub-regional level (e.g. United Nations Economic Commission for Africa). In addition, several states have bilateral agreements and treaties in place that date back to the colonial era.

The Nile Basin Initiative (1999) is the first time that an all-inclusive basin-wide institution has been set up. It is an attempt to implement a legal and cooperative framework for sustainable management of the Nile Basin. As water resources are squeezed, it faces real challenges in ensuring each riverside land user receives its 'fair share'.

While populations of these regions will continue to increase and need feeding, there will be less water for irrigation and grain production, which is predicted to decrease. With less supply, prices are set to rise and the two largest nations on the planet, India and China, may be forced to become net food grain importers for the first time. The knock-on effects will be global. The poorest could starve, food riots could become more common-place and fears over *food security* could lead to international conflict.

Nevertheless, reduced water availability does not have to lead to conflict. Where countries plan ahead and cooperate across international boundaries, such as in Singapore or Senegal, **water security** can still be guaranteed.

Did you know?
Three American TNCs (Cargill, ADM and Bunge) control 90 per cent of the global grain trade.

Did you know?
Rising grain prices were a significant causal factor in the Arab Spring uprisings of 2010.

ACTIVITIES

1 Study Figures **1** and **2**.
   a Compare and contrast UK coal net imports and exports from the time of the British Empire (Figure **1**) to today (Figure **2**).
   b Suggest reasons for any differences.

2 What is the link between water and energy resources, and geopolitics?

3 Read the example of Gazprom in 5.14. How has geopolitics influenced the export of natural gas from Russia?

4 Study Figure **5**.
   a Describe the pattern of water insecurity on the map.
   b Using evidence from the map and an atlas, suggest reasons why this should not necessarily be the case.

5 Forty per cent of Africa's population lie within the Nile Basin. Egypt has traditionally benefited most from the Nile even though, for example, 85 per cent of the flow originates in Ethiopia. Using the internet, research the US$70 billion Toshka Project. How is this an example of water geopolitics?

Figure 6 *Despite a recent international tribunal (UNCLOS) ruling against it having any rights in the South China Sea, China still claims almost all of the area, including the potentially oil- and gas-rich Spratly and Paracel Islands*

STRETCH YOURSELF

Look at Figure **6**. Research the UNCLOS ruling (July 2016) and the South China Seas dispute. Why should the Philippines, Japan and other south-east Asian countries be concerned over China's increasing interests in the South and East China Seas? Consider issues of energy security, potential maritime energy reserves and also the increased importance of the region's sea lanes for transportation of energy reserves from Russia and the Middle East into Asia.

In this section you will learn about sources of water, water demand and water stress

As with other natural resources, the Earth's water is distributed unevenly and access to it is determined by a range of socio-economic and political factors such as governance, development of infrastructure and environmental legislation. *Water security* is far more than just living somewhere where it 'rains a bit' (Figure **1**)!

Sources of water

Look at Figure **2**. Although 96.5 per cent of all water on Earth is salt water, the remaining water sources should still be enough to meet our needs. The majority of freshwater is in the icecaps of Antarctica and Greenland. Less than one per cent of the world's freshwater is accessible for direct human use. This is the water from lakes, rivers, surface-water stores (reservoirs) and groundwater stores that are economically accessible (also see *AQA Geography A Level & AS Physical Geography*, 1.2).

▲ **Figure 1** *Each year, World Water Day raises awareness of a different aspect of the importance of water; it is supported by bodies such as UNICEF and UN-Water*

Water source	Water vol (km³)	% of freshwater	% of total water
Oceans, seas and bays	1 338 000 000	–	96.54
Ice caps, glaciers and permanent snow	24 060 000	68.6	1.74
Groundwater (total)	23 400 000	–	1.69
Fresh	10 530 000	30.1	0.76
Saline	12 870 000	–	0.93
Soil moisture	16 500	0.05	0.001
Ground ice/permafrost	300 000	0.86	0.022
Lakes (total)	176 400	–	0.013
Fresh	91 000	0.26	0.007
Saline	85 400	–	0.007
Atmosphere	12 900	0.04	0.001
Swamp water	11 470	0.03	0.0008
Rivers	2120	0.006	0.0002
Biological water	1120	0.003	0.0001

Did you know?

Agriculture accounts for 70 per cent of all water used. When used efficiently, water improves productivity and reduces poverty for millions.

◀ **Figure 2** *Estimate of global sources of water*

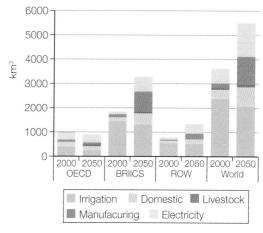

OECD (Organisation for Economic Cooperation and Development)
BRIICS (Brazil, Russia, India, Indonesia, China, South Africa)
ROW (rest of the world)

Figure 3 *Global water demand in 2000 and 2050*

Think about

Water resource optimism?

Neo-malthusianism suggests that the human relationship with water is that of consumers exploiting a fragile nature (e.g. the use of irrigation within agribusiness). Against this view, some suggest that any natural 'resource' is determined by both current needs and technological capability. New technologies bring new resources into being and may even change what we consider to be 'nature'. Water optimists suggest that humans are producers and not just consumers – deep-lying aquifers were not considered a water resource 100 years ago, yet they are commonly referred to as 'reserves' today.

Components of water demand

Look at Figure **3**. The demand for water is predicted to increase by around 55 per cent by 2050 due to population increase, urbanisation and improving standards of living. In particular, emerging LDE and EME economies will cause demand to soar. This is largely a result of their rapid expansion of water-hungry economic activities such as manufacturing, intensive agriculture and coal-fired power stations, as well as from the demands of a growing and more affluent population. **Water stress** and *water scarcity* are inevitable.

Water stress

Water stress is when renewable water in a country falls below 1700 m^3 per person. It occurs when the demand for water exceeds available water reserves or when poor quality restricts its use. Water scarcity is when annual water supplies fall below 1000m^3 per person. Despite these terms being used frequently, there is no consensus on how these terms should be defined or indeed measured.

As the largest centres of population, urban areas are facing the greatest challenges posed by water stress. Easily available surface water and groundwater sources have already been depleted, so cities are going further (such as Los Angeles – see 5.22) or digging deeper (e.g. London – see 5.6) to access water. In the future, it is probable that cities will have to depend on innovative solutions (e.g. by collecting far more rainwater in urban water catchments) or use of more advanced technologies such as desalination or reclaimed water to meet their water demands.

The Organisation for Economic Cooperation and Development (OECD)

Established in 1961, the OECD represents 34 of the world's richest nations. It aims to promote and support policies that will improve the economic and social well-being of all people. The OECD is an important independent 'gatherer' of global socio-economic data and their website, particularly the excellent *iLibrary*, should be on every Geography A-Level reading list!

STRETCH YOURSELF

In response to severe drought and climate change in Australia (see 5.6), several cities have constructed large scale desalination plants. Using the internet, write a report evaluating the impact of these desalination plants. Look at 5.7 Another View for a starting point.

ACTIVITIES

1 Study Figure **1**.

  a What do you understand by the term 'water security'? Write your answer in no more than 50 words.

  b Why is water security a difficult term to define in just a few words? (Hint: consider the contrasting needs of African women, who may walk over eight hours each day to collect water, to thirsty urban populations in HDEs – clean water is a basic right and fundamental need for all of the world's population.)

2 Do HDEs *really* need more water than the rest of the world?

3 Should we be optimistic or pessimistic about the future availability of freshwater resources?

In this section you will learn about the relationship of water supply to climate, geology and drainage

Access to safe water is seen as a fundamental need and basic human right. So what happens when the taps or bore holes run dry? In 1976, Britain experienced the worst drought in 150 years (Figure **1**). Rivers and **aquifers** dried up, soil cracked and blew away, heathland caught fire and the unthinkable nearly happened – water supplies almost ran out. Standpipes were introduced on streets, water was rationed to schools and businesses (Figure **1**), and people were even encouraged to share baths! Will this ever happen again? If so, a plan is needed…

⚫ **Figure 1** *Residents collect water from a standpipe in North Devon during the infamous 1976 drought*

Water supply and physical geography

Climate is the major geographical control on water supply. Low air pressure is associated with precipitation, although intense low pressure weather systems, such as tropical storms, are likely to result in excessive overland flow rather than infiltration, which would recharge groundwater stores or aquifers. Summer precipitation is less effective at recharging water sources as the ground tends to be harder, which discourages infiltration. There is also increased water loss from evapotranspiration. Spring may bring increased snow melt and most storms occur in autumn.

Impermeable surfaces such as clay will act as water-shedding surfaces, while permeable and porous rocks such as chalk will act as aquifers. Alternating bands of hard and soft rock may lead to natural water gathering basins or **synclines**. When groundwater is trapped beneath such synclinal layers of rock it is called an **artesian basin** (see 5.7). Hard, resistant rock might lead to areas of upland and subsequent relief rainfall but, on the opposite slope, a rain shadow might reduce precipitation totals. River systems that have a higher drainage density and a range of inputs, such as from groundwater flows, are more efficient water gatherers than those that rely on one water source alone. Without monitoring and controls, any water supply can become polluted from industry, agriculture or domestic uses (see 5.10). *Over-abstraction* from groundwater and rivers can also cause ecological damage, including intrusion of saltwater into aquifers at the coast and subsidence in cities as aquifers become depleted.

Did you know?
In 1976, Britain appointed its first ever minister for drought – the day after he was installed, it rained for the next two months!

The Murray–Darling Basin

So, what is the plan to tackle water shortages? The UK has largely rejected the idea of a national water grid that would connect areas of water surplus to areas of shortage – it is simply too expensive and cost-prohibitive for the now privatised water companies. In contrast, the Australian Government announced a National Plan for Water Security in January 2007 which focused on the Murray–Darling Basin (MDB).

QUEENSLAND

Brisbane

SOUTH AUSTRALIA

Murray–Darling Basin watershed

NEW SOUTH WALES

Darling R.

Adelaide

Murray R.

Sydney

Canberra

VICTORIA

Melbourne

Southern Ocean

N

0   400
km

▶ **Figure 2** *The Murray–Darling Basin*

Look at Figure **2**. The Murray–Darling Basin:

◆ is the size of France and Spain combined

◆ covers 14 per cent of the Australian land mass

◆ provides 75 per cent of Australia's water (85 per cent of the country's irrigation water)

◆ provides 40 per cent of the nation's farm produce, worth A$13.6 billion

◆ is home to two million people, with millions more outside the MDB – in the cities of Sydney, Melbourne, Brisbane and Adelaide – who all depend on it for food and water.

There has been a five-fold increase in water extraction from the MDB since the 1920s – matching the increase in Australia's population. Average annual rainfall across the MDB is 480 mm but this hides significant geographical differences (Figure **3**). Additional inputs into the MDB include groundwater extraction from the Great Artesian Basin and the Murray Groundwater Basin. Outputs from the MDB include high evaporation rates on the eastern edges of South Australia in the west. The climate also varies over time. Look at Figure **4**. Aside from the impacts of climate change, El Niño brings significant rainfall differences that have knock-on effects on the management of the whole MDB area.

With competing demands on water at a record high, managing the dwindling water resource of the MDB to meet the needs of contrasting stakeholders (particularly farmers) is a real challenge. Already, some areas face almost continual drought, while water abstraction upstream has permanently damaged wetland ecosystems near the mouth of the Murray River (Figure **5**). All of this makes one certainty – the recently formed Murray–Darling Basin Authority (2008) will continue to face a huge challenge balancing economic, social, environmental and political priorities.

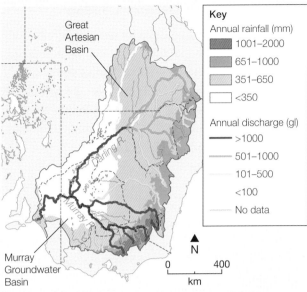

◆ **Figure 3** *The Murray–Darling Basin; average rainfall and river flow and the two main groundwater sources*

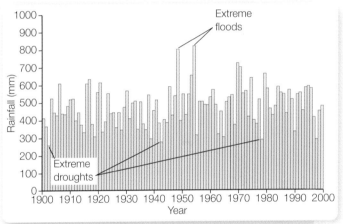

◆ **Figure 4** *The variability of rainfall in the Murray-Darling Basin, 1900–2000*

▶ **Figure 5** *Likelihood of drought in the Murray–Darling Basin*

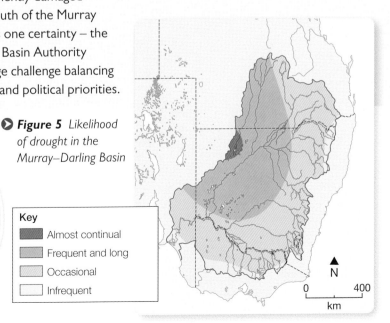

Did you know?

Since 1980, 6500 climate-related disasters have been recorded. Today, an average 375 million people are affected by extreme weather events caused by climate change.

285

Water supply to Greater London

The Thames (drainage) Basin is classed as a region suffering 'severe water stress'. This means that the majority of accessible water resources are already used to supply water for homes and businesses – there is little spare capacity. Thames Water supplies 2.6 billion litres of drinking water to nine million people across London and the Thames Valley every day and demand is increasing as the population continues to grow. Geographical factors influence supply.

Climate

With an average annual rainfall of 690 mm (the national average is nearly 900 mm), the region is one of the driest areas in the UK. Only 250 mm of this annual rainfall is available for use by the environment and population as the remainder is lost through evapotranspiration. Frequent winter rainfall, rather than occasional storms, is particularly important as it allows groundwater stores to be recharged. Typically, it takes two winters of below average rainfall to lead to drought conditions.

Geology

Look at Figures **6** and **7**. The geology of the Thames Basin is particularly significant. The chalk acts as an aquifer and provide base flow into the tributaries of the River Thames. Additionally, around one-third of the water supply is pumped directly from the aquifers via bore holes. Many of London's fountains once used natural pressure to propel the water upwards (Figure **8**).
As water levels have dropped, bore holes are now as deep as 200 m, meaning that hydrostatic pressure cannot be relied upon to power the fountains.

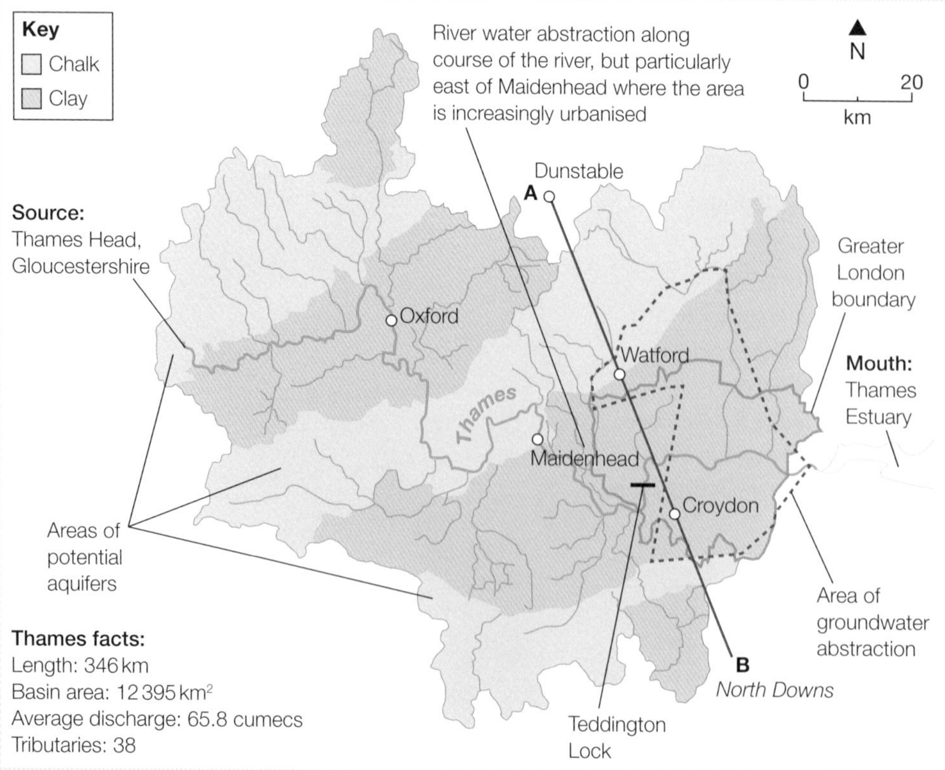

> **Did you know?**
> The Thames (drainage) Basin accounts for 10 per cent of the land area of England and Wales yet is home to almost 25 per cent of the population.

> **Did you know?**
> Londoners use 10 per cent more water a day than anywhere else in the UK.

Key
☐ Chalk
☐ Clay

River water abstraction along course of the river, but particularly east of Maidenhead where the area is increasingly urbanised

N
0 — 20
km

Source:
Thames Head,
Gloucestershire

Dunstable
A

Oxford

Watford

Greater
London
boundary

Thames

Maidenhead

Mouth:
Thames
Estuary

Croydon

Areas of
potential
aquifers

Area of
groundwater
abstraction

B
North Downs

Thames facts:
Length: 346 km
Basin area: 12 395 km²
Average discharge: 65.8 cumecs
Tributaries: 38

Teddington
Lock

◁ **Figure 6** *Water sources and geology of the Thames Basin (see Figure **7** for cross section between Dunstable (**A**) and the North Downs (**B**))*

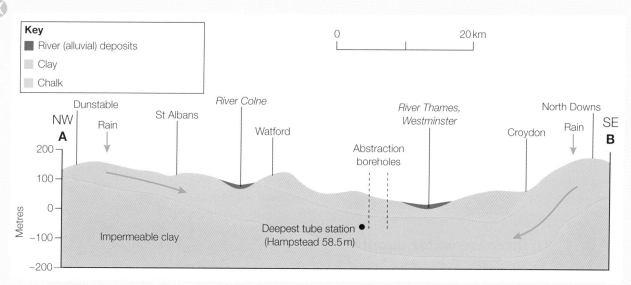

Key
■ River (alluvial) deposits
▨ Clay
▦ Chalk

0   20 km

⬛ **Figure 7** *Cross-section of the Thames Basin (between Dunstable (**A**) and The North Downs (**B**) on Figure **6**)*

Drainage

Around 80 per cent of London's water supply is taken from the River Thames upstream of Teddington Weir. As the majority of this river water is supplied by aquifers, ground water is the most important source of water. The other source is from the tributaries of the Thames that flow over impermeable clay outcrops where rates of overland flow are higher. The remaining 20 per cent of London's water supply is from groundwater abstraction adjacent to Greater London.

⬛ **Figure 8** *London lies within the centre of an artesian basin. Water flows down into the bottom of the chalk basin where it is confined under hydrostatic pressure. The fountains of Trafalgar Square once used this natural pressure for flow.*

ACTIVITIES

1 Study Figures **2** to **5**.
   a Using an atlas, identify areas of upland, tropical rainforest and semi-desert within the area of the MDB. How will each natural environment impact on water availability?
   b Study Figure **5**. Suggest reasons for the pattern of drought shown.
   c Read sections 2.8 and 3.7 again. What is a GIS? How might this tool help in the analysis of spatial data such as those shown in Figures **2** to **5**?

2 Does Greater London have a sustainable fresh water source? Suggest reasons for your answer with reference to the region's physical geography.

STRETCH YOURSELF

What caused the 1976 UK drought? Was it caused by physical and/or human factors? Are we ready for the next severe drought? Try to include specific references from news sites on the internet.

In this section you will learn about strategies to increase water supply including catchment, diversion, storage and water transfers and desalination

California's Central Valley is sinking by as much as 30 cm each year! Farmers are pumping water from groundwater stores faster than they can be replaced. The pore spaces in the aquifer are now empty which is causing the clay to collapse and the ground above to sink. Ironically, this subsidence is damaging the wider water supply network that the farmers and communities rely on – irrigation channel walls are buckling and pipes cracking. Californians are digging deep in their search for new water – some wells are over 760 m deep! (See 5.22.) But they are not alone. All over the world, populations are searching for new water supplies.

Did you know?
The water from your tap could contain molecules that dinosaurs drank!

Strategies to increase water supply

Catchment

Construction of more **water catchment** areas, for example, wetland restoration or reafforestation, allows more rainwater to be collected. However, it is not just about more capacity – management schemes work with different users of the land. For example, water companies can block drainage channels to help to retain water, while pastoral farmers are encouraged to shift to arable farming where runoff (and so silting up of channels) is less, especially when overgrazing has previously taken place.

Storage

Surface water storage can be increased by the construction of dams that then facilitate water diversion. Dams do not necessarily have to be permanent features or only constructed in isolation. For example, during the wet season, inflatable dams made of thick, laminated rubber and nylon tubes, block stream flow and raise the water level (Figure 1). In drier months, water may then be diverted onto fields for flood irrigation. These dams are clamped securely to concrete foundations across a streambed and can be quickly filled with air or water to create a barrier and then subsequently deflated to lie flat.

▼ **Figure 1** A rubber dam stretches across the Tankabati River in Bangladesh; the inflatable dam is 5 m high and helps deliver water to irrigate rice fields

▼ **Figure 2** Pont du Gard aqueduct, France; the Romans built long stone channels (aqueduct) aqua = water and ductus = channel), to carry clean water from nearby hills to the towns

Diversion

Water can be removed, or abstracted from its natural course or location by canal, pipe or other conduit. In California alone (see 5.22), there are more than 25 000 points where water is diverted – at scales ranging from the irrigation of a single field – to providing drinking water for an entire urban area. Any water abstraction should be carefully managed to avoid long-term environmental damage.

Water transfer

Transporting water from areas of excess to areas of supply is not a new strategy (Figure **2**) but schemes such as those in California (5.22), China (pages 290–1) and Lesotho (5.8) illustrate the scale and extent of current water transfer projects.

Desalination

Technologies such as **thermal distillation** and **reverse osmosis** allow large-scale desalination of sea water or brackish groundwater (Figure **3**). However, this comes with significant economic and environmental costs. Desalination has high capital costs (machinery is expensive), is energy-intensive and must dispose of highly concentrated saline by-products. Despite dividing opinion (see Another View), desalination plants have been built in several drought prone areas including the Middle East and United States (Figure **4**).

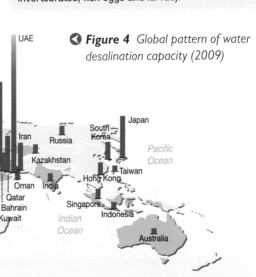

⬥ **Figure 3** *Sorek is the world's largest seawater desalination plant near Tel Aviv, Israel; it became operational in 2013 and provides 20 per cent of the water consumed by the country's households and cost US$500 million*

Did you know?
Desalination provides 70 per cent of drinking water in Saudi Arabia!

ANOTHER VIEW

Desalination – is it worth its salt?

Desalination divides opinion. Supporters point out that, in areas suffering from drought, there are few other sustainable options. It reduces reliance on aquifers and the source of brackish water is likely to be closer than freshwater, thus reducing the need for building an expensive and potentially environmentally harmful pipeline. Opponents explain that, aside from the huge economic costs, there are environmental risks. These include contamination of freshwater aquifers by brine water, high energy (running) costs of the plant and damage to fisheries (the ocean water intake pipes can suck in and kill marine invertebrates, fish eggs and larvae).

◀ **Figure 4** *Global pattern of water desalination capacity (2009)*

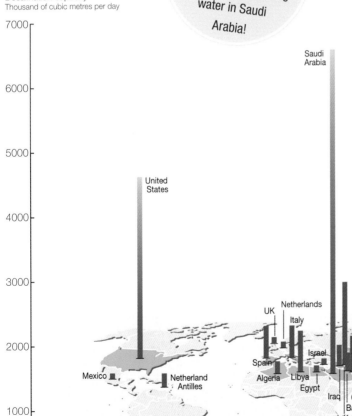

Desalination capacity
Thousand of cubic metres per day

Note: only countries with more than 70 000 cubic metres per day are shown.

Sources: Pacific Institute, The World's Water, 2009.

A large-scale water transfer project – China

The Yangtze is the third longest river in the world and its river basin in central southern China produces half of all the country's grain. Look at Figure **7**. In 2007, the almost unthinkable happened – the Yangtze almost ran dry. Drought is increasingly prevalent in China. The effects of drought are greatly increased as a result of overabstraction, particularly for industrial uses.

As part of China's industrial growth, some of the country's most productive farmland in the southern regions has been converted to industrial zones (see Figure **5**). Not only has this caused high levels of water pollution (e.g. heavy metals and toxic wastes) and increased demand for water but it has also forced the drier northern regions (see Figure **6**) to grow more. Today, these regions face unprecedented water shortages. One result is increasing reliance on groundwater stores. However, this is unsustainable as the aquifers are being used at a quicker rate than they are recharged.

Look at Figure **8**. The South–North Water Transfer Project is one of China's so-called **megaprojects** (see 5.11). It is a massive integrated and ambitious project that started construction in 2002 and is planned to continue until 2050. It aims to abstract almost 45 billion m³ of water a year from the Yangtze River and its basin, and move this to the drought affected north using a 2400 km network of tunnels and canals.

Did you know?

China has 20 per cent of the world's population but only an estimated 7 per cent of its freshwater reserves.

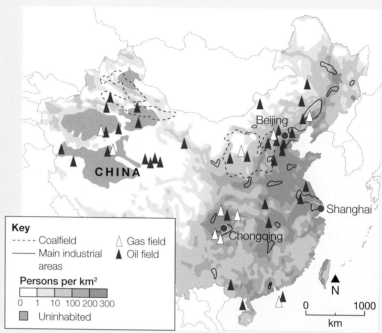

Figure 5 China's population and industry are also unevenly distributed; both tend to be concentrated on the coast and/or on gas, coal and oil fields.

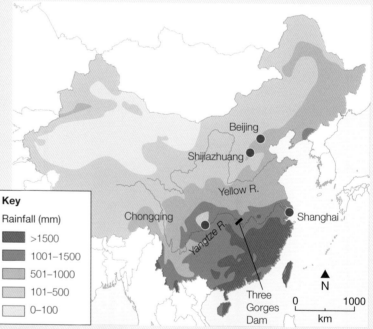

Figure 6 China's rainfall is unevenly distributed; the drier north, including megacities such as Beijing, rely on water transfers from the wetter southern regions

The project will help to alleviate a drying north but critics argue that it is making a poisoned south. The dramatic decline in discharge has affected the Yangtze River ecosystem that is already negatively affected by the Three Gorges Dam. Socio-economic impacts include the loss of livelihoods as some residents have been forced to relocate. At a cost of US$62 billion, the project is hugely expensive and it has been described as a 'white elephant' that benefits the urban rich and further exacerbates existing regional inequalities.

▲ **Figure 7** *The Yangtze River 'flowing' through Chongqing during the historical drought of 2011*

▼ **Figure 8** *China's South–North water project*

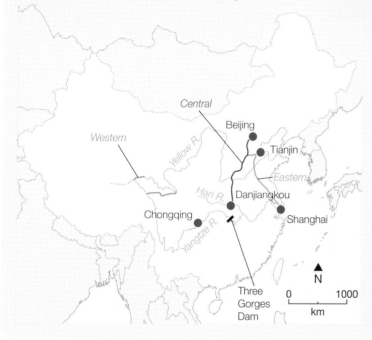

——— Western (under review): If built, this scheme would divert 20 billion m^3 of water from three tributaries of the upper Yangtze. Such is the scale and engineering required (300 km of tunnels, 200 m high dams, permafrost, tectonic activity) that it may never be built.

——— Eastern (1155 km): Completed in 2013. Includes two 9.3 m diameter tunnels 70 m under the Yellow River.

——— Central (1267 km): Completed in 2014. The expansion of the reservoir at Danjiangkou has forced around 350 000 people to be resettled.

ACTIVITIES

1. Choose *three* strategies for increasing water supply. List possible advantages and disadvantages of each (e.g. economic cost, environmental impact and issues of sustainability – particularly for LDEs).

2. Complete the summary table of strategies used to increase water supply in California (look at 5.22, California case study).

Strategy to increase water supply	Example from California
Catchment	
Storage	
Diversion	
Water transfer	
Desalination	

3. Study Figure **4**. Explain the global pattern of desalination capacity?

4. Compare and contrast Figures **5**, **6** and **8**. Suggest reasons for any relationships.

5. Study Figure **8**. To what extent is China's mega water transfer project 'necessary'?

Did you know?
More than half of China's 50 000 rivers have dried up over the past two decades.

STRETCH YOURSELF

Read sections 5.4 and 5.5 again. How do China's attempts to increase water supply illustrate the ideas of 'water geopolitics' and 'water security'?

In this section you will learn about environmental impacts of a major water supply scheme incorporating a major dam and/or barrage and associated distribution networks

Where does your drinking water come from? In Oceania, fewer people have access to piped water than in sub-Saharan Africa. The problem for many of these island communities is not a lack of rainwater but capturing it across the many low-lying atolls. Desalination plants and other large-scale infrastructure projects can be found in the Pacific islands.

However, many local communities cannot afford the capital investment nor the fuel to run them. Consequently, Pacific island states such as Tuvalu are dependent on rainwater harvesting, but this does not meet supply and imported bottled water is the primary source for many of the island's 11 000 residents.

The Lesotho Highlands Water Project (LHWP)

Lesotho is a poor country that is completely surrounded by the much wealthier and more powerful South Africa. However, despite an almost complete lack of natural resources, Lesotho has managed to trade the one resource that it has in abundance but its neighbour is lacking – water. Today, the Lesotho Highlands Water Project (LHWP) is one of the world's largest water supply/transfer schemes (Figure 1) but it has not been delivered without significant environmental impacts.

Background

Look at Figure 1. Lesotho is a mountain kingdom with an abundance of water. Rainfall is variable across the country and also from one year to the next, although most of the precipitation falls within the seven month wet summer season from October to April (Figure 2). The Maluti Mountains cover an area exceeding 18 000 km^2 of the Drakensberg range and are the source of the Senqu (Orange) River and its tributaries. Together with the widespread mountain wetlands, the highlands region forms an important part of the southern African region's water resources. Lesotho's water resource far exceeds current and future requirements – present patterns of consumption are about 2 cumecs, while the total available is around 1500 cumecs. The steep slopes, fragile soils and unreliable rainfall make the dominant economic activity of subsistence farming vulnerable. Consequently, when the LHWP was finally signed by Lesotho and South Africa in October 1986, it was widely welcomed. In this binational agreement, South Africa agreed to pay the water transfer costs and Lesotho agreed to finance the hydroelectric power projects. The ambitious project is planned over several phases and involves up to five dams, 200 km of tunnels and water pumping and transfer stations.

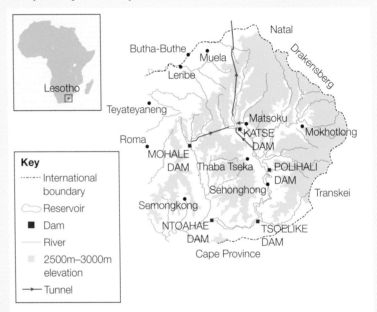

Key
- ⸱⸱⸱⸱⸱ International boundary
- ⬭ Reservoir
- ■ Dam
- — River
- ▨ 2500m–3000m elevation
- ↦ Tunnel

	Phase IA	Phase IB	Phase II	Phase III	Phase IV
Dam name	Katse	Mohale	Polihali	Tsoelike	Ntoahae
Dam height (m)	185	145	164	158	
Storage capacity (million m³)	1950	958	2200	2223	
Water transfer capacity (cumecs)	16.9	10.1	28.0	8.6	
Completion date	1998	2004	2024	On hold	On hold

🔺 **Figure 1** *The Lesotho Highlands Water Project (LHWP)*

Environmental Impacts

Lesotho has a highly vulnerable environment. It is characterised by steep slopes, fragile soils, soil erosion (including gullying) and climate variability. There is already significant pressure on the land from (largely subsistence) farming and unplanned settlements. Nevertheless, Phase 1A of the LHWP began in 1986 without an environmental impact assessment (EIA – see 5.2) for the overall project. Latterly, environmental concerns have been taken more seriously and EIAs have been in place since Phase 1B. However, significant environmental damage has already occurred and the impacts of such a large-scale water project can never be fully mitigated.

Loss of farm land

The project is taking over Lesotho's most fertile land in the mountains, including the Mohale area, which was the only region in the country that produced an agricultural surplus. Less than 10 per cent of Lesotho is suitable for farming and the loss of farmland is placing increased strains on **food security**, particularly as two-thirds of the people living in affected areas depend on locally produced food crops. New settlements to house those displaced from the flooding have far less access to natural resources.

Reduced access to natural resources

Reduced access to woodland, herbal medicines, wild vegetables and grazing lands is not just as a direct result of flooding – the reservoirs also act as barrier to movement for both people and cattle and exacerbate the problem of poverty-related environmental activities.

Habitat destruction

Habitats of endangered species such as the Maluti minnow and bearded vulture are being destroyed. A threatened endemic plant, the Spiral Aloe, is affected by the project, as too are rare bird species that nest in the area. Ecosystems are threatened downstream.

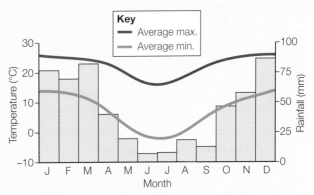

⬤ *Figure 2* *Annual temprature and rainfall graphs for Lesotho*

Downstream damage

There have been reductions in wetlands habitat, less water available downstream for people and wildlife, reductions in fisheries and cessation of flooding. Natural flooding is important because it increases sediment, oxygen and nutrient levels and changes water temperature. These seasonal changes are critical to the survival of many species found in ecosystems downstream of the LHWP.

Soil erosion

An increase in soil erosion has taken place as result of construction of not only the dams, but also the associated infrastructure such as roads and power lines. For example, poorly designed culverts on LHWP roads funnel runoff downslope and have exacerbated gullying. This forces some farmers to plough against the contours of the slope, which accelerates erosion. The problem is aggravated further as displaced communities are forced to farm on more marginal land, such as on steeper slopes, and they may lack the necessary knowledge of sustainable farming practices.

ACTIVITIES

1 In HDEs, we now drink as much bottled water as milk. What are the possible environmental impacts of this?

   a What do you understand by the term 'poverty related environmental activities'?

   b Give examples of the vicious cycle (or multiplier effects) of the LHWP on people and their interaction with the environment.

2 What role has the World Bank in large-scale water supply projects?

STRETCH YOURSELF

The Lesotho Highlands Development Agency (LHDA) is responsible for the implementation, operation and maintenance of the LHWP. Since it started in 1986, the five phases of the project have been subject to much discussion and some change with still no completion date set for Phases III and IV. Look at the website of the LHDA (www.lhda.org. ls). Suggest reasons why the LHWP is still not finished. Are such large-scale water supply schemes 'worth it' for LDEs such as Lesotho?

In this section you will learn about:
- impacts of climate change on UK water supplies and demand
- our water footprint and strategies to manage water consumption

In 1830, each person in the UK managed with around 18 litres of water a day. Today, this figure is around 150 litres, rising to a staggering 4645 litres when 'hidden levels' of water use (**virtual water**) are considered – this is our *water footprint*. Increased demands, as well as the impacts of climate change, are having a significant and long-term impact on the water cycle.

Many suggest that our water-intensive lifestyle is unsustainable (Figure **1**), especially as emerging economies also aspire to a similar extravagant use of water resources. To meet demand, any increase in supply must work alongside conservation measures (5.7 and 5.10) and strategies to manage water consumption.

UK Climate Impact Programme predictions

In general, the UK climate is expected to become hotter and drier in summer, and warmer and wetter in winter. Only Scotland is likely to retain precipitation all year round. However, increased precipitation in winter is not predicted to offset decreases in precipitation in summer and increased losses through evapotranspiration. For example, Wales and Cornwall are likely to be much drier than at present and East Anglia alone could have up to 60 per cent less summer rainfall – with potentially devastating impact on its agricultural productivity.

Whilst climate modelling can never be accurate, by 2080 UK climate overall is expected to be:

- Hotter – annual UK temperatures may rise by between 2 and 3.5 °C.

- Drier – annual precipitation across the UK is likely to decrease slightly. Precipitation in summer may decrease by as much as 50 per cent in the south-east but increase by up to 30 per cent in winter. Snowfall could also decrease significantly.

- More extreme – high summer temperatures and dry conditions are likely to become more common, while extreme autumn and winter precipitation could become more frequent.

Activity	Water used
Running the tap	8–12 litres/min
Washing up in the sink	6–8 litres
Washing hands and face	3–9 litres
Taking a normal shower	6–12 litres/min
Taking a power shower	13–22 litres/min
Flushing the toilet	5–12 litres
Running a dishwasher	15 litres
Running a washing machine	60–80 litres
Having a bath	75–90 litres
Using a hosepipe	550–1000 litres/hr
Making food and drink	6–10 litres

Figure 1 *Lifestyles of HDEs are considered to be increasingly water intensive*

The consequences of all this could be bleak – and not just regarding water management. Rising sea levels, increased storm surges and rain falling increasingly as storm events could double the number of homes currently at risk from flooding. Risks from skin cancer and insect-borne diseases could increase. Furthermore, droughts and life-threatening heatwaves might be particularly acute in urban areas, with London most at risk. Summer temperatures in the capital could reach a searing 45 °C – the equivalent of Marrakesh, Morocco, today!

Strategies to manage water consumption

Food consumption

Responsible food retailers are starting to consider water footprint in the same manner as the more established carbon footprint – food is increasingly sourced from growing regions that do not suffer water stress. This may necessitate changes to our diet. So rather than eating fruit and vegetables that are out of season (such as Spanish strawberries in winter), we may need to rely more on home-grown vegetables (such as winter season sprouts) or those that are more drought or salt tolerant, such as broccoli.

Water conservation

Look at Figure **2**. Emerging technologies, combined with traditional water saving methods and good sense, offer significant opportunities to use water more efficiently in the home or workplace. For example, energy and water efficient dishwashers and washing machines use less water than washing by hand. Other methods to save water include lifestyle changes such as showering rather than bathing or practical decision-making such as planting drought resistant plants in the garden.

Water meters

Fitting a water meter means you only pay for the water you use. This is a positive step towards reducing a water footprint because it forces the user to think more about management of water consumption. The average family in the UK reduces their water usage by 10–15 per cent after a water meter is fitted.

Water footprint

'Water, water everywhere. Nor any drop to drink…'! This certainly applies to many daily items included within our *water footprint*. We may consume 150 litres of water in our day-to-day existence of, for example, drinking, cooking and washing but far more is used for growing food, making our clothes and even our favourite digital gadgets! The water footprint measures the amount of virtual water used to produce each of the goods and services that we use. It can be expressed per person, per company or even for an entire nation or continent.

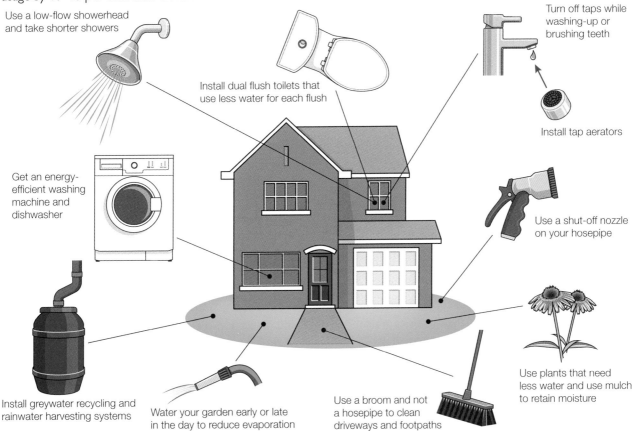

Use a low-flow showerhead and take shorter showers

Install dual flush toilets that use less water for each flush

Turn off taps while washing-up or brushing teeth

Install tap aerators

Get an energy-efficient washing machine and dishwasher

Use a shut-off nozzle on your hosepipe

Install greywater recycling and rainwater harvesting systems

Water your garden early or late in the day to reduce evaporation

Use a broom and not a hosepipe to clean driveways and footpaths

Use plants that need less water and use mulch to retain moisture

🔺 *Figure 2* *Water conservation methods in the home*

ACTIVITIES

1 How are the effects of climate change in the UK likely to impact on both the supply and demand for water? (Think about how our water usage changes when it is hotter.)

2 Study Figure **1**. Populations of HDEs lead a water-intensive lifestyle. What do you understand by this phrase?

3 Study Figure **2**. Is it optimistic to think families will adopt water conservation measures at home? Explain your answer.

4 Suggest how changes in land use affect levels of water consumption.

5 'About 25 per cent of all fresh water consumed in the US is associated with wasted food.' Discuss this statement with reference to water consumption.

STRETCH YOURSELF

Research green water, blue water and grey water footprints.

In this section you will learn about sustainability issues associated with diffuse pollution and water management

Successful water management is more than water conservation strategies (see 5.9). It is about changing mind-sets towards sustainable use of finite water resources, whether in a four-bed house in Middlesex or a mud and thatched hut in Mozambique.

Sustainable water management is, in part, about addressing the needs of all different water users while ensuring monitoring and other controls are in place to maintain water quality and flow rates.

Diffuse pollution

Look at Figure **1**. Diffuse pollution occurs when small amounts of pollutants, often from many different sources, are washed into a water catchment across a wide area. The effects of a pollution cocktail like this are more significant than an individual pollutant. This is because the sources of pollution are often widespread, and so hard to spot.

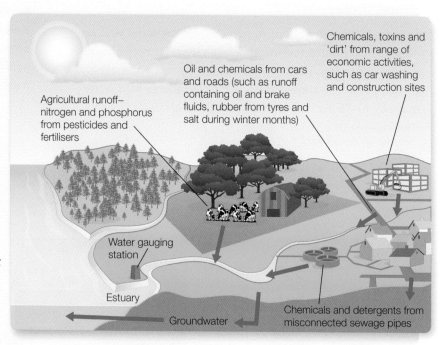

Chemicals, toxins and 'dirt' from range of economic activities, such as car washing and construction sites

Oil and chemicals from cars and roads (such as runoff containing oil and brake fluids, rubber from tyres and salt during winter months)

Agricultural runoff– nitrogen and phosphorus from pesticides and fertilisers

Water gauging station

Estuary

Groundwater

Chemicals and detergents from misconnected sewage pipes

▶ *Figure 1 Diffuse pollution includes pollutants from many different sources; this is in contrast to point source pollution which comes from a specific, identifiable source such as a factory effluent pipe*

Examples of water management

Greywater recycling

Don't throw it away! Women and children in countries such as those in sub-Saharan Africa walk long distances to collect water from muddied boreholes. In HDEs, we let around one-third of recyclable water, such as from sinks, showers, washing machines, go down the plughole. This domestic 'dirty' water, excluding sewage, is referred to as *greywater*.

Greywater recycling systems clean the greywater and then plumb it back into your toilet, washing machine or outside tap. However, treatment and storage of greywater is costly and additional energy has to be used to pump the water to where it is needed. As a result, the main users of such systems are high water consumers such as hotels and leisure companies.

Rainwater collection recycling

Rainwater harvesting systems (RHS) collect water from the roofs and divert it into existing pipework to flush toilets and wash clothes. The main advantage of RHS over greywater recycling is that rainwater is likely to contain less bacteria and fewer contaminants with less need for expensive cleaning.

Groundwater management

Groundwater supplies 30 per cent of all available freshwater (*AQA Geography A Level & AS Physical Geography*, 1.2) but this figure hides significant regional variations (see 5.6). It is therefore important to manage groundwater to maintain sustainable water reserves and prevent overabstraction. Water abstraction licences are issued in many countries and consumption levels are monitored. In England, the Environment Agency charges licence holders to ensure water resources are managed effectively and sustainably.

WaterAid

WaterAid is a UK-based international NGO that aims to improve access to safe water, sanitation and hygiene for the poorest and most disadvantaged. WaterAid believes that the success of their work is dependent on a number of interconnected factors (Figure **2**). For example, if good practice is not followed in the exploitation of groundwater resources, such as using qualified supervisors to spot-check work, there is a good chance that boreholes will fail and the investment will be wasted.

WaterAid has also been undertaking a sustainable 'water recycling' experiment – the use of composting toilets and urine reuse (Figure **3**). Two of the three main nutrients used in fertilisers, nitrogen and phosphorous, are found in urine. The Kubasisa Muganga Project in Mozambique has succeeded in changing perceptions towards use of urine as fertilisers. Urine is separated from faeces and collected into containers. Wearing masks and gloves, locals then pour this directly over the seedbeds. After 30 days, the crops or vegetables are perfectly safe to harvest and eat. The project is sustainable on a number of levels:

◆ Water is conserved as no water is used in toilet flushing.

◆ Pollution of groundwater (an important drinking water supply) is prevented.

◆ It is hygienic and does not attract flies (and so disease).

Indeed, the project has been so successful that urine is now viewed as an economic commodity that can be traded – the so called 'gold' of agriculture.

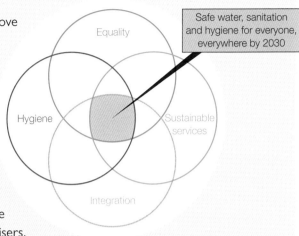

▲ **Figure 2** *WaterAid's global strategy*

❯ **Figure 3** *The composting toilet uses no water to flush. Urine and composted solid waste is recycled as fertiliser*

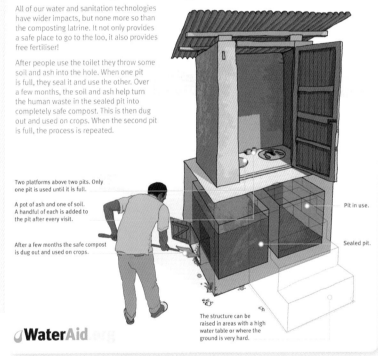

The composting latrine

All of our water and sanitation technologies have wider impacts, but none more so than the composting latrine. It not only provides a safe place to go to the loo, it also provides free fertiliser!

After people use the toilet they throw some soil and ash into the hole. When one pit is full, they seal it and use the other. Over a few months, the soil and ash help turn the human waste in the sealed pit into completely safe compost. This is then dug out and used on crops. When the second pit is full, the process is repeated.

Two platforms above two pits. Only one pit is used until it is full.

A pot of ash and one of soil. A handful of each is added to the pit after every visit.

After a few months the safe compost is dug out and used on crops.

Pit in use.

Sealed pit.

The structure can be raised in areas with a high water table or where the ground is very hard.

WaterAid

ACTIVITIES

1 Study Figure **1**.
   a What is meant by diffuse pollution?
   b Compare sources of diffuse pollution in an urban drainage basin to those in a rural drainage basin?

2 Work with a partner to brainstorm as many different examples of water conservation and recycling (including greywater and rainwater harvesting) in the home. These might be changes in behaviour (such as showering rather than a bath) or using alternative technologies (such as a water meter). It may help to read section 5.9 again.

3 Study Figure **2**.
   a What do you understand by the term 'water sustainability'?
   b To what extent is WaterAid's Kubasisa Muganga Project achieving economic sustainability?

STRETCH YOURSELF

Read 'Everyone, Everywhere 2030: WaterAid Global Strategy 2015–20'. Is there reason to be optimistic that WaterAid will achieve their goal of reaching 'everyone, everywhere with safe water, sanitation and hygiene by 2030'?

Water conflict

In this section you will learn about water conflicts at local, national and international scales

The politics of water

In Iraq in 2500 BCE, the King of Lagash diverted water from the rival region of Umma in a dispute over the edge of paradise region around the Tigris River. *Water conflict* is not new but as competition for water increases, combined with increasing political instabilities, poor management and the impacts of climate change, water conflict is becoming more frequent at local, national and international scales (Figures **1**, **2** and **3**).

Did you know?
In a 2012 report, the US Director of National Intelligence warned that overuse of water was a source of conflict that could potentially compromise US national security.

▲ **Figure 1** *A disputed freshwater source near the village of Xalatlaco on the outskirts of Mexico City led to civil unrest among the villagers*

▼ **Figure 2** *Examples of water conflicts in 2014 at local, national and international scales*

Parties involved	Scale	Violent conflict or in context of violence	Basis of conflict	Description
South Africa	Local	Yes	Development dispute	At least four people killed during protests over water shortages in the northern town of Brits
Libya	Local	Property damage	Development dispute	Disputes between the Zwai and Tebu tribes, included attacks on pumps that operate a pipeline system, leading to water shortages
Venezuela	Local	Property damage	Terrorism	Unidentified groups attacked the Bourgin water treatment plant in Merida and contaminate the local supply with diesel fuel
Darfur, Sudan	Local	Yes	Military target; Military tool; Terrorism	Armed Janjaweed militia burn and loot a South Darfur refugee camp and destroy all water wells
Nigeria	Local	Yes	Development dispute	Tiv/Agatu farmers and Fulani herdsmen clash over access to grazing land and water points
India	Local	No	Development dispute	Upper caste women in Kashmir restrict access to higher quality water resources
Mexico	Local	Yes	Development dispute	Confrontation over a water spring between 1500 police and residents of a village on Mexico City's outskirts

Parties involved	Scale	Violent conflict or in context of violence	Basis of conflict	Description
USA	National	No	Terrorism	Three men in the state of Georgia are arrested for planning to attack water treatment plants, power grids, and other infrastructure
South Sudan	National	Yes	Military target; Military tool	A military checkpoint and water pipeline to the United Nations Mission are targeted and destroyed
Syria	National	Yes	Political tool; Military tool	Water supply for the city of Aleppo is cut off; pumping stations and water distribution networks are bombed
Kyrgyzstan/ Tajikistan	International	Yes	Development dispute; Military target	Border dispute over access to pasture and water resources
Russia and Georgia	International	Yes	Military target; Development dispute	Since the 2008 war, continued reports from Georgia that water and gas supplies are regularly cut off by Russia
Ukraine and Crimea	International	No	Political tool; Development dispute	After Russia annexed Crimea, Ukraine accused of cutting water supply in the North Crimea Canal, leading to a water shortage for Crimea's agricultural fields
Iraq	International	Yes	Military tool; Military target	Insurgents from self-styled ISIL (Islamic State in Iraq and the Levant) seize the Falluja Dam in Iraq and close the floodgates to cause upstream flooding and to cut downstream water supply
Iraq	International	Yes	Military target; Military tool	Heavy fighting around Haditha Dam and Euphrates River Dam between ISIL and government forces

⑤ Being 'place specific'

Geography is all about place – knowing specifically where areas of study are located raises all sorts of questions that geographers are best placed to answer. Always ask yourself the question 'could my answer be about any place on the planet or have I linked my answer to named places that provide the reader with no doubts as to location'.

🔽 Figure 3 *Mosul Dam, northern Iraq (2014); water is increasingly being used as a tool in conflicts around the world*

Will the Tibetan Plateau determine the course of Asia's future?

Known as the 'Roof of the World', the Tibetan Plateau covers an area of more than 6 million km² and is an environmentally strategic area and critical to the health of the planet (Figures **4** and **5**).

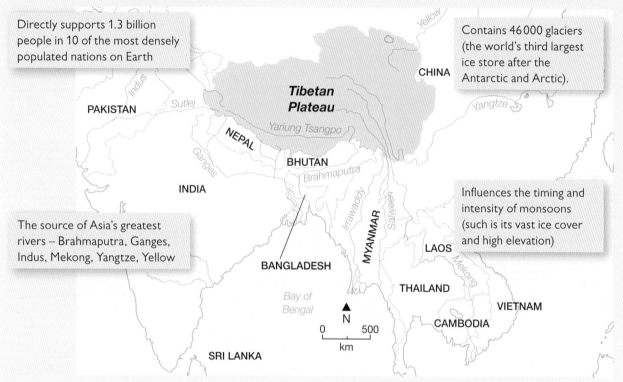

Directly supports 1.3 billion people in 10 of the most densely populated nations on Earth

Contains 46 000 glaciers (the world's third largest ice store after the Antarctic and Arctic).

The source of Asia's greatest rivers – Brahmaputra, Ganges, Indus, Mekong, Yangtze, Yellow

Influences the timing and intensity of monsoons (such is its vast ice cover and high elevation)

The Tibetan Plateau is also one of the fastest-warming areas on Earth. The snowline has risen from 4600 to 5300 m and Tibet is now hotter than at any time in the past 50 years. This alone represents a potential global environmental catastrophe. Look at Figure **6**. China's current and future plans to dam or divert water from five of the large rivers coming out of the Tibetan plateau is not only exacerbating the environmental problem but also creating a political storm with nearby states.

The geopolitical significance of all this should not be underestimated:

◆ The headwaters of the Yangtze, Yellow and Mekong rivers originate in this area and are already running weak, threatening water supplies, irrigation and HEP potential for hundreds of millions of people.

◆ The Yellow River alone supplies water to one-fifth of China's 1.3 billion people and directly serves 50 major cities along its 5430 km. It has sometimes slowed to a trickle causing urban water pipes to run dry.

◆ The Indus, Brahmaputra and Ganges rivers supply India and Bangladesh.

◆ Melting permafrost could change the pattern of summer monsoons, causing more droughts in northern India and in the wheat-growing north of China. More intense and frequent floods in southern China are likely. The resulting mudslides and desertification of grasslands, vital for sheep and yak grazing, have already displaced over 30 000 herders in the plateau.

▲ **Figure 4** *The sources of many major rivers are found on the Tibetan Plateau*

▼ **Figure 5** *The geopolitically significant Tibetan Plateau*

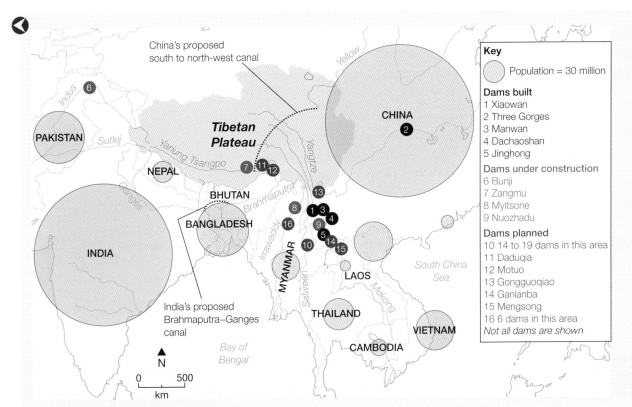

In 2009, China attempted to block a US$2.9 billion Asian Development Bank loan because it believed it would help finance Indian water projects in a disputed region of Tibet. China's proposed western route of the South–North Water Transfer Project (see 5.7) threatens the peace and security of the entire region. In turn, India is on a dam-building spree of its own, including construction of mega dams as well as micro-hydro projects on the Brahmaputra River (that has its source in the Tibetan Plateau) in Arunachal Pradesh. All of these dams, barrages, canals and irrigation systems can be wielded overtly in a water conflict or more subtly as a political bargaining tool. As data on river projects is already rarely shared in full with other states, it seems that 'trust' is now at a trickle. Is a water war inevitable?

Figure 6 *Turning the tap off – plans to manage the flow of water from the Tibetan Plateau will affect huge numbers of people (national populations shown by proportional circles)*

ACTIVITIES

1 Study Figure **2**.
  a On an atlas map of the world, locate each example of water conflict in 2014.
  b Analyse the distribution and scale of water conflict in 2014.

2 Study Figure **8** and refer to the case study on the Tibetan Plateau.
  a Explain the causes of water conflict in the Tibetan Plateau.
  b Should we be optimistic about a sustainable solution to the water conflict? Explain your answer.

STRETCH YOURSELF

Research water conflict in the Middle East and summarise the key issues. You may wish to use the headings from Figure **2** as a guide. Areas of study might include Israel (Golan Heights is a crucial source of water but it is an occupied territory), Turkey (the controversial GAP scheme) or Egypt (pressures on the Nile, particularly for irrigated farming). Be place-specific in your answer.

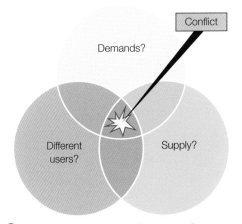

Figure 8 *The causes of water conflict. Water conflict is most likely where causes overlap*

Sources of energy and the energy mix

In this section you will learn about:
- sources of primary and secondary energy
- demand and energy mixes in contrasting settings

If water is our most vital resource, then energy must be our most useful and valuable. The twentieth century saw an enormous increase in global energy production and consumption – a trend continued into the twenty-first century. However, the relative importance and the balance of energy sources changed – and continues to do so. Furthermore, there has always been an uneven distribution of energy sources on a global scale and a significant difference in demand between rich and poor nations.

Primary and secondary sources of energy

Look at Figure **1**. Energy sources are categorised as *stock* or *flow*, and *primary* or *secondary* within each of these categories. Primary sources are raw materials used in their natural form such as coal, petroleum (oil), uranium and wood. They are often converted into more practical and convenient secondary energy sources such as petrol, diesel and electricity.

Did you know?

One tonne of uranium ore produces as much electricity as 25 000 tonnes of coal!

▼ **Figure 1** *Primary and secondary energy sources*

STOCK			
Type	**Key characteristics and uses**	**Key locations**	**Key trends**
Coal (Primary)	• Dirty, bulky and difficult to transport • Traditional markets – railways and domestic heating • Now used for electricity generation in thermal power stations, the production of coke for metal smelting and as a raw material in the heavy chemical industry	• The belt between 30°N and 60°N that stretches across North America and Eurasia contains 70% of the world's coal • Relatively little is mined in the southern hemisphere • China is the biggest producer and consumer	• Declining use in older industrial countries of Europe, and Japan • Increasing use in Eastern Europe and Russia • Dramatic increase in use in India and China
Petroleum (oil – Primary)	• 'Black gold' is relatively cheap and flexible • Refined and used for transport, industry and electricity generation • Key raw material of petrochemical industry • Submarine oil fields now contribute a large proportion of the world's supply	• OPEC members produce 75% – the Middle East (chiefly Saudi Arabia, Iran, Iraq and Kuwait), Venezuela, Ecuador, Nigeria, Gabon and Indonesia • Caspian Sea region has enormous reserves – second only to Saudi Arabia • 1973 OPEC oil crisis made previously uneconomic sources viable in environmentally difficult areas such as the North Sea and Alaska • Very large reserves in mainland and offshore Mexico	• Caspian Sea region production currently 'significant but not major' • Notwithstanding the collapse of oil prices, the Arctic Ocean and South Atlantic could be an exploration bonanza of the future • Oil shale 'fracking' (hydraulic fracturing) a new potentially abundant source once natural gas is condensed into oil shale (see 5.4)
Natural gas (Primary)	• The most flexible and 'clean' fossil fuel – domestic, industrial and electricity generation uses • Most transported direct to consumers by pipeline • Some transported by sea in a liquefied form (LPG)	• Leading producers include the USA, Russia, Canada and China • Caspian Sea region – presently 'significant but not major'	• Production of natural gas has increased very quickly • European production notable but declining (such as the North Sea reserves of the Netherlands and UK) • 'Fracking' is potentially an abundant source
Uranium (Nuclear fission – Primary)	• Contentious, 'clean' and extraordinarily powerful – used for electricity generation • Most associated with HDEs • Expensive – in part to ensure safety • Huge potential in combating climate change	• Kazakhstan produces over 40% of the world's uranium • USA produces the most nuclear energy (one-fifth of its electricity)	• Nuclear electricity generation increased tenfold globally between the mid-1960s and mid-1970s • EMEs are rapidly developing the technology • China alone could build 300 more stations by 2050

FLOW			
Type	Key characteristics and uses	Key locations	Key trends
Biomass (Primary and secondary)	• Fuelwood used for heating, lighting and cooking • Biogas (biomethane) produced from organic decomposition of plants, animal and human waste – used to generate electricity • Alcohol (bioethanol) produced from fermenting cane sugar, root crops (such as cassava) and cereals (such as maize and soghum) • Biodiesel produced from oilseed rape	• Main source of energy in LDEs – particularly in rural areas • Biomethane production notable in Nepal • Brazil – main producer of bioethanol	• 40% of world's trees removed for fuelwood – supplying energy for over 2 billion people • Ethanol is a cheap alternative to petrol in suitably modified vehicles • Biodiesel can be mixed with (petroleum) diesel • Drax Power Station in North Yorkshire, the last coal-fired power station built in the UK, is now Europe's largest biomass-fuelled station (burning imported wood pellets)
Hydro-electric power (HEP) (Primary)	• Produced by running or falling water – no fuel so low operating costs but high installation and transmission costs • Produces small proportion of the world's electricity supply • Dams and reservoirs can also control floods and supply domestic, industrial and irrigation water	• Important in HDEs with mountainous relief, such as Norway, Sweden, Austria and Switzerland • Huge potential throughout South America, south-east Asia and, especially, Africa • LDEs and EMEs dependent on HEP tend to go for multi-purpose large-scale schemes	• Frequent call for alternative type of HEP station using the energy of incoming and outgoing tides. The Rance Estuary in France is still the world's only tidal power station. Potential for the Severn, Dee, Solway and Humber in UK.
Other renewables (Secondary)	• Solar, wind, wave, tidal and geothermal energy are examined in section 5.16		

Global demands and the energy mix

Look at the compound graph (Figure **2**). The world today sees global demands and an energy mix still dominated by fossil fuels, but alternative flow resources are growing in significance. Most of the world's energy is still consumed by a fairly small number of HDEs where living standards are high.

The increasing global demand for energy means that stock resources are being used up more quickly and the threat of shortages loom. Furthermore, as LDEs develop and their populations aspire to HDE living standards, their demand for energy will increase. However, prices rise as reserves dwindle, prompting the search for cheaper and/or more accessible alternatives. The different primary energy sources used to meet demand make up the energy mix – this will usually change over time. For example, the last 50 years in the UK has seen a progressive decline of coal, the so-called 'dash for gas', increasing proportions of nuclear and renewables (including HEP and wind) and the rise and fall of North Sea oil.

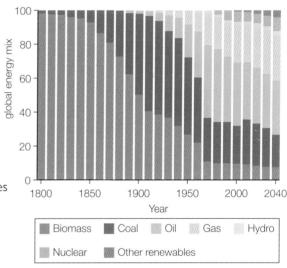

▲ **Figure 2** *Global energy mix, 1800–2040. Note that these are proportional figures which hide large increases in total energy consumption.*

ACTIVITIES

1 a List every use and source of energy in your typical day – think of lighting, heating, transport and the items you used.

b Is energy the most useful and valuable resource? Could you live without it?

S

2 a Study Figure **2**. Describe the trends shown in the compound graph.

b Using a table, state the percentage of energy provided by each primary energy source in 1800, 1850, 1900, 1950, 2000 and projected for 2040.

i Which energy sources have increased in importance over the time-span of the compound graph? Explain your answer.

ii Which energy sources have decreased in importance over the time-span of the compound graph? Explain your answer.

In this section you will learn about:
◆ the relationship of energy supply to geology, climate and drainage
◆ China's energy demands, mix and security

Energy and geography

For any country, access to reliable and affordable sources of energy is essential. Unfortunately, few countries have abundant reserves, such as oil in Saudi Arabia or natural gas in Russia. Physical geography can determine a country's energy mix. For example:

◆ Geology might yield stock energy resources of fossil fuels.

◆ Climate may give opportunities for the development of flow resources such as solar or wind energy.

◆ Drainage could be harnessed for HEP (Figure **1**).

◆ All three combine in creating potential for the production of biomass.

Human geography is also a factor, not least in determining how much energy and in what forms. It is also relevant in quality of life issues, such as the environmental impact associated with energy supply and political commitment to conservation and addressing climate change. All these interrelationships amount to a challenging test for any country trying to achieve **energy security**.

△ **Figure 1** China's Three Gorges Dam was completed in 2012

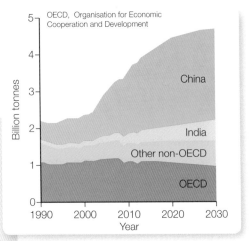

△ **Figure 2** Coal demand by region

China's energy demands, mix and security

Everything geographical about China is big – land area, population, economic growth or resource consumption. All raw statistics continually stagger. So it may not come as much of a surprise that, since 2010, China has been the world's biggest consumer and producer of energy. To put this in perspective:

◆ China is the world's top consumer, producer and importer of coal. Around half of all coal burned worldwide is consumed in China (Figure **2**).

◆ China is the world's largest net importer of oil.

◆ China is the world's biggest producer of greenhouse gases – especially CO_2.

China's energy mix is more varied than these headline figures suggest (Figure **3**) and, as with all countries, it is projected to change in future as social, economic and environmental circumstances dictate.

China's energy demands are so great that they already threaten other countries' supplies. However, because it is still largely an agricultural country and energy demand per person is still low, energy demands are projected to accelerate as it continues to economically develop.

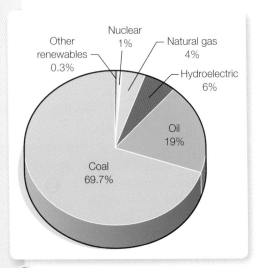

△ **Figure 3** China's energy mix (2012). Take care not to interpret this as electricity generation

Why is increased energy demand important?

This acceleration in demand is not just because China is reinforcing its contemporary status as 'the world's workshop' and so meeting the demands of new industry. It is also due to rapid urbanisation, increasing personal wealth and growing car ownership. By 2030, China is expected to account for nearly 40 per cent of the world's total energy consumption.

Look at Figure **4**. China's large area and variable physical geography ensures wide-ranging domestic energy sources – it was self-sufficient in oil until 1993. Many stock deposits of oil, coal and gas, however, are in remote locations and are not economically viable. Furthermore, over two-thirds of China's electricity generation is still being produced by coal-fired power stations resulting in serious pollution problems including poor air quality and acid rain.

Natural gas stocks are high, but at best could only provide one-fifth of the demand and this assumes that costly pipelines could be extended to the gas fields in western China.

Renewable energy in China

Such is the geographical diversity of China that wind, solar and biomass energy sources are slowly adding renewable capacity to its energy demands and mix. Its HEP potential is far greater, already accounting for 22 per cent of China's electricity generation. The Three Gorges Dam (Figure **1**) is the biggest HEP scheme in the world producing the energy equivalent of over 20 nuclear power stations. The government has long-term ambitions for new projects on all its major rivers which address environmental concerns related to the country's over-dependence on fossil fuels, especially coal. However, China's water security is far from assured (see 5.7) and HEP projects raise their own social and environmental concerns:

◆ flooding of established settlement, infrastructure and productive agricultural land

◆ displacement of population

◆ reservoir silting and pollution

◆ vulnerability to major earthquakes and landslides.

In conclusion, China's energy mix remains heavily dependent on fossil fuels, especially highly-polluting coal. Despite a generous range of domestic energy sources, such are its extraordinary demands that 90 per cent has to be sourced from foreign suppliers. China is **energy dependent** rather than energy secure.

▲ **Figure 4** *China's energy resources*

ACTIVITIES

1 List the reasons for China's growing energy insecurity.

2 Describe and explain the short- and long-term problems associated with China's increasing energy demands.

S 3 With reference to China, analyse the compound line graph in Figure **2**. A good answer would:

  • outline the past patterns and predicted trends

  • suggest what these predictions imply about alternative energy and environmental issues, such as climate change.

STRETCH YOURSELF

China has published several five-year plans, each one covering key national issues. The twelfth and most recent Five-Year Plan: Energy (2011) included a pledge to boost the country's share of non-fossil fuel energy to around 20 per cent by 2030. Outline ways in which such a target could be achieved.

In this section you will learn about energy supplies in a globalising world

Global energy supplies in a globalising world

Look again at Figures **2** and **3** in 5.3 and Figure **1** in 5.12. Energy reserves are known resources that could be accessed and used. Physical geography dictates the abundance or otherwise of stock energy resources in different global regions. Global energy sources are distributed unevenly. For example, solar energy potential is greatest in regions with clearest skies, such as the tropics with descending high pressure air.

The human geography of energy supplies is also important. For example, note how many of the largest reserves of stock energy resources are concentrated in politically unstable parts of the world, such as the Middle East. This can lead to disruptions in supply. The OPEC oil crisis in 1973 led to a complete rethinking of energy supplies across Europe – including sourcing new supplies (the North Sea) and diversification.

Different countries have varying demands, sources and energy mixes, yet compete in a globalising world where energy is vital. Political friendships and differences can therefore be crucial to ensuring energy security. Just as geographical regions, such as the Middle East and the Caspian Sea region, can become dominant as energy suppliers, so too can TNCs such as Gazprom and Royal Dutch Shell (Figure **1**). For example, the geopolitics of energy security is particularly well illustrated by Gazprom and its close links with the Russian government.

🔻 **Figure 1** *The world's biggest TNCs (2015); whether measured in total assets, wealth, production or sales, it should come as no surprise that seven of the top ten TNCs in the world are energy producers, processors and distributors*

TNC	Country of origin	Main business	Total revenue (2015) US$ (billion)
Wal-Mart	USA	Retail	421.8
Royal Dutch Shell	UK/ Netherlands	Oil, gas and alternative energy	378.2
Exxon Mobil	USA	Oil and gas	354.7
BP	UK	Oil and gas	308.9
Sinopec Group	China	Oil, gas and chemicals	273.4
China National Petroleum Corporation	China	Oil and gas	240.2
State Grid Corporation of China	China	Electricity generation	226.3
Toyota Motor Corporation	Japan	Automobiles	221.8
Japan Post Holdings	Japan	Mail, banking and insurance	204.0
Chevron	USA	Oil and gas	196.3

Gazprom and the Trans-Siberian Pipeline

Moscow-based Gazprom is the world's largest natural gas supply company controlling 17 per cent of the Earth's reserves (Figures **2** and **3**). It provides all of the gas for the bordering countries of Estonia, Finland and Latvia, and one quarter of the EU's!

Gas exports are controlled through pipelines and the Trans-Siberian is the most important at 4500 km in length, with a capacity of 32 billion m³ per year. It was funded by German, French and Japanese banks and completed in 1984. The pipeline is now partially owned and operated by Ukraine which is a 'transit state' because 80 per cent of the gas that Russia exports to Western Europe crosses it.

The pipeline has always been controversial. The original Soviet plans to build it were considered a threat to the balance of energy trade in Europe. It was strongly opposed by President Reagan's administration in the USA which prevented American companies from selling supplies to the Soviets for its construction.

⬆ **Figure 2** *Gazprom's headquarters in Moscow*

However, in recent years, the Trans-Siberian pipeline has really made the headlines. In late 2004, in a peaceful democratic revolution, Ukraine replaced its pro-Russian government with one led by pro-western reformers. Russia retaliated by quadrupling the price of its natural gas to Ukraine and then cutting it off completely early in 2006 when the new government refused to pay! This cut European supplies by 40 per cent in some areas.

By 2008, Ukraine was seeking to join both **NATO** and the EU, prompting Russia's Gazprom to cut the supplies again over so-called disputed debts. Further disputes in the winter of 2009–10 led Gazprom to cut supplies to Bulgaria, Hungary, Poland and Romania.

In 2014, Gazprom again shut down supplies to Ukraine over non-payment of huge debts running into billions of dollars. This was followed a day later by an explosion blamed on depressurisation by some, sabotage by others (Figure **4**). All such events are alarming for many European countries because they depend on Russia for so much of their natural gas. However, Gazprom's European energy markets are too valuable to lose and so it has been helping to secure Europe's energy supplies with the construction of a new pipeline bypassing Ukraine and Belarus. The Nord Stream pipeline running across the bed of the Baltic Sea opened in 2012 and may be extended further (Figure **3**). The South Stream pipeline that was planned to run across the bed of the Black Sea was cancelled in 2014.

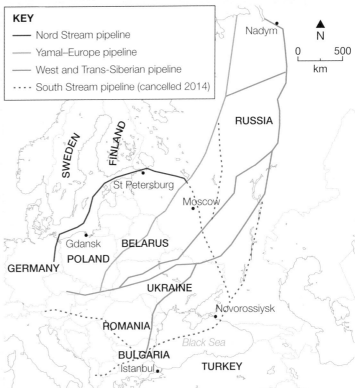

Figure 3 *Gazprom's natural gas pipelines*

KEY
— Nord Stream pipeline
— Yamal–Europe pipeline
— West and Trans-Siberian pipeline
···· South Stream pipeline (cancelled 2014)

Figure 4 *An explosion on the Trans-Siberian pipeline in Ukraine in June 2014 resulted in a fireball rising 200 metres into the sky*

ACTIVITIES

1 'The geopolitics of energy security is particularly well illustrated by Gazprom and its close links with the Russian government.' Explain how this relationship poses a threat to other countries dependent on Russian gas.

2 In 2015, Gazprom started work building a 4000 km pipeline, following the signing in May 2014 of a 30-year gas-supply contract with China. The US$8.8 billion project, called the 'Power of Siberia', should be completed in 2017. Outline the economic, environmental and political advantages and disadvantages of such a scheme to both Gazprom and China.

Did you know?
Russia and China share a 4300 km border, so a good relationship between them is vital. But they are suspicious of each other, with Russia seeing China as a rival power and potential threat.

In this section you will learn about environmental impacts of a major energy resource development and associated distribution networks

'There's gold in them hills. Black gold!'

In popular consciousness, the oil business is all about wealth and power – and that without it, industrial civilization would shudder to a stand-still. Furthermore, too often the history of 'black gold' is perceived as if through a Hollywood lens – misleadingly quintessentially American images of Texan rigs ('nodding donkeys') and a rugged workforce who are a law unto themselves … It is portrayed as an exciting and even romanticised industry (Figure **1**).

Certainly, the immense value of oil has long been recognised. As early as 2600 BCE, ancient peoples were using the so-called 'fire water' for attack and defence. Indeed, the first oil extraction dates back to the seventh and eighth centuries, and the first exports into (not out of!) the Middle East took place in the fourteenth century. Even its first, mid-nineteenth century large-scale commercialisation – the birth of the modern-day oil industry – was not American. All these milestones relate to the western shores of the Caspian Sea – to Baku, Azerbaijan.

△ **Figure 1** The rugged romance of 'black gold'

Baku – the start of the modern-day oil industry

Look at Figure **2**. Contrary to popular belief, the origins of the modern-day oil industry are not American, but Azerbaijani. Large-scale petroleum production started around Baku (now the capital city) in the mid-nineteenth century. Despite Azerbaijan's vast and easily exploitable reserves, and skilled workers attracted to the region, transporting the oil to markets was difficult and expensive. Yet, as is so often the case throughout the history of this remarkable industry, necessity proved to be the mother of invention. By 1890, Baku was the busiest port in the world, operating fleets of tankers. Further international investment in pipelines, sealed storage reservoirs and railway networks followed. Distribution of Azerbaijan's 'black gold' to the western world and beyond was assured.

Following the 1917 Russian revolution and also Communist nationalisation (1921), decades of Soviet mismanagement threatened the very existence of the oil industry. The reality of Communism proved to be overproduction, with no incentives to boost efficiency or to invest in better technology. This all led to falling productivity and environmental mayhem.

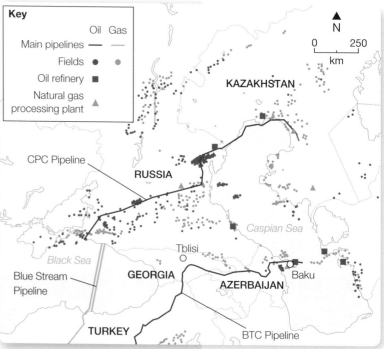

△ **Figure 2** Oil and natural gas infrastructure in the Caspian Sea region

The environmental legacy

Following the Soviet Union's collapse in 1991, western scientists were finally able to study the environmental impact in the region and measure the magnitude of damage caused:

◆ The Caspian Sea had been polluted by leakages from outdated offshore drilling equipment, and runoff from chemically polluted rivers.

◆ Along with the rusty derricks and pools of oil scum, well fires had been burning for years, releasing toxins into the air.

◆ Natural gas flares from Soviet drilling equipment added to the existing air pollution problem (4.5 million m^3 of harmful gases were being released from the flares every day).

◆ Drilling on the Absheron Peninsula of Azerbaijan had left much of the peninsula a barren wasteland, riddled with standing ponds of oil, high concentrations of sulphurous residue and a shoreline blackened by oily tides.

The wildlife and human costs

Serious consequences of such prolonged environmental degradation include:

◆ a decline in Caspian Sea fish stocks, especially sturgeon, threatening cavier production

◆ plummeting seal populations – in 2000 around 11 000 dead seals washed up on the Caspian shoreline

◆ most forms of agriculture became unproductive due to poor soil conditions

◆ safe drinking water supplies vastly reduced

◆ average life expectancies decreased by as much as four to five years

◆ rising infant mortality and death rates in pregnant women.

Figure 3 *The environmental impact of drilling on the Absheron Peninsula*

Hopes for the post-Soviet era

Renewed independence for Azerbaijan and four other Caspian basin states may have started in environmental and economic misery, but hope was never in short supply. This hope was multiplied in the mid-1990s with the discovery that remaining Caspian oil reserves were so vast that they were second only to Saudi Arabia! No longer discouraged by Communist rule, almost every oil-based TNC, including BP, rushed to invest in both exploitation and associated infrastructure for distribution. All drilling and pipeline technologies would be state-of-the-art and so present a markedly reduced threat to the environment. Improving infrastructure, such as roads, airports, electricity, communications and services, could also begin as well as the cleaning up of the existing environmental mess. A vast array of job opportunities would be created giving locals the chance to markedly improve their way of life. However, politically, proposed routes for new pipelines would always be contentious because of the huge potential revenues earnable in transportation fees.

So where are we now?

Political instability, ethnic conflicts, marked disparities between the living conditions of local people and expatriate oil workers, and also companies such as BP not living up to their environmental promises, are issues as disappointing as they are alarming. Reports on everything from onshore pollution, to wildlife decline and human health deterioration show that the problems that were endemic during the rule of the former Soviet Union are still apparent today.

ACTIVITIES

1   Look again at 5.12, 5.14 and forward to 5.16. With particular reference to the Caspian Sea region, assess:

   a   the environmental and social costs of the oil industry

   b   whether these costs outweigh the benefits and are too high a price to pay.

In this section you will learn about strategies to increase energy supply

Why is there a need to increase energy supply?

Globally, future population growth and economic development dictate an inevitable increase in energy demands. The exact scale of this increase is difficult to predict with any certainty but by 2035:

◆ most demand is still likely to be met by fossil fuels

◆ of the fossil fuels, natural gas will grow fastest – up to 2 per cent per annum

◆ nuclear power might double, at most, in EMEs and HDEs

◆ renewable alternatives will more than double in the global energy mix.

Furthermore, by 2035 vehicle numbers are anticipated to double, but against a backdrop of slowing oil consumption. This clearly highlights the importance of alternative technologies and advances in energy efficiency (Figure **1**).

But all this needs to be considered within the wider context of **energy poverty**. For example, about 1.2 billion people, an equivalent of the entire population of India, have no access to electricity and 2.8 billion rely on fuelwood, crop waste, dung and other biomass to cook and heat their homes. The need to increase energy supply is unarguable.

Environmental issues come into play too – climate change, air quality and acid rain can hardly be ignored.

Oil and gas exploration

For many, finding new oil and gas resources is essential to secure sustainable access to energy for everyone. To others, it is quite simply environmental madness. But, given the global economic significance of oil and gas, and also that six of the world's biggest TNCs are petrochemical companies, exploration is ongoing. Current notable advances are:

◆ the deep offshore oil potential in environmentally challenging regions such as the South Atlantic and Arctic Ocean (see 5.23)

◆ shale oil and gas 'fracking' (see 5.4) demonstrates enormous potential but it is proving to be very controversial (Figure **2**).

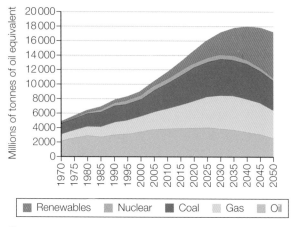

⬤ **Figure 1** Long-term global energy demand predictions

Think about

Fracking

In May 2016, North Yorkshire County Councillors voted to approve Third Energy's application to frack in Ryedale. The decision flies in the face of the wishes of thousands of local residents, businesses and Ryedale District Council. It also caused concern to anti-fracking protestors in neighbouring East Yorkshire where widespread applications to frack are still pending (Figure **2**). The environmental pressure group Friends of the Earth both supports and promotes local petitions against fracking, stating 'the fight isn't over... we know the tide is turning against fracking, public support is vanishing. It has been suspended in Wales, Scotland and in countries across Europe as the unacceptable risks to people's health and the environment become clearer.'

◗ **Figure 2** Fracking is a relatively new source of oil and gas, although controversial

Nuclear power – keeping an open mind

Increasing energy supplies through further development of nuclear power is also controversial. Concerns over nuclear waste and its close association with weapons of mass destruction are often expressed by those opposed to nuclear power. Many also quote the 1986 Chernobyl disaster which contaminated thousands of square kilometres of Ukrainian farmland. It will continue to be contaminated for many years to come. Radioactive rain affected every European country and fall-out was detected as far away as the USA and Japan.

Furthermore, after the 2011 Japanese earthquake and tsunami (see *AQA Geography A Level and AS Physical Geography*, 5.13), explosions and mass evacuations were shared globally in real time through social media. Social media also allowed the struggle to prevent meltdown at the Fukushima Daiichi nuclear power plant to be followed. Power to the reactor's cooling system was disabled by the earthquake and the back-up generators destroyed in the tsunami that followed (Figure **3**). This caused panic selling across stock markets. The explosions also prompted a government-ordered shutdown of the majority of Japanese nuclear power plants, responsible for 30 per cent of Japan's electricity generation. Since then, stringent safety checks and a progressive softening of public opinion have allowed progressive nuclear restarts since 2015.

Yet nuclear power is an extraordinarily powerful, efficient, albeit expensive technology. Those in favour, for example, argue that Chernobyl was a ghastly exception caused by a combination of unforgivable cost-cutting and human error. Current reactor designs have passive 'walk-away' safety systems, which, Fukushima Daiichi notwithstanding, make the chance of an accident extremely unlikely. Nuclear power is one of the very few low-carbon energy sources already developed. Indeed, it is easy to argue that although it might not be an ideal way to meet future energy demands, the dangers of climate change are certainly far worse. (Also see 5.18.)

⬆ **Figure 3** *Japan's Fukushima Daiichi nuclear power plant before and after the 2011 Tōhoku earthquake and tsunami*

Renewable alternatives – our sustainable future?

We've established the scale of increasing energy demands and also the undeniable need – up to 85 per cent of the increase by 2035 is likely to be in LDEs and EMEs. Fossil fuel will still dominate but it is, in the opinion of many, environmentally unsustainable. Nuclear power is as controversial as it is expensive.

Are renewable alternatives the answer?

Once again, it is not quite an open and shut case. Few, if any, would disagree that the potential of renewable alternatives must be developed at a much greater rate if future energy needs are to be met sustainably – all future projections suggest that the greatest relative growth is in this sector.

Considered individually, each has distinct qualities and limitations which make them no less controversial than fossil fuels and nuclear power (Figure 4).

> **Did you know?**
> A key subsidy scheme that helped the spread of UK wind turbines – the Renewables Obligation – closed in April 2016.

🔻 **Figure 4** Solar, wind, wave and geothermal alternatives: advantages and disadvantages

Solar

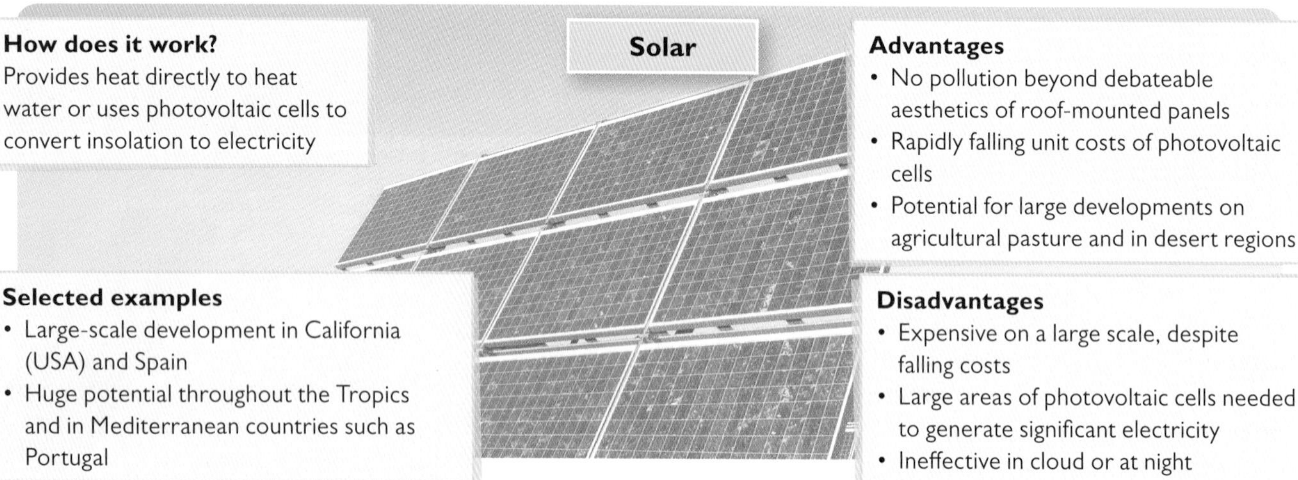

How does it work?
Provides heat directly to heat water or uses photovoltaic cells to convert insolation to electricity

Advantages
- No pollution beyond debateable aesthetics of roof-mounted panels
- Rapidly falling unit costs of photovoltaic cells
- Potential for large developments on agricultural pasture and in desert regions

Selected examples
- Large-scale development in California (USA) and Spain
- Huge potential throughout the Tropics and in Mediterranean countries such as Portugal

Disadvantages
- Expensive on a large scale, despite falling costs
- Large areas of photovoltaic cells needed to generate significant electricity
- Ineffective in cloud or at night

Wind

How does it work?
Generates electricity from a rotating turbine – either a single turbine or many in wind farms

Advantages
- No atmospheric pollution
- Abundant exposed sites
- UK is one of Europe's windiest countries
- Proven technology
- Low running costs

Selected examples
- *Onshore*; Whitelee, Glasgow *Offshore*; London Array, Kent Burbo Bank, Liverpool
- Significant potential in Spain, India, China, USA, Germany, and in the UK and the North Sea

Disadvantages
- Weather dependent
- Noise and visual pollution
- Expensive to build, despite falling costs – especially marine wind farms
- Bird and marine life affected

Wave

How does it work?
Oscillating waves force air into a chamber to rotate the turbine

Advantages
• No atmospheric pollution
• Largest storm waves during winter high demand

Disadvantages
• Wind and weather dependent
• High costs of construction and maintenance
• Not yet commercially viable
• Marine life affected

Selected examples
Coastal regions with large fetches, such as the Western Isles and Orkney, Scotland and Cornwall

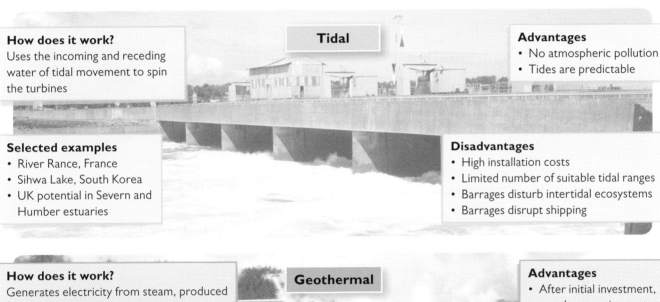

Tidal

How does it work?
Uses the incoming and receding water of tidal movement to spin the turbines

Advantages
• No atmospheric pollution
• Tides are predictable

Selected examples
• River Rance, France
• Sihwa Lake, South Korea
• UK potential in Severn and Humber estuaries

Disadvantages
• High installation costs
• Limited number of suitable tidal ranges
• Barrages disturb intertidal ecosystems
• Barrages disrupt shipping

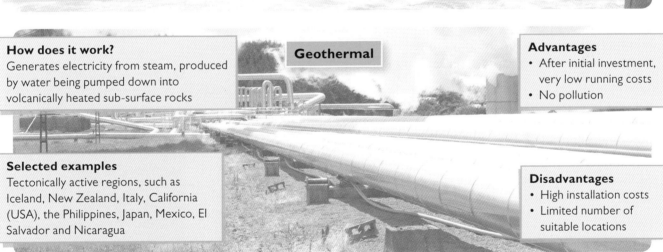

Geothermal

How does it work?
Generates electricity from steam, produced by water being pumped down into volcanically heated sub-surface rocks

Advantages
• After initial investment, very low running costs
• No pollution

Selected examples
Tectonically active regions, such as Iceland, New Zealand, Italy, California (USA), the Philippines, Japan, Mexico, El Salvador and Nicaragua

Disadvantages
• High installation costs
• Limited number of suitable locations

ACTIVITIES

1 Study Figure **1**.
   a Describe the global energy demand trends predicted.
   b Suggest reasons for the trends predicted beyond 2035.

2 Look again at Figure **1**, 5.12.
   a Outline the case for and against nuclear power.
   b To what extent do biomass and HEP (including tidal) represent viable energy sources to help meet future demands?

3 Explain why a reliance on fuelwood, crop waste, dung and other biomass for cooking and heating in LDEs is unsustainable?

STRETCH YOURSELF

Working in small groups or in pairs, prepare a debate arguing the cases for and against fracking in the UK.

In this section you will learn about strategies to manage energy consumption

Think about energy wastage in your home today. Count how many standby lights are always glowing red and how many gadget chargers have been left warming in a live socket. It has been suggested that not switching electrical devices off at the wall costs the average UK household £200 a year! – economic and environmental insanity. That is just one of numerous energy-saving measures each household could achieve by adopting domestic energy conservation examined in section 5.19.

Energy conservation is just one example of *demand management* – adopting techniques to reduce energy consumption, and therefore demand, rather than focusing on generating more and more electricity.

Managing energy consumption

All over the world, vast amounts of energy are wasted – in homes, industries, offices, schools, farms and transport. Electricity generation itself is wasteful of primary stock reserves – how wasteful depends on the method used. For example, electricity generation in thermal coal-fired stations (Figure **1**) accounts for almost 41 per cent of the world's output – but their efficiency only ranges from 32 to 42 per cent compared to 85 to 90 per cent for HEP.

⬆ **Figure 1** *Coal-fired power station at Eggborough, North Yorkshire*

An important aspect, therefore, of managing energy consumption is managing electricity generation. To some extent, most countries are already doing this by changing their energy mixes. As we have seen earlier in this chapter, improvements in energy efficiency and sustainability can be made by considering options such as nuclear power and renewables like HEP, wind and solar power.

Governments can also effectively encourage change through taxation. For example, the UK Climate Change Levy was introduced in 2001 to 'encourage improved energy efficiency and reduce greenhouse gas emissions'. All businesses have to pay this tax, but employers receive tax incentives if they adopt energy-efficient schemes and other good practices. However, removing the exemption for renewable energy companies in 2015 did raise the question of whether it is worth continuing when it penalises clean and dirty power plants alike!

⬇ **Figure 2** *The German Reichstag – the greenest parliament building in the world*

Refurbishing the Reichstag

Energy efficiency is increasingly central to an architect's brief – particularly when designing or refurbishing commercial or public buildings.

Look at Figure **2**. The German parliament building, the Reichstag, was extensively refurbished in the late 1990s to be more energy efficient – it is now the greenest parliament building in the world. Its new energy system is based on a mixture of solar energy, geothermal power, *combined heat and power* (CHP), biomethane generators, and innovative ventilation. More than 80 per cent of the electricity it uses is generated internally and a geothermal installation cools the building in summer and provides heat in the winter. Special insulation limits heat loss. This has led to a 94 per cent cut in the Reichstag's carbon emissions!

Decentralising energy generation

Decentralising energy (DE) generation by developing CHP systems can be cheap, clean and efficient. A DE network using CHP involves the local generation of electricity and, where appropriate, the recovery of the surplus heat generated as a by-product. This is then pumped into homes, industries, offices and so on, either as hot water or as steam, through District Heating (DH) networks of reinforced, insulated pipes.

Examples of CHP and District Heating systems are also found in the UK, albeit rarely on the scale of continental European examples. A CHP and DH system on the Byker Wall estate in Newcastle includes three efficient gas-fired boilers and a biomass boiler that burns wood chippings. Residents enjoy remarkably stable heating charges and surplus electricity is sold back to the grid.

DE – the 'Danish model'

District Heating networks provides Denmark with 60 per cent of its space and water heating. In Copenhagen, the figure is 98 per cent (Figure **3**). This so-called 'Danish model' is much admired. For example, Copenhagen's remarkable District Heating network has steadily evolved since 1984. It now:

♦ delivers energy to households and businesses via 50 000 km of pipework

♦ halves energy costs for 'typical' home owners and businesses

♦ cuts carbon emissions from heating – 40 per cent lower than gas boiler systems, and 50 per cent lower than oil boilers.

Nationally, the Danish approach to managing energy consumption is even more wide-ranging. Initiatives include:

♦ laws requiring utilities to provide capacity for 450 MW of electricity via decentralised CHP systems

♦ high taxation applied to fossil fuels

♦ banning electrical heating in new buildings in 1988, and in all existing buildings in 1994

♦ converting fossil fuel CHP systems to biomass (wood pellets and straw) and *Municipal Solid Waste* (MSW). MSW incineration now meets approximately 30 per cent of Copenhagen's heat demand.

Such are the successes of this national approach that Denmark is the only net exporter of energy in the EU.

▶ **Figure 3**
Combined heat and power (CHP) in the Copenhagen area

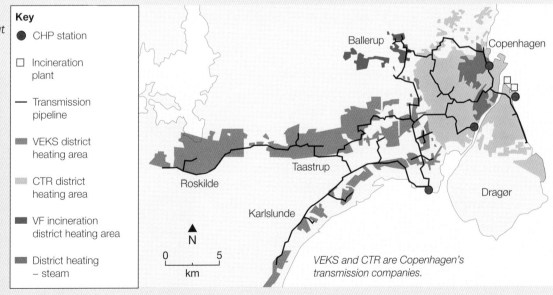

Key
● CHP station
□ Incineration plant
— Transmission pipeline
▨ VEKS district heating area
▨ CTR district heating area
▨ VF incineration district heating area
▨ District heating – steam

N
0   5
km

VEKS and CTR are Copenhagen's transmission companies.

ACTIVITIES

1 Between 1 and 2 per cent of UK heating (5 per cent in London) comes from CHP systems. Outline the lessons to be learnt from the 'Danish model' in managing UK energy consumption in future.

2 Why are District Heating Networks best suited to urban areas?

STRETCH YOURSELF

Which is easier for a government – managing energy supply or managing energy demand? Explain your reasoning. (You will find section 5.16 useful in your decision-making.)

In this section you will learn about sustainability issues associated with energy production, trade and consumption: the enhanced greenhouse effect

What is the enhanced greenhouse effect?

The greenhouse effect is much misunderstood (Figure **1**). Without it our planet would be frozen and lifeless! It is climate change resulting from the *enhanced* greenhouse effect that is the problem – it is arguably the greatest environmental challenge of our time where sustainable approaches to energy production, trade and consumption are key elements to avoiding the worst consequences.

Figure 1 *The what, why and where of the greenhouse effect*

What are greenhouse gases?	• CO_2 is the principal greenhouse gas • Other greenhouse gases include water vapour, methane and nitrous oxides
What do they do?	• They allow short-wave radiation from the Sun through to the Earth, and then trap some of the longer wavelength radiation that would otherwise be radiated back into the atmosphere • They act like a chemical duvet within our atmosphere • They keep the Earth much warmer than its distance from the Sun should allow
Why key trends show that the effect is being 'enhanced'?	• Our atmospheric duvet's 'tog' rating is rising • Since the Industrial Revolution, human activities have caused CO_2 levels to rise by a third • Methane levels have also doubled – and methane is 20 times more potent than CO_2 • CO_2 could, theoretically, double during this century • Nitrous oxides are steadily increasing
Where do the greenhouse gases come from?	• Most CO_2 comes from burning fossil fuels. Burning tropical rainforests is another major source (Figure **2**) – trees take carbon dioxide from the air, and lock up the carbon as they grow (see *AQA Geography A Level & AS Physical Geography*, 1.9) • Methane emissions are growing even faster than CO_2. Much is belched by ever-increasing numbers of cattle. Microbe activity in (rice) padi fields, the burning of MSW and emissions from landfill sites, coal mines, natural gas pipelines and melting permafrost are also significant factors • Nitrous oxides are most associated with road and air transport but they are also released by fertilisers

Why does it matter?

The Earth is warming abnormally affecting climates, weather patterns and sea levels with potentially alarming consequences. Ice is being lost rapidly – melting sea ice, continental ice sheets and mountain glaciers (Figure **3**). But this has impacts far beyond changing landscapes and ecosystems. Millions of people in countries like China, India, Bolivia and Peru already depend on meltwater from mountain glaciers for irrigation, domestic and industrial water supplies, and HEP. Longer-term concerns tend to focus on:

- ◆ projected future sea-level rise threatening low-lying regions such as Bangladesh, the Nile Delta, the Netherlands and Florida, USA, and also world cities including London, Bangkok, Singapore, Kolkata (Calcutta) and Tokyo

- ◆ more intense rainfall events causing localised flooding

- ◆ more droughts, heatwaves, forest and bush fires

- ◆ stronger hurricanes, cyclones and typhoons

- ◆ major environmentally driven migrations (with increased potential for conflict).

Figure 2 *The rate of Amazon rainforest destruction is now slowing following an initiative whereby the international community pays governments not to carry out deforestation. These Climate Protection Payments are reducing emissions and protecting natural habitats.*

What can be done?

Most of the technologies required to curb greenhouse gas emissions are already mastered but further development just needs sufficient political will and financial investment. With the exception of stopping all tropical deforestation (and planting 300 million hectares of new trees), it is notable that all technological 'solutions' involve energy production, trade and consumption. For example:

◆ reducing distances travelled by cars and increasing engine efficiency

◆ converting coal-fired power stations to natural gas, with **carbon capture and storage (CCS)** for all (see *AQA Geography A Level & AS Physical Geography*, 1.9)

◆ increasing the use of biofuels, as long as virgin rainforest isn't cleared to grow them

◆ doubling the production of nuclear power

◆ increasing the area of solar panels by a factor of seven hundred

◆ reducing carbon emissions from buildings and appliances by a quarter, through more efficient insulation, lighting and appliances.

The Kyoto Protocol

The economic and political aspects of combating climate change are often more difficult to be positive about. From the 1992 United Nations Framework Convention on Climate Change (UNFCCC), the hugely influential Kyoto Protocol of 1997 followed. This was the first agreement between nations committing them to reducing greenhouse gas emissions. The Protocol's first commitment period ended in 2012, the second will end in 2020. Crucially, Kyoto has not been an end point, with 195 countries adopting the first-ever legally binding global climate deal at the Paris Climate Conference in 2015. This agreement, due to come into force in 2020, sets out a global action plan to limit global warming to 'well below' 2 °C (Figure **4**).

Emissions (carbon) trading

One particularly influential outcome of the Kyoto Protocol was **emissions (carbon) trading**. These 'cap and trade' schemes operate by setting strict emission targets to reduce the amount of pollutants released into the atmosphere. Those factories or utilities that can reduce their emissions to come under their current cap can then 'trade' their remaining allowance by selling to others who do not meet their targets. So trading lets organisations buy and sell allowances while acting as a huge incentive to invest in cleaner technology. Each year the cap is lowered gradually, so organisations can plan well in advance to be allowed fewer and fewer permits.

The UK set up the world's first carbon trading scheme which was subsequently adopted throughout the EU. The Climate Change Act also commits the UK to reducing emissions by 2050 by at least 80 per cent of 1990 levels.

STRETCH YOURSELF

Permafrost melting is the most dangerous feedback associated with climate change. Furthermore, there are as yet no agreed estimates of the potential effect of the methane that would be released. Explain this worry.

🔺 **Figure 3** *Beneath the melting Aletsch Glacier, Valais, Switzerland*

1 °C warmer	The world is already 1 °C warmer than in pre-industrial times
2 °C warmer	The longer-term concerns already outlined
3 °C warmer	The threats start to ratchet up – 80% of Arctic sea ice could melt, there would be major species extinctions and sea levels could be up to 25 m higher than at present
4 °C warmer	Apocalyptic scenarios resulting from the irreversible thaw of permafrost – plummeting food production, hundreds of millions hungry, a fifth of humanity affected by flooding, Amazonian rainforest collapse

🔺 **Figure 4** *Global warming – key average temperature increase consequences*

ACTIVITIES

1 Cap and trade has been described as 'the most environmentally and economically sensible approach to controlling greenhouse gas emissions.'
   a Why is such emission (carbon) trading so effective?
   b What socio-economic and political considerations influence governments in making decisions about annual reductions in 'caps'?

2 To what extent does emission (carbon) trading give wealthy organisations and countries the excuse to do nothing about climate change?

In this section you will learn about energy conservation, acid rain and nuclear waste

Energy conservation

Energy conservation is a key issue in reducing greenhouse gas emissions, addressing energy poverty, ensuring sustainability and saving money. Indeed, it is potentially big business given the promise of financial savings to energy consumers.

A focus on domestic consumption alone demonstrates just what is needed and possible, not least given that more than a quarter of UK CO_2 emissions come from energy used in the home.

For a long time, UK building regulations have encouraged energy efficiency, for example through the promotion of double glazing and government grants for fitting cavity wall and loft insulation in existing buildings. Electrical appliances now carry energy-use ratings in order to encourage consumers to think about their energy demands. 'Smart' electricity meters using real-time displays of consumption are now available. Tungsten light-bulbs have been banned and replaced with low-energy alternatives (which also last much longer).

Look at Figure 1. In 2008, Woking Borough Council in Surrey converted a double-glazed detached house to show homeowners what further steps they could take for themselves to conserve more energy and reduce their carbon footprint. The UK government's Code for Sustainable Homes, introduced in 2006 pushed this further by:

- improving thermal efficiency of (triple-glazed) windows, walls and roofing
- installing high-efficiency condensing boilers
- installing energy-efficient domestic appliances and light bulbs
- using environmentally-friendly building materials
- adopting solar water heating and electricity-generating photovoltaic panels, and wind turbines where practical

The code was withdrawn for new developments in 2015 with some of the measures introduced into building regulations for all new builds.

▼ **Figure 1** *Woking Borough Council's conversions to conserve energy and make a low-carbon home*

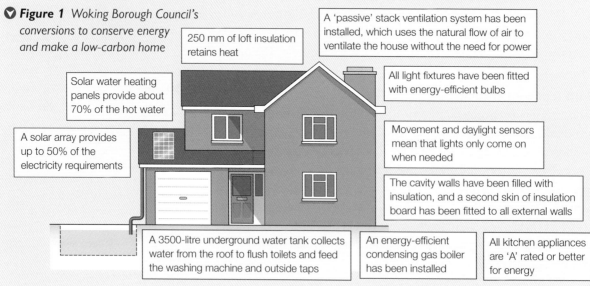

250 mm of loft insulation retains heat

A 'passive' stack ventilation system has been installed, which uses the natural flow of air to ventilate the house without the need for power

Solar water heating panels provide about 70% of the hot water

All light fixtures have been fitted with energy-efficient bulbs

A solar array provides up to 50% of the electricity requirements

Movement and daylight sensors mean that lights only come on when needed

The cavity walls have been filled with insulation, and a second skin of insulation board has been fitted to all external walls

A 3500-litre underground water tank collects water from the roof to flush toilets and feed the washing machine and outside taps

An energy-efficient condensing gas boiler has been installed

All kitchen appliances are 'A' rated or better for energy

What can we do?

Individual positive actions can become an energy-saving revolution when collectively multiplied:

- calculating our carbon footprint enables us to understand and change behaviours which have most impact
- driving less by sharing journeys, using public transport or walking (which has health benefits)

- reducing our 'food miles' by sourcing locally produced, seasonal produce wherever possible
- switching off electrical devices at the wall
- recycling and reusing
- buying energy-efficient appliances
- flying less or choosing an energy-efficient airline.

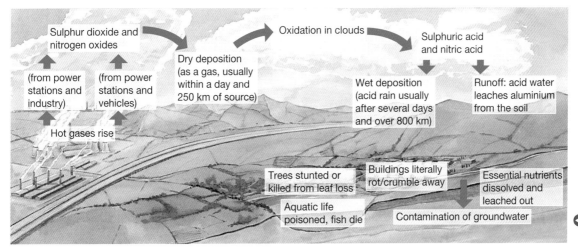

Sulphur dioxide and nitrogen oxides

(from power stations and industry)

(from power stations and vehicles)

Hot gases rise

Dry deposition (as a gas, usually within a day and 250 km of source)

Oxidation in clouds

Sulphuric acid and nitric acid

Wet deposition (acid rain usually after several days and over 800 km)

Runoff: acid water leaches aluminium from the soil

Trees stunted or killed from leaf loss

Buildings literally rot/crumble away

Essential nutrients dissolved and leached out

Aquatic life poisoned, fish die

Contamination of groundwater

◀ **Figure 2**
Acid rain issues

Acid rain

Greenhouse gas emissions dominate our thinking now, but in the 1980s and 1990s fossil fuel air pollution was more commonly associated with acid rain (Figure **2**). The environmental impact of dying forests, dead lakes and corroded buildings could not be ignored and so followed air quality concerns (discussed in 5.13) as another **trans-boundary pollution** event requiring international cooperation. Emission targets set at the 1979 Geneva Convention, Clean Air Acts and technical innovations such as sulphur dioxide scrubbers in coal-fired power stations have reduced the problem in Europe and the USA. However, Asia is the new acid rain hotspot with China, India, Thailand and South Korea now having to address the same issues.

Nuclear waste

We have already established that nuclear power has much credibility as a 'clean', sustainable, albeit expensive solution to energy emission and security issues, particularly in EMEs and HDEs (see 5.12, 5.16). But nuclear power stations produce high-level radioactive waste (in the form of used fuel rods), which has to be managed.

Look at Figure **3**. The THORP reprocessing plant at Sellafield in Cumbria takes used fuel rods from all over the world, extracts reusable uranium and plutonium and stores the remaining waste in steel-clad or concrete and lead-lined vitrified glass containers. This waste has a long half-life and so safe transport and storage challenges are a major issue. Certainly, waste stockpiles at Sellafield cannot increase indefinitely so the need for geologically stable, accessible, safe and secure long-term storage sites is pressing. Such sites would need to consider:

◆ disturbance during construction

◆ short- and long-term safety concerns

◆ accident and terrorism risks

◆ effects on the local economy.

Management of nuclear waste, high research and development costs, past accidents and excessive decommissioning costs are widespread concerns that this remarkable energy source is more of a curse than a blessing.

▲ **Figure 3** *Storage pond for used fuel rods at the THORP nuclear waste reprocessing facility at Sellafield, Cumbria, UK*

ACTIVITIES

1 Calculate your carbon footprint (search the internet for ways to do this).

   a List five simple measures to reduce your carbon footprint.

   b Choose one best suited to saving your family money and implement it.

   c What is the relationship between carbon footprint and water footprint? (see 5.9)

2 What is meant by 'trans-boundary' pollution?

3 Why has acid rain re-emerged as an environmental problem?

4 Is nuclear power a curse or a blessing? Explain your answer.

In this section you will learn about:

◆ sources, demand, distribution and uses of copper

◆ aspects of physical geography and geological conditions associated with ore occurrence and working

◆ environmental impacts and sustainability of extraction, processing and distribution

Copper – one of the world's most important metals

Imagine a world without electronics… without electricity! Copper conducts heat exceptionally well, but more importantly it plays a crucial role in all modern technology because of its excellent ability to conduct electricity. This is why copper is one of the world's most important metals. Copper is a key raw material for electricity transmission cables, wiring, telecommunications and electronic components such as semi-conductors (Figure **1**). Other important uses are in:

◆ construction, such as water piping, tanks and taps, roofing, lightning rods and decorative facing (Figure **2**)

◆ transport – train and vehicle radiators, bearings, brakes, relays and switches

◆ refrigerators, air-conditioning systems and microwave oven magnetrons

◆ combined with nickel to make a corrosion-resistant material used in shipbuilding, oil rigs and desalination plants

◆ fungicides and nutritional supplements.

Copper's value should never be underestimated. It has been used for over 7000 years and is the third most widely used metal in the world (behind steel and aluminium). It is durable, malleable and oxidises (corrodes) very slowly. (This is why copper cylinders can be used to store spent nuclear fuel. For example, the integrity of a copper cylinder with 10 cm thick walls would remain for one million years!) Copper is also totally recyclable, with nearly 90 per cent of the available scrap currently recycled. It is also very useful when blended with zinc to make the alloy brass, used in many musical instruments.

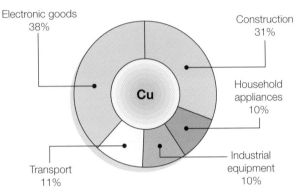

Electronic goods 38%
Construction 31%
Household appliances 10%
Cu
Industrial equipment 10%
Transport 11%

🔺 **Figure 1** *Uses of copper worldwide*

Did you know?

New York harbour's iconic Statue of Liberty was a gift from France, designed by the French sculptor Bartholdi and built by Gustave Eiffel (of Eiffel Tower fame). Dedicated in 1886, it was an inspiring sight to new immigrants because it represented the freedom and hope America promised.

▶ **Figure 2** *The figure of the iconic Statue of Liberty is constructed of shaped copper plates, oxidised to a distinctive green patina and fixed to an iron frame; the flame of the torch is gold leaf*

Sources, distribution and trade

Look at Figure **3**. Both recycling and copper mining is needed to meet an annual consumption that exceeded 21 million tonnes in 2015. Ore deposits are highly concentrated (unlike other basic metals like iron ore and bauxite, which are dispersed). Latin America accounts for almost half of global copper ores, with Chile having the highest reserves, followed by Peru (Figure **4**). Geology accounts for the wide scatter across a large number of countries because the ore is only found in veins and cavities of basic igneous rocks. In almost every case, the metal content is very low with mined ores containing less than 0.5 per cent copper. Consequently, only open pit mining and primary processing on site is economic.

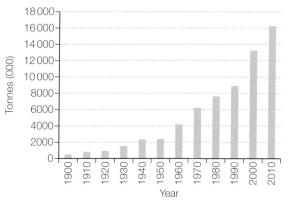

⬣ **Figure 3** *World copper mining production*

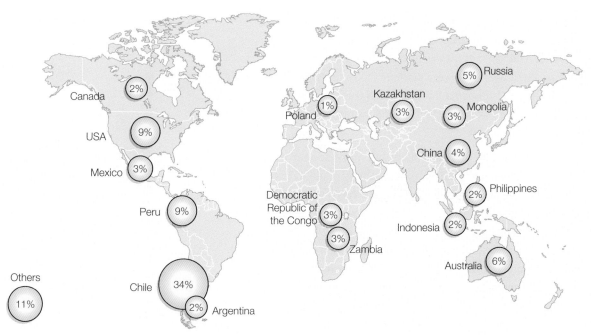

⬣ **Figure 4** *The global distribution of copper ore reserves*

While copper mining is concentrated in Latin America, more than 60 per cent of global copper consumption is in Asia. This necessitates a global copper trade, whether as copper concentrates, anodes, cathodes, ingots, semi-finished products or scrap for recycling. Over one-third of all copper now used has been recycled.

Yet, while trade has long been necessary, the patterns have changed markedly over recent decades. For example, in 1990 the USA was the world's biggest copper consumer, with Japan second. By 2015, China had overtaken the USA as the world's leading manufacturer and also accounted for 42 per cent of global copper consumption. Despite having only 4 per cent of global copper ore reserves, it had become the second largest miner by 2013 and made up the shortfall by becoming the key driver of global trade.

The new world dynamics dictated by China's economic boom years have been explored widely throughout this book. Inevitably, therefore, with the Chinese economy now growing at a slower pace, demand for all commodities, including copper, has fallen and with it the price. But analysts know that copper is always amongst the first to rebound because it has such a wide range of uses, including home appliances.

The Kennecott Bingham Canyon Mine, Utah, USA

Look at Figures **5** and **6**. If you're into supersizing, biggest ever lists and mega this and that, then this copper mine in Utah, USA has it all! The Kennecott Bingham Canyon Mine, 30 miles south-west of Salt Lake City, is the largest copper mine in the USA and has been nicknamed 'the granddaddy of all copper mines' and the 'richest hole on Earth'. What was a mountain over 100 years ago is primarily a copper mine but also produces gold, silver, molybdenum, platinum and palladium. It is the world's deepest man-made open pit excavation – 4.5 km across and 1.2 km deep.

Scale is everything in this facility: the mine employs 2000 workers, has 500 miles of internal roads and 450 000 tonnes of rock are extracted daily. The value of metals produced annually exceeds US$1.8 billion and if the mine was a stadium, it could seat nine million people. Its 320-tonne haul trucks, each the size of a two-story house, need new tyres every nine months at a cost of US$25 000 each (Figure **7**)!

Since opening in 1906, the mine has produced more copper than any other in history – more than 19 million tonnes. The UK's Rio Tinto Group bought Kennecott in 1989 and has invested nearly US$2.5 billion in modernising the mine and addressing environmental problems. For example, water trucks with 50 000-gallon tanks constantly spray water onto roads to reduce dust. From an in-pit crusher, the ore is moved by conveyer belt, partly through a tunnel, to the processing plant. A high-grade copper-gold 'skarn' deposit has been discovered approximately 300 metres below the open pit. This has led to Rio Tinto looking at the possibility of extending the life of the mine to 2029.

▲ **Figure 5** *Aerial view of the Kennecott Bingham Canyon Mine; this open pit copper mine is the largest man-made excavation on Earth*

▲ **Figure 6** *Location of Kennecott Bingham Canyon Mine, Utah, USA*

◀ **Figure 7** *A new 320-tonne haul truck costs about US$3.5 million*

Record-breaking disasters

Even potential disasters break records. For example, on the morning of 10 April 2013, all employees were evacuated from the bottom of the pit after ground probes and radar monitoring equipment detected slope deformation increasing five-fold (from 1 mm to 5 mm a day). Two landslides of both bedrock and mine waste followed that same evening – the second described as 'an avalanche' of 160 million tonnes of material moving at estimated speeds of 70–100 miles per hour. It proved to be the largest non-volcanic landslide ever recorded (Figure **8**). The two slides caused the equivalent of magnitude 5.1 and 4.9 earthquakes followed by 16 smaller aftershocks. The truck maintenance building (described as more of an aircraft hanger than a work shop) was sliced nearly in half, 13 vehicles were buried and the pit access road too. Debris over 60 metres deep covered the pit floor.

▲ **Figure 8** *The 2013 Kennecott Bingham Canyon Mine landslide*

The human impacts of this event were far less serious than might be expected. As a result of effective site monitoring and the resulting evacuation order, no fatalities or injuries occurred. Crews were back on site three days after the slide to start the removal of six million tonnes of waste rock and to build a new access road so that mining could be resumed. The task was completed seven months ahead of schedule and limited output losses to 50 per cent of capacity. The mine's visitors centre and outlook point had to be closed while parts of the structure were moved ahead of the slide. But job losses were inevitable. Around 170 workers took early retirement and another 40 were made redundant or transferred elsewhere in Rio Tinto's operations.

Did you know?

The Kennecott Bingham Canyon Mine Visitor Centre is dedicated to educating the public about mining practices, sustainable development and the importance of mining in modern life. Since opening in 1992, more than three million visitors have generated US$2.8 million for Rio Tinto's charitable foundation.

Think about

The copper production process

Copper ore has a very low metal content producing vast quantities of waste rock. An ore containing 2 per cent copper is regarded as rich! The copper must then be treated further to make it usable. The first step after extraction is called *beneficiation* where the copper-bearing rocks are converted into more usable shapes for smelting which increases the copper content beyond 50 per cent. After this, the copper is melted and cast as anodes, which are then electro-refined to produce high-purity cathodes containing more than 99.99 per cent copper content. The refined copper is then melted and cast by fabricators into different shapes, including rods and wire (Figure **9**) depending on the demands of end users. Most fabricators nowadays are concentrated in China – the world's largest copper consumer.

▼ **Figure 9** *Copper wire bundled and ready for transport*

Environmental impacts of copper mining

Look again at section 5.2. Copper mining is not untypical of all mineral extraction in that mining operations, whether surface or underground, can have severe environmental impacts by:

◆ disrupting the landscape

◆ removing vegetation and topsoil

◆ contaminating the air with dust and toxic substances

◆ causing toxic compounds (including acids) in mine *tailings* to percolate into the groundwater (Figure **10**).

Mineral processing also has environmental impacts. Most deposits generally contain less than 30 per cent of the desired mineral, some, such as copper, often much less than this. This means large quantities of rock must be dug up and processed to yield a much smaller quantity of mineral. Inevitably, there are large amounts of both *overburden* (waste material lying over the ore) and *tailings* (the materials left over after extracting the valuable mineral from the ore) to be managed. Furthermore, as minerals are used up, miners have to resort to lower grades of ore – and the lower the grade of ore, the worse the environmental impact. For example, 400 years ago copper ores typically contained about 8 per cent metal. Now it is less than 1 per cent.

This waste can cause environmental problems if simply dumped on spoil heaps or back-filled into the pit. If the tailings are not covered and stabilised, dust and water leaching through the waste can carry toxic materials into the environment. Consequently, dams, embankments and other types of surface impoundments, known as Tailings Management Facilities (TMFs) are by far the most common storage methods used today. Air pollution from both smelting gases and dust also require management (Figure **11**).

▲ *Figure 10 Copper tailings ponds are sequentially covered and landscaped; this one is in Utah, USA*

Restoration and rehabilitation

Fortunately, restoration and rehabilitation of sites is becoming more common, but it can be expensive. For example, Kennecott Bingham Canyon mine tailings are sent through a pipeline from the processing plant to a tailings impoundment north of the town of Magna where they are stored but not left unmanaged. Since the Rio Tinto Group bought the mine in 1989, it has spent US$450 million in cleaning up historic workings and groundwater.

The success of any environmental restoration depends on the:

◆ characteristics of the site

◆ quantity of material removed

◆ depth of the deposit

◆ composition of the surrounding rocks, soil and ore

◆ type of mineral being mined.

Success of course also depends upon political conditions in the country or region, not least environmental regulations and how well they are enforced, and the power and/ or influence of environmental pressure groups. Furthermore, the mining industry and conservationists often disagree over whether rehabilitation can restore ecosystems to their original state and even whether that is necessary in particular cases.

◆ **Figure 11** Managing the impacts of copper mining

Impact	Steps taken to reduce the impact
Dust is emitted from open pit blasting and also trucks carrying ore, tailings and stockpiles	Water trucks (using recycled water whenever possible) are used to spray mine sites and roads to suppress the dust
Vast quantities of overburden is generated from open pit mining that is dumped in spoil heaps	Monitoring of spoil heaps to limit size, maintain stability and control water content; eventually they can be landscaped by recontouring, covering with topsoil and revegetating
The land is changed significantly as a result of open pit mining	Top soil is removed before mining and reused later in rehabilitation. Landforms are recontoured to resemble the natural landscape
Tailings from processing are impounded in tailings dams	Tailings dams can eventually be covered with clay, topsoil and revegetated
Large pits are created from open pit mining	Exhausted pits can be filled in with rock waste, used for MSW landfill (see 3.20) or left to fill with groundwater and used as recreational lakes
Copper smelting produces sulphur dioxide gas	Most of the gas can be collected and used to make sulphuric acid for use in fertiliser

Sustainability issues associated with copper extraction, trade and processing

Section 5.2 examines sustainability issues associated with mineral ore extraction and processing in another of the Rio Tinto Group's operations – the Rössing Uranium Mine in Namibia. The knee-jerk assumption that all mineral ore extraction is as environmentally destructive as it is unsustainable (in the widest sense of the word) is crude, dated and therefore rightly challenged. Very often, sustainability is economically sensible. For example, copper recycling is now the norm and uses only 15 per cent of the energy that is required to extract and purify copper from its naturally occurring ore.

Furthermore, where copper is extracted, it is usually refined at source and, as mentioned earlier, bulk transport efficiently distributes copper concentrates, anodes, cathodes, ingots, semi-finished products or scrap. Nowadays copper extraction, processing and trade are typical of other major mining operations with an emphasis on environmental stewardship, energy conservation and health and safety.

Community and well-being

The development of sustainable communities for workers and their families is a feature of many mining operations. For example, just as Rio Tinto Group developed the new town of Arandis in support of its Rössing Uranium Mine operations in Namibia, so it conceived the Daybreak community in South Jordan, Utah to serve the Kennecott Bingham Canyon Mine (Figure **6**). The development was planned as a 'mixed-use, walkable community that incorporates quality education, a healthy and renewable environment, and a vibrant local economy'. Development is still ongoing and has distinct liveability characteristics (see 3.22). Rio Tinto Group sold it in 2016 to the global alternative investment adviser Värde Partners in order to 'move the Daybreak vision forward' and concentrate on its mining operations.

ACTIVITIES

1 Study Figure **4**.
   a Describe the distribution of copper ore reserves.
   b Plot the information on a bar graph.
   c Evaluate the usefulness of this representation in comparison to the original map.

2 Explain the economic necessity for copper smelting to be located at a raw material location.

3 a Describe and explain the difference between overburden and tailings.
   b Explain the paradox that both are increasing in volume, yet decreasing in their long-term environmental impact.

4 What evidence could be presented in support of an argument that the Kennecott Bingham Canyon Mine is environmentally, economically and socially sustainable?

Did you know?

The effect of copper exposure upon mine workers has been studied widely. Exposure to high levels of copper can increase the risk of both lung cancer and coronary heart disease. To reduce the chance of long-term exposure, copper mine workers are moved between operations such as smelting, electro-refining and fabricating.

STRETCH YOURSELF

'The assumption that all mineral ore extraction is as environmentally destructive as it is unsustainable is crude and also dated.' Critically evaluate this contention with respect to the copper industry.

In this section you will learn about:
* the factors influencing future water and energy demands and supplies
* water and energy futures in India

Water and energy – the future

It will rain at the weekend – at least that's what the weather forecast predicted! But are forecasts always accurate? The further we look into the future, the less certain our predictions become. (This is one argument that climate change scientists have grappled with for years – see 5.18.) However, the more information that we have available to us, the clearer our vision of the future becomes.

Supply and demand

How nations respond to future challenges of supply and demand may reflect one of three scenarios:

* Business as usual – there will not be a crisis and therefore no changes are needed.
* Technology, economics and **privatisation** – accepting that problems exist but the **market economy** will solve them using technological fixes.
* Values and lifestyles – dramatic shift in attitudes, requiring education, international cooperation and behavioural changes.

Look at Figure **1**. It is already clear that the providers of global water and energy supplies are struggling to meet demands – the futures of both are anything but certain.

Did you know?
Drilling in the North Sea Culzean oil and gas fields is possible because of not only new high pressure and high temperature drilling techniques but also tax incentives!

Factors influencing water and energy futures	Effect on water supply and demand	Effect on energy supply and demand
Technological	GM crops are being designed so they need less water to grow. Improved desalinisation plants increase fresh water supply and are becoming cheaper to run.	New energy reserves become exploitable with application of new technologies (e.g. directional oil drilling in the 1930s, fracking in USA in 2000s)
Economic	Increasing costs of storage and transfer might force end users to reduce their water footprint. Conversely, our need for water might result in further expensive engineering solutions	Stock energy resources will become more expensive as supplies become exhausted – which may force reduced consumption. Alternatively, previously uneconomically recoverable resources may become viable (e.g. oil in the Arctic Ocean)
Environmental	Improvements in efficiency and waste reduction, e.g. rainwater harvesting or use of greywater, might lead to reductions in demand for freshwater	As reserves become exhausted, environmental taboos may 'have' to be crossed (e.g. oil in Alaska or even Antarctica. Delayed effects of climate change might force reduction in use of hydrocarbons/fossil fuels)
Political	Fewer dams will be built if environmental and social opinions influence policy. Attempts to reduce water footprints might result in less food miles and local produce being consumed instead	**NIMBYism** is likely to continue to influence policy-making at all levels. Nuclear energy remains politically contentious but governments are increasingly worried over meeting future energy gaps

Figure 1 Factors affecting future water and energy demand and supplies

India's water and energy future crises

The Oxford Dictionary defines *crisis* as 'a time of intense difficulty or danger… when a difficult or important decision must be made'. India is facing a water and energy crisis (Figure **2**). This has resulted in significant economic and environmental challenges with many asking if India is able to provide sustainable responses.

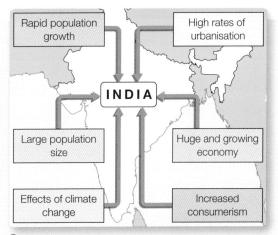

▲ **Figure 2** Factors in India's water and energy crisis

Water futures

Demand for water is increasing across India – from rural farmers, industry and urban consumers. These factors are linked. For example, a growing middle class (Figure **3**) demand more 'western' clothing and diets including dairy products and meat. This causes a shift in agriculture to cotton and dairy cattle, both of which use more water. Climate change is making rainfall less predictable – failed monsoons occur more often which impacts in particular already marginal semi-arid regions. Economic growth has resulted in over 70 per cent of surface water becoming polluted and, despite government controls, the sheer daily volume of waste materials is a huge management task.

Look at Figure **4**. India has the engineering capacity to match China's water megaprojects (see 5.13) but, almost inevitably, this will lead to future conflict.

Energy futures

Widespread blackouts in northern India in July 2012 were a stark reminder that energy demand may be reaching a point when it outstrips supply. Since the 1990s, India's energy sector has no longer been solely owned by the state but this privatisation policy has only been partly successful. The result of this has been under investment and an increasing reliance on energy imports which, along with dwindling domestic reserves, have resulted in a fall in domestic production. This has left India exposed to fluctuating energy prices and the uncertainties of energy geopolitics. India now imports over one-third of its gas, yet geopolitical concerns have meant that pipelines have not been built with close neighbours such as Pakistan and Iran.

▲ **Figure 3** New housing in Kerala, India; an increasingly affluent and growing middle class are enjoying benefits of an improved lifestyle

▲ **Figure 4** Thenmala Dam, Western Ghats, India; there are already over 400 large dams over 15 m high and construction of further large-scale water engineering projects is inevitable

ACTIVITIES

1 Explain the relationship between energy usage and water supplies.

2 Why do the poor rarely benefit from any developments that address national water or energy shortages?

3 Study Figure **3**. Why is India's growing middle class placing increased demands on water and energy resources?

STRETCH YOURSELF

Study the 'Supply and demand' text. Consider the scenarios outlined. Explain how each one might be achieved. What are the particular challenges for EMEs?

In this section you will learn about California's water crisis at regional and local scales

California – The Golden State

Where else can you surf in the morning and snowboard in the afternoon? Lifestyle and affluence – California screams 'America' to the world! Yet the assumption that earthquakes are California's greatest geographical threat is not shared by its growing population. Their worry is water – an unquenchable thirst that is increasingly difficult to meet.

Crisis … what crisis?

California's water crisis is not sudden. Forty years of population growth shows no sign of slowing – and the demand for water already exceeds natural supplies. Furthermore, there is a **spatial imbalance** – three-quarters of the population lives to the south of Sacramento, yet three-quarters of the precipitation falls to the north.

Look at Figure **1**. See how low those precipitation figures are and 65 per cent is lost to evapotranspiration. Climate change is increasing the variability and unreliability even more! Persistent high pressure in the western Pacific Ocean stops storms reaching California between November and March. It is during this period that the state typically gets 50 per cent of its precipitation. This means that droughts are now frequent. There was protracted drought between 2000 and 2007, and 2015 was the fourth consecutive year of yet another. It was estimated that drought cost the California economy US$2.7 billion in 2015. Much of this cost was in the huge irrigation-dependent farming industry as fields were left fallow and valuable crops such as grape vines died.

Keeping California's taps running

The majority of California's supply normally comes from surface water, with 30 per cent from groundwater aquifers and a very small proportion from desalinated sea water. However, groundwater demands rise by up to 60 per cent in drought years. This puts huge strains on unsustainable aquifers where recharge since 1999 has not kept pace with extraction.

Six main local, state and federal funded water systems now manage and redistribute water from ten drainage basins across California (Figure **2**). However, given the shared infrastructure (such as the State Water Project using the Central Valley Project's California Aqueduct) it has become an interconnected, productive but controversial system.

Did you know?

With a population of almost 39 million, California is the most populated state in the USA. If it was a country, it would be the seventh largest economy in the world!

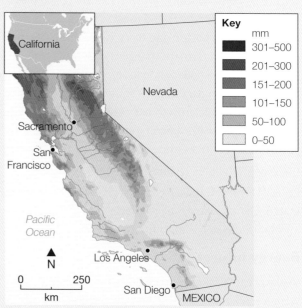

Key

mm
301–500
201–300
151–200
101–150
50–100
0–50

⬆ **Figure 1** *California's annual precipitation. Sacramento, the state capital, separates northern and southern California*

Key

Height (m)
>1000
501–1000
0–500
Local
State
Federal

⬆ **Figure 2** *California's water supply system*

Crucially the SWP combines with the CVP to provide water to both central and southern California, with the Colorado River Water Scheme providing further water to southern regions and neighbouring states, such as Nevada and Arizona (Figure **3**).

The State Water Project (SWP)	• 29 dams, 5 HEP projects and 970 km of canals and pipelines • Supplies nearly half of California's drinking water • Irrigates 300 000 hectares of farmland
The Central Valley Project (CVP)	• originally intended for irrigation and control of flooding • 22 dams and reservoirs provide HEP • Principal use is irrigation for 1.2 million hectares and drinking water for 2 million consumers
The Colorado River Water Scheme	• Supplies half of all southern California's water via a system of dams and aqueducts • Irrigates 1.4 million hectares of farmland • Supplies drinking water to major population centres such as Los Angeles and San Diego • Supplies HEP and water beyond California to such as Nevada and Arizona (e.g. Hoover Dam (Figure **4**), Glen Canyon Dam

🔺 **Figure 3** *Summary of California's water supply systems*

Challenges, conflicts and controversies

Ensuring water security in California raises significant challenges. Conflicting demands, environmental consequences, conservation issues and even questions of 'ownership' all complicate a management problem that must result in long-term sustainability. The battles between environmentalists and farmers, the south and the north, rural and urban give some sense of the conflicts that underlie this issue:

◆ Of all California's water, 80 per cent is used for farming – albeit increasingly efficiently. Nearly half of farmers now use advanced scientific methods that determine when, how much to and how best to irrigate – sprinkler, drip, and so on. But 1.6 million hectares of land are still flood irrigated, suggesting that there is further progress to be achieved.

◆ Environmentalists are concerned that wetlands have already been drained, natural habitats altered and river fish stocks depleted. For example, the Colorado River delta at the Gulf of California is massively degraded – the flow is little more than a trickle. This means that any industrial and agricultural effluent is more concentrated.

◆ Water rights are continually disputed. Native American tribes have longstanding legal rights. Ownership of the Colorado River has long been disputed by USA and Mexico. Political arguments on water shares and quality are complicated to say the least!

Did you know?
Each of the 860 golf courses in California uses almost 410 million litres of water annually which would fill 136 Olympic-size swimming pools!

🔺 **Figure 4** *The Hoover Dam and the Colorado River Bridge, Nevada/Arizona, USA*

◆ The spatial imbalance between distribution of precipitation and population was addressed by the State Water Project (SWP). However, northern Californians feared 'ownership' of their water by the south, while southerners worried about continuity of supply. Meanwhile, the demands of the Bay–Delta transfer region between the two increases the controversy of who exactly has 'rights' to what water and why?

◆ In 2015, mandatory restrictions to cut demand by 25 per cent were introduced. Water waste is discouraged through higher rates and fees, state-sponsored aid for homeowners who install more efficient garden watering systems, conservation through the use of greywater (water recycling), drought-tolerant landscaping, and so on.

Good enough to drink?

Look at Figure **5**. Industrial pollution and dirty agricultural and hazardous household waste runoff has been seeping into groundwater wells throughout Los Angeles County for decades – the most polluted wells had to be closed. Water transfer schemes (see Figure **3**) have, until recently, met demand, but as the Californian drought continues and water becomes scarcer, water agencies are increasingly having to use groundwater sources again. Indeed, most communities in southern LA County now rely, at least partially, on contaminated groundwater sources. The contaminated water has to be treated and purified, which is expensive. But is it safe? In January 2016, treatment failures led to 20 000 residents in Watts, a poor multicultural community in inner city Los Angeles, drinking untreated well water for six hours. Nevertheless, abstraction and subsequent treatment from polluted aquifers is becoming a cheaper option than imported water – despite any potential health risks.

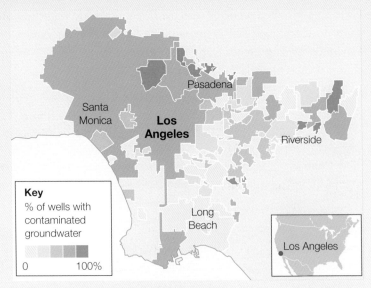

⚫ **Figure 5** *Percentage of water wells with contaminated groundwater (prior to purification), Los Angeles County*

Clean water – it's a steal!

In 2016, bottled water was donated to the many hundreds in drought-affected Tulare County (Figures **2** and **6**) who were without drinkable mains water. Such was the desperation that residents stole bottled water from neighbours. All this in the richest state in the USA. What is happening?

Look at Figures **6**, **7** and **8**. Following successive years of drought, the Tule River that once supplied the wells of Tulare County has long since run dry. Many local residents depend on a private water well for supply. As these wells dry up, whole communities have been forced to change their attitudes towards water conservation and also to make fundamental changes to their lifestyle. Porterville is one of the worst-affected communities.

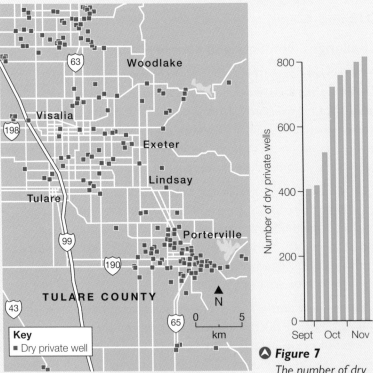

⚫ **Figure 6** *Tulare County had the highest percentage of dry water wells of any county in California*

⚫ **Figure 7** *The number of dry private wells doubled between September and December 2015*

Finding water in Porterville

In 2015, you were one of the more fortunate households in Porterville if mains water was still on tap. For many of the 55 000 residents that are reliant on private water wells, the daily routine involved pumping water from containers and then transporting this back home. Water containers that served the community became commonplace at the roadside and non-potable (non-drinking) water was available at fire stations and community centres (Figure **9**).

In the absence of any mains water, such pumped water (often by hand) had to serve all domestic needs – including flushing the toilet and washing clothes.

Aid programmes were a lifeline to thousands of the most disadvantaged households – for example, in the first six months of 2016, the county-wide Bottled Drinking Water Programme delivered water to nearly 2000 qualifying households and FoodLink distributed 268 000 food packages.

Socio-economic costs

Tulare County lies in the heart of California's rich agricultural belt. Labour-intensive crops, such as fruits and vegetables (not including wine grapes) have traditionally employed many farm workers and been a mainstay of the local economy. This is no longer the case. The prolonged drought has reduced water allocation to farmers who, in turn, have chosen to grow smaller quantities of less water-intensive crops. In practice, this means that some fruits and vegetables (such as 'delicate' melons and lettuces that are hand-picked) are being replaced by higher value crops such as almonds and pistachios that are subsequently machine harvested. In other cases, land is simply being left fallow.

Consequently, there is less need for, in particular seasonal, farm workers. In Tulare County, there were 1500 fewer farm workers in 2015 than the previous year – a drop of nearly 19 per cent. As agriculture is the main industry, this is all causing a negative socio-economic multiplier effect for whole communities. In 2015, one in four Tulare County families lived below the federal poverty level and were reliant on food stamps and other forms of aid. Families were being forced to relocate to find work. This impacts on everything from purchases at the local grocery store to attendance at school (fewer students means less state funding, fewer teachers and staff and so on). Even when and if the rains return, the shift to more mechanised forms of farming may mean recent changes to agriculture become permanent.

▲ **Figure 8** *The dry riverbed of the Tule near the town of Porterville, October 2014*

▲ **Figure 9** *Water tank near the fire station in East Porterville, California, September 2014*

ACTIVITIES

1 Explain why demand management is the only realistic 'solution' to California's water crisis.

Ⓢ
2 Study Figures **5** and **6**.
   a Describe the distribution of contaminated water wells in Los Angeles county (prior to purification)
   b Describe the distribution and number of dry private water wells in Tulare county
   c Comment on the socio-economic impacts of these patterns.

3 Give examples of how water shortages can cause socio-economic multiplier effects at both regional and local scales.

4 Study Figure **9**. Making reference to security, language and health issues, comment on the design and labelling of the water tank.

STRETCH YOURSELF

Recent droughts in California, USA, may prove to be a long-term blessing in disguise. Discuss.

In this section you will learn about the effects of the physical environment on the oil exploration industry

Alaska – physical geography

Tectonically active, and divided by mountain ranges into markedly contrasting coastal and interior regions, Alaska has a diverse natural resource base including major fisheries, gold and other minerals (Figure **1**). It also demonstrates remarkable glaciated and periglaciated landscapes, ecologically diverse wilderness, and protected environments such as reserves and wildlife refuges. Its extreme climatic variability is complicated further by significant warming (up to 1 °C per decade has been recorded) resulting in melting of glaciers, thawing of permafrost and associated environmental impacts. Alaska's socio-economic geography is much more straightforward, dominated by exploitation of its vast oil reserves.

Oil exploitation in Alaska, USA

An oil boom that began in the 1970s with the development of the 25 billion barrel Prudhoe Bay oilfield and Trans-Alaska pipeline (Figure **2**) has generated wealth that funds up to 90 per cent of the state's operating budget and now accounts for one-third of all jobs! Alaska is a classic **single-product economy**. Furthermore, Alaska residents do not pay any state income or sales taxes. In fact, since 1982 they have collected annual dividend cheques from the Alaska Permanent Fund (the state's oil-wealth savings account) valued at more than US$50 billion.

However, oil production has been on a slow decline since production from the Prudhoe Bay region peaked in 1988 at about two million barrels a day. Volumes running through the Trans-Alaska pipeline are now down by 75 per cent from peak flows.

Oil security and human welfare are inextricably linked in Alaska. So it could be argued that Alaska has to keep the oil flowing – whether through more onshore development or by increased offshore drilling. This would push the technology beyond anything yet known, given the extreme cold, ice and darkness of Arctic waters in winter.

Look again at Figure **1**. The Western Arctic Reserve, with its extensive wetlands and large population of threatened species, is home to groups of Native Americans who depend on the wildlife for their food, clothing and shelter. Almost 500 000 caribou live and migrate through the reserve. However, the coastal plain of the Arctic National Wildlife Refuge (ANWR) was test

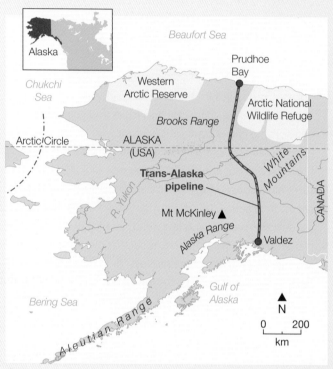

▲ **Figure 1** Trans-Alaska pipeline, Alaska, USA

▼ **Figure 2** The 1300 km Trans-Alaska pipeline took two years to construct and links the Prudhoe Bay oilfield to the ice-free port of Valdez. In a tectonically active region, where temperatures drop as low as minus 50 °C, it crosses 800 rivers, two mountain ranges, permafrost, fragile tundra fauna and flora, and wildlife migration routes.

drilled in the 1980s and proved to have reserves estimated at over ten billion barrels, provoking emotive arguments as to whether or not oil exploitation should be allowed. Environmentalists argue fiercely that the fragile tundra soils and vegetation would never recover from oil exploitation – with all the implications for biodiversity – whilst economists highlight socio-economic arguments including:

◆ furthering US energy security

◆ securing and creating jobs

◆ lowering oil prices for consumers

◆ increasing federal, state and local tax revenues

◆ reducing the US trade deficit.

For Alaskan politicians, the failure to exploit the ANWR elevated the importance of exploration offshore. Royal Dutch Shell holds Arctic Ocean offshore licences and in 2008 started to explore for oil in the Chukchi Sea off Alaska's north coast (Figure **1**). The US Bureau of Ocean Energy Management estimated that the oilfield could hold nearly 30 billion barrels of oil, development of which would provide jobs and much-needed oil for the Trans-Alaska pipeline. But in September 2015 the US$7 billion plan was abandoned. Although environmentalists had been concerned about the impact of exploitation in these pristine Arctic waters, the decision was largely driven by economics, with the price of oil having plummeted to US$50 per barrel.

Think about

Ironies of the Alaskan oil story

• Alaska is undergoing extraordinary climate change – and currently warming at one of the fastest rates on Earth. A warmer, wetter climate, particularly in the south-east and interior, is now believed to be inevitable. This would lead to glacial retreat, thawing of permafrost, a shorter snow season and reduced sea ice – much of which will make oil exploration, exploitation and transportation easier, but it could also render the Trans-Alaska pipeline redundant.

• President Barack Obama, a Democrat, opposed any drilling in the ANWR and extended the ban to further areas on the coastal plain east of Prudhoe Bay. Yet he sanctioned offshore drilling.

• Obama's Republican predecessor, President George W. Bush, continues to argue that it would be possible to recover oil from the ANWR – using directional drilling techniques – whilst simultaneously protecting the fragile environment. Before becoming a professional politician, he worked in the oil industry.

The Arctic National Wildlife Refuge (ANWR)

The ANWR was established in remote north-eastern Alaska to protect a virtually untouched wilderness of extraordinary wildlife and recreational qualities (Figures **1** and **3**). The 7.7 million hectare refuge includes diverse habitats such as treeless tundra, boreal forest, barrier islands and coastal lagoons. It runs from the mountains of the Brooks Range north to the Beaufort Sea.

Described as one of the last intact landscapes in America – where natural processes remain little influenced by the modern world – it contains an abundance of wildlife, including polar bears, oxen, caribou and millions of migratory birds. The ANWR is also the home Alaska's Gwich'in Indians, whose hunting grounds are protected by law.

⬢ **Figure 3** *Caribou calf crossing the Kongakut River in the Arctic National Wildlife Refuge*

ACTIVITIES

1 Why has the Trans-Alaska pipeline been described as one of the world's greatest engineering feats?

2 Outline the arguments for and against oil exploitation in the ANWR.

3 Why do politicians, oil companies and environmentalists hold mixed views as to the future viability and sustainability of oil exploitation in Alaska?

STRETCH YOURSELF

'Oil companies have looked longingly at the Arctic for years, but its often icebound seas and treacherous weather make exploring expensive and dangerous. Shell's decision could spell the end of Arctic drilling for some time, although low oil prices and geopolitics – not environmental concerns – are the main reason.' (*Wall Street Journal*, 29 Sept 2015) Critically evaluate this statement.

Now practise...

The following are sample practice questions for the Resource security chapter.

They have been written to reflect the assessment objectives of Component 2: Human geography Section C of your A Level.

These questions will test your ability to:

◆ demonstrate knowledge and understanding of places, environments, concepts, processes, interactions and change, at a variety of scales [AO1]

◆ apply knowledge and understanding in different contexts to interpret, analyse and evaluate geographical information and issues [AO2]

◆ use a variety of quantitative, qualitative and fieldwork skills [AO3].

1 Outline the different components of water demand. (4)

2 Study Figure 1. How might the concept of virtual water be helpful in addressing the issue of the sustainability of water management at a global scale? (9)

Figure 1 *The virtual water content of a product is the volume of freshwater used to produce the product, measured at the place where the product was actually produced*

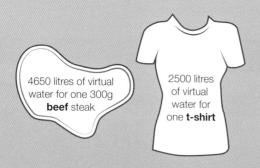
4650 litres of virtual water for one 300g **beef** steak

2500 litres of virtual water for one **t-shirt**

1000 litres of virtual water for one litre of **milk**

650 litres of virtual water for one loaf of **bread**

70 litres of virtual water for one **apple**

3 With reference to a named country or region, what strategies have been successfully employed to increase energy supply in recent years? (9)

4 With reference to a metal ore, compare the global distribution of reserves/resources with the geography of its end users. (6)

5 'The global trade in energy resources defines geopolitics more widely.' To what extent would you agree with this statement? (9)

6 Assess the environmental impact of a major energy resource development and its associated distribution networks. (9)

7 What type of techniques or technologies can be used to manage energy consumption in HDEs? (6)

8 Evaluate the success of different strategies used to increase water supply to a location or locations you have studied. (9)

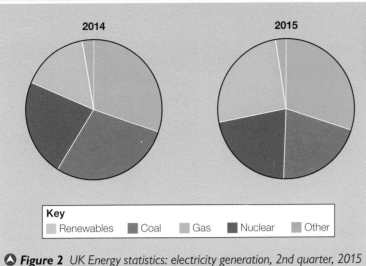

	TWh (Terrawatt hours)	% change from 2014
Electricity generated from		
Coal	16.14	−27.4
Nuclear	16.92	−3.3
Gas	23.79	0
Renewables	19.94	+51.4
Total	78.75	−0.1
Electricity supplied to		
Industry	22.69	+0.8
Domestic	24.16	−0.6
Other final consmers	25.00	+2.8
All	71.85	+1.0

Key
Renewables Coal Gas Nuclear Other

⊘ **Figure 2** *UK Energy statistics: electricity generation, 2nd quarter, 2015*

◁ **Figure 3**

Renewable energy outstrips coal for first time in UK electricity mix

In the second quarter of 2015, renewable energy contributed a record 25 per cent of the UK's electricity. This was the first time that renewables had overtaken coal in electricity generation for a sustained period. This announcement came in the same week that the government was widely criticised for cutting support for clean energy.

The increase in the production of electricity from renewable sources between April and June was, in part, because of the high winds and long sunshine hours during this period, but also because more turbines and solar panels had been installed since the same quarter in 2014, when renewables contributed just 16.7 per cent of electricity.

The rate of construction of new sites for renewable energy production has been rapidly increasing in recent years, while aging coal and nuclear plants have been closing down.

9 Study Figures **2** and **3**. Using these sources, assess the relative contribution of renewable energy to the UK's energy mix. (9)

10 Explain the relationship between energy supply and the physical geography of a named country or region. (6)

11 Outline some of the sustainability issues associated with the development of a mineral resource you have studied. (9)

12 What challenges do water companies and authorities face trying to manage the consumption of a resource viewed less as a commodity and more often as a common good? (20)

13 'When it comes to government approval of major resource development projects, environmental impact assessments are often no more than a "fig leaf" for decisions that have already been taken.' To what extent do you agree with this statement? (20)

AS Geography fieldwork investigation

Stages in a geographical investigation

Like all other scientific investigations, a good fieldwork investigation follows a series of logical stages. This helps to ensure that the investigation is accurate, thorough and stands as a valid piece of research.

In *Component 2: Human geography and geography fieldwork investigation*, you could be asked questions on any aspect of a fieldwork enquiry, as described in Figure **2**, which you may find useful as a checklist when you are revising. Would you be able to answer a question on every aspect?

With regard to fieldwork, get involved at every stage! At AS you should experience a minimum of two days of fieldwork. Over the course of these two days you should undertake fieldwork relating to both physical and human processes in geography. Always be clear on the theory or idea your investigation sets out to test. You will need to go back to it towards the end of your analysis and consider how your results relate to it.

▲ *Figure 1* *Taking in the street scene in Sion, Switzerland*

❏ Preparation for fieldwork, including:
  o background reading
  o drawing up aims and objectives for the enquiry
  o planning research in the field and from secondary sources
  o using data-sampling techniques
  o carrying out health and safety procedures.

❏ Collection of primary data in the field and using secondary data sources.

❏ Processing and presenting data using relevant graphical and cartographical techniques.

❏ Analysing data, including using statistical techniques where relevant.

❏ Drawing conclusions related back to the original aims and objectives and linking these conclusions to both the place studied and the general ideas forming the basis of the enquiry.

❏ Reviewing the success, or otherwise, of all stages of the enquiry.

❏ Considering how the enquiry could be further developed.

▲ *Figure 2* *AS Geography: stages of a fieldwork-based enquiry*

Using different kinds of data

You need to be aware of the pros and cons of selecting qualitative or quantitative methods of data collection. How do you go about analysing quantitative and qualitative data differently? (See 2.10, 2.11.) may give you some ideas.

Statistical skills

At AS, your study on the use of statistical techniques could include:

- measures of central tendency (mean, mode, median)

- measures of dispersion (range, interquartile range and standard deviation)

- inferential and relational statistical techniques, to include Spearman's rank correlation and application of significance tests.

Think also about your secondary sources. How would you justify your choice of sources? How useful and how reliable are they?

If your study 'doesn't work'

Remember, there is always room for improvement in any investigation – so reflect on and recognise this when evaluating your methods or analysing your results.

A Level Geography fieldwork investigation

Stage 1: Identifying an appropriate question or issue

The specification requires that you undertake:

'an independent investigation [with] a significant element of fieldwork … based on either human or physical aspects of geography, or a combination of both.'

During your A Level course you will be introduced to a range of different fieldwork techniques. Your time 'in the field' will cover techniques that relate to processes in both physical and human geography and should help you to make a decision about what you want to find out more about in your independent investigation.

Once you have chosen a topic, you will need to define your key question, or identify the specific issue that your investigation will focus on and, within that, a number of hypotheses you want to test. This will require some further reading on your part. Take a look at more than one textbook, so go to the library and also look online.

Choosing your title

In choosing the title of your independent investigation you should consider the following points.

- Your title must be geographical, linked to the specification and written by you.

- It could be the statement of a hypothesis or, alternatively, a question you would like to answer.

- There must be one or more clear connections to sound geographical theory, concept or process (such as Agnew's definition of place (2.2). You'll need to set your key question in the context of relevant literature in your write-up.)

- You will need to incorporate primary data (field data collected first-hand through observations or measurements) and/or good-quality secondary data (evidence from other people's field investigations) – either way you will have to demonstrate an understanding of methods used in the field

- The investigation must be based on a small, manageable area of study such as a small town or an area of woodland. Consider that a maximum of 1 to 2 days should be needed for collecting the primary data.

- The investigation title must lend itself to the full development of a geographical investigation. For example, it should prompt the collection of plenty of data for interpretation and statistical analysis.

- You must be able to conduct the investigation safely – no abseiling, wading up to your waist in fast-flowing rivers, crossing motorways, etc.

Some examples of appropriate titles

How has counter-urbanisation changed the characteristics of the village of Trull?

Media representations of Great Missenden affect the sense of place felt by residents

Do 'insiders' and 'outsiders' have different experiences of Dovedale, in the Peak District National Park?

Stage 2: Planning

Many of the common problems students find when conducting an independent investigation come down to lack of proper planning. Thorough, careful planning will reduce problems later on and help you to achieve a successful outcome. If you have time, consider a small pilot study (a few hours of fieldwork to try things out) or take time to reflect on past fieldwork experiences; what worked and what didn't?

When planning your investigation you need to consider the following, presented in Figures **3** and **4**, with regard to your data collection:

▼ *Figure 3* *Planning your data collection*

Data collection

Where are you going to collect your data and when?

- The choice of site(s) is very important and must be justified.
- You should also consider when to carry out your data collection: early morning or late afternoon; on a weekday or weekend? Figure **4** lists some factors you should consider when conducting a coastal or an urban study.

What kind of data are you going to collect, how and why?

- You need to carefully consider what data you need to fully test your hypotheses and answer your question.
- Would qualitative or quantitative data be more useful, and what is the most appropriate technique to employ? (E.g. questionnaire or interviews, see 2.11)
- What sort of secondary data would be helpful, in addition to primary data?
- Be careful not to collect so much data that you simply become overwhelmed.

What sampling strategies are you going to employ and why?

- Make sure you choose an appropriate strategy for your question or issue.
- Consider whether you need to use *point*, *line* (transect) or *area* (e.g. quadrat) sampling.
- Also consider what kind of sampling you will use – *random* (chance), *systematic* (at regular intervals, e.g. every 10th person or every 10 metres) or *stratified* (biased, e.g. 25 per cent sample points from each of four areas). If sampling the population of a place, you might question a predetermined percentage of people from three or four different age groups, to ensure your sample reflects the age makeup of the community. Check the population data from the 2011 Census.

What equipment is needed and how does it work?

- You need to select appropriate equipment for an A Level investigation. For example, the use of a decibel meter to accurately measure noise levels in different locations at different times of day.
- The equipment should be reasonably easy to use, not too expensive and should be of good enough quality to provide you with accurate and reliable results.
- Take time to practise using the equipment before you start your investigation.
- Consider having 'Plan B' equipment in case your phone or camera does not work (e.g. a pad and pencil) and carrying spare batteries if necessary.

▼ *Figure 4* *The type of environment in which you collect your data will affect your planning*

Considerations	Coast	Urban
Where?	• Is the length/area of coastline an appropriate size to enable me to collect the data I need? • How can I get there? • Does it have the appropriate characteristics for my study, such as exhibiting marked changes east to west, or with distance from the foreshore? • Can I gain safe (and legal!) access to an appropriate number of sites (e.g. about 8 to 10 if conducting a transect across coastal dunes)?	• Where do I go to collect my data and why? • Do I need to get permission to talk to people in certain places? How do I get it? • Is the settlement big enough? Is it too big? • How safe is the urban area? Are some areas too risky for me to go to alone? • Will my presence and data collection cause any problems for the local people?
When?	• Should I collect data at high tide or low tide? Or both? Why? • Is access available year-round? • Will the time of year affect biological indicators, such as vegetation? • Will the weather be ok?	• Will the time of year affect my results (e.g. tourism)? • Will the time of day (e.g. rush hour) affect my results? • Is the day of the week important (e.g. Sunday compared with Monday)? Do I need to be there on both days? • Will the weather affect footfall or visitor numbers?

Risk assessment

It is a legal requirement to carry out a risk assessment. A risk assessment aims to identify the potential risks of your fieldwork investigation and then minimise their likelihood of occurring, particularly for those risks that are potentially most likely or the impacts most severe.

To start, think about the risks that could occur on the day you conduct your fieldwork investigation (ideally, you should pay a preliminary visit to your study site to identify any risks that may be likely). Then, identify ways of minimising the risks, for example, wearing the correct footwear – Figure **5** below shows a typical urban fieldwork site with the associated risks.

It is also important that you consider what to do if something does happen if, for example, a member of your group becomes lost or unwell, or the weather turns wet and foggy.

There is a huge amount of helpful advice on the internet, and you can check the latest information from your phone (e.g. tide times and the weather forecast). Your school or college will also be able to help.

Crowded streets: people and vehicles

Risk: accident, causing an obstruction or nuisance

Solution: observe your location before you decide where to stand; look at where people walk or gather and whether vehicles also use the same space

Noise and atmospheric pollution

Risk: poor sound recordings and/or headaches

Solution: undertake a mini-investigation; try a few trial recordings of interviews at different locations and avoid areas with heavy traffic or fieldwork during the rush-hour

Weather

Risk: warm weather increases the risk of dehydration and sunburn

Solution: take plenty of water, sunscreen, hat, etc.

General public, safety and equipment

Risk: equipment gets damaged/stolen or is faulty or inaccurate

Solution: carry spares or alternatives; stay vigilant about keeping equipment safe, for example, in a safe place out of sight if not in use

▲ **Figure 5** *The potential risks and some possible solutions in the field*

Data presentation and analysis: before you go

It might seem strange to consider how you are going to present and analyse your data at the start of your investigation, but it is very important that you do so. In order for you to present your data in different diagrams and interpret it using statistical techniques, you need to have the appropriate type and amount of data. For example, if you plan to use Spearman's rank correlation test, you will need to collect two sets of data that can be ranked (see *AQA Geography A Level & AS Physical Geography*, 6.9). So you might collect data on the cost of housing (data set 1) with distance from the station (data set 2). Make sure you have enough observations to make your result meaningful (statistically significant); you must have at least five pairs of data (more than ten will give more reliable results).

Stage 3: Data collection

Be methodical

Be methodical when collecting data in the field, you need to try to ensure that the results you obtain are accurate and reliable. For example, make sure that you take several readings at a particular site, and check the equipment regularly. On the day of the data collection, you may make small amendments to the method you use. The need for these tweaks may only be evident at your first site, once you are knee-deep in the river! As far as possible, keep your method consistent between different sites.

Keep a record

Remember to make a note of any changes made to your methods so that you can describe and justify your approach in your report. You need to be clear on what you did and why you did it, so talk to other members of your team if you are collecting data in a group.

Do not forget to take photos and draw field sketches. These provide a good record of what you did and also help to put your fieldwork sites in geographical context. For example, the photograph in Figure **5** could be used to help explain certain trends or anomalies in the results of urban land use.

Other people's data

Try to include secondary data if possible. This might include maps, extracts from books, articles in magazines or online, official statistics or other students' work. Having another study to compare with your own is an excellent strategy for your write-up.

Stage 4: Data presentation

Raw data is of very little value when it comes to data interpretation. It needs to be processed and displayed so that it makes sense and reveals any trends and patterns. Look online or in any fieldwork textbook for presentation methods available to you, such as line, bar and scattergraphs, pie charts, triangular graphs and dispersion diagrams – many are included in this book. At A Level you need to select your method of presentation with care, make sure they are drawn accurately and are appropriate to your study. For example:

- If you carry out a transect in a town, data should be displayed in a linear manner along a scaled line as opposed to a single pie chart.

- Points on a scattergraph should not be joined up to form a line graph but should have a best-fit line to show the trend (drawing the line is a form of analysis).

- Spatial data (e.g. noise level) is most effectively shown by an overlay that uses colour to distinguish between different zones within an area.

Whether drawn by hand or on computer, make sure that all diagrams, are complete. Do not forget to include:

- a clear and focused title
- north arrow and scale bar on a map
- a key, if appropriate
- labelled axes on graphs
- annotations on field sketches, explanatory photos and what you observed on the ground.

Remember that any annotation of figures, content of textboxes and other text linked to presentation techniques all count towards the total word count of your final 3000–4000 word report.

Try to present your various datasets clearly and logically (e.g, by date of collection, sub-theme, or location) to guide the reader through your report.

Ensure you have included a range of presentation techniques – demonstrate the breadth of your abilities.

Which presentation technique? Ask yourself is your data continuous or discrete (non-continuous)? For discrete data use a bar graph, for continuous try a histogram.

Stage 5: Interpretation and analysis

In this section of your investigation you need to discuss, describe and, where possible, explain your results. It is also important to critically examine your data. To do this effectively you should:

- make use of appropriate statistical techniques to analyse your data. Consider using measures of central tendency (e.g. mean) or dispersion. If you are examining a relationship between two variables then you should consider using Spearman's rank correlation test Book 1, (6.9). If you are examining associations or looking to identify significant differences between data sets, you should consider using the Chi-squared or Mann–Whitney U tests (see below).

- consider patterns and trends rather than simply quoting lots of numbers. But be sure to use numbers to support any points you make. Translate raw data into percentages, describe the range.

- look for any exceptions (anomalies) to the general trends and try to explain why they have occurred.

- attempt to explain the patterns and trends you have observed using geographical terminology and referring to concepts and processes as appropriate.

- try to make statements that link your data sets together.

- make connections back to your title/hypothesis/ underlying theory.

Significance levels – don't forget!

With all statistical tests, the calculated result is only as good as the interpretation. Significance tables should be used to assess the probability that the result was due to chance. For most geographical investigations a significance level of 0.5 per cent is reasonable, i.e. there is a 95 per cent likelihood that the result was not caused by chance.

Any test of association or correlation (e.g. Spearman's rank correlation test) must be based on a sound geographical concept or theory.

▲ **Figure 6** *Data analysis*

Chi-squared test

The Chi-squared test is used to test the difference between grouped data sets. The two sets of data can be either two sets of raw data or a comparison between actual observed values and theoretical 'expected' values representing equal chance.

There are a number of important aspects of this test:

- Data must be in the form of frequencies, grouped or categorised.

- The total number of observations must exceed 20.

- The expected frequency for any one group must exceed 5.

- The categories should not have a directly causal link.

Mann–Whitney U test

The Mann–Whitney U test is used to test the difference between two sets of data. It is often carried out after drawing a dispersion graph that might suggest a possible difference between two sets of data, for example, the sizes of pebbles collected on two beaches.

There are a number of important aspects of this test:

- It measures the difference around median values.

- Raw individual data must be used rather than grouped data.

- There should be between 5 and 20 values, although it is possible to use more than 20.

- The two data sets do not need to have an equal number of values.

Qualitative data

Not all data sets are suitable for statistical analysis. Old photographs, your own photographs, field sketches and observation notes all provide valid data for analysis, and should be included, but these qualitative sources may require a different approach. Use overlays or online annotation that demonstrates your interpretation of these sources. You might wish to present your images on a map ('geolocate' them) or organise them by the date they were taken, to allow the reader to observe the patterns or changes you describe. If you have a lot of this form of data you will need to be selective, presenting those images or sources that best illustrate physical phenomena, the interaction between different groups of people and places, or between people and the environment.

The use of coding to analyse interview transcripts involves the careful categorisation of this kind of data (see 2.11). It is an iterative, stage-by-stage process. For an approach to textual analysis, including poetry, painting and advertisements, see 2.7.

Relating your results to the wider context

Within your report you are expected to set your key question(s) for investigation in context. This means that early on you need to outline the theory or concept that prompted you to design your investigation as you did, demonstrating further reading on your part. Later on, as part of your analysis, you should attempt to explain any patterns and trends, linking back to relevant geographical theory.

Extending geographical understanding

The experience of completing your independent investigation will deepen your understanding of the theme you choose – you might suggest how and in what ways, as part of your conclusion or evaluation. Furthermore, in the local context, your study will provide colour and detail that overarching theories from textbooks cannot. You can comment on how your specific study develops the theory or process(es) as they manifest themselves in your local or study area. Just like any other researcher, your work will have 'pushed back the boundaries' of geographical understanding – congratulations!

Stages 6 and 7: Drawing conclusions and your evaluation

Your conclusion should pull together all of your results so that you can address the question or hypothesis in your title. To do this, briefly run through the main thrust of your results, but make sure that you focus on the title.

On reflection, you should also be realistic about how your investigation went. Consider what you would do differently if you began again, knowing what you know now. To evaluate your investigation you should:

- consider whether there were any inaccuracies in the data collection

- suggest how data collection could have been improved (e.g. more sites, more data)

- consider how representative your sample was and suggest whether or not you would take a different approach to sampling next time

- discuss the accuracy of your results, consider any problems with the data collection

- assess the validity of the conclusion, i.e. would a different person doing the same investigation get exactly the same results as you did?

Your written report should make reference to fieldwork ethics and how they applied to your investigation, either in the justification of your methods, your evaluation, or both.

Fieldwork ethics

Over the course of all geographical investigations you should strive to maintain the highest standards of scientific rigour, to protect the environment (Figure **7**) and treat people that you meet with respect.

The Code of Practice for researchers applying for funding from the Royal Geographical Society (with the Institute of British Geographers) is a useful guide on fieldwork ethics. It requires:

- accurate reporting of findings, and a commitment to enabling others to replicate results where possible
- fair dealing in respect of other researchers and their intellectual property
- confidentiality of information supplied by research subjects and anonymity of respondents (unless otherwise agreed with research subjects and respondents)
- independence and impartiality of researchers to the subject of the research.

It is science, be it physical or human

Do not exaggerate your results or leave data out that does not fit the wider trend or suit your hypothesis. You can discuss any anomalies ('outliers'), exceptions and possible inaccuracies in your evaluation.

In addition, you should give proper credit to other researchers. If you have used their research work to inform your understanding of a place or theme and you have included their data, you must cite this. You usually do this at the foot of the page or in a bibliography, as you would with any other source of secondary data.

In the field, in the street

Still photography is a great way to capture the street scene; but take care. People might like to be asked before they become the central focus of your observational photography – although it is okay to take quick snaps of a busy high street. Do not trespass on private property, even to get the 'ultimate angle' on the juxtaposition of old and new. And do consider the feelings of others when using images in your report or sharing your findings in class.

Do not get so absorbed in the use of your camera or survey that you forget to look both ways before crossing the road. You also should not become an obstacle that shoppers or workers have to navigate around!

When you try to stop and talk to people in the street, you will find that many say they do not have the time to stop. Do not be offended or harass reluctant participants – they are just busy (Figure **9**). Also do think about how you present yourself on the day. Do you look smart and well organised to a distracted stranger walking past? Would your style of dress make your approach off-putting for some groups within society? While you might not want to compromise your style, the practical need to collect some data may prompt some amendments depending on the sector of the community you would like to talk to.

⬆ **Figure 7** *Beach sign on the Gower Coast, South Wales*

⬆ **Figure 8** *An environmental quality survey in Buckinghamshire*

⬆ **Figure 9** *Interviewing in Zermatt, Switzerland*

343

The role of researcher

If you are investigating the impact on people of an issue or major change in the environment, ensure you approach the subject with sensitivity – it may be something that keeps them up at night! Take care with question design and avoid leading questions that might suggest your views and/or introduce bias to your data. Your role is to remain as neutral as possible and record what is said.

When recruiting people to answer your questionnaire or take part in a group discussion or interview, briefly explain the purpose of your research and how the results will be used. You will need to get their permission to record what they say or permission from their parents, perhaps, if they are not old enough to give their consent. And thank people for their time afterwards; being polite will always pay off. For a full discussion of your role as a researcher in the interview process, see 2.11.

Naming names: don't do it!

The results of any questionnaire or survey should make no mention of real names and the reader should not be able to identify individual participants from the data you present. Giving your interviewees the guarantee of anonymity allows them to talk freely, so you are much less likely to hear what they think you want to hear, and more likely to get their real views.

⬣ **Figure 10** *Your role as a researcher is to remain neutral and record what is said*

What does a successful independent investigation look like?

3.3.4.1 Introduction and preliminary research
Level 4: 9–10 marks

- A research question(s) is effectively identified and is completely referenced to the specification.

- Well supported by thorough use of relevant literature sources.

- Theoretical and comparative contexts are well understood and well stated.

⬣ **Figure 11** *St Brelade's Bay, Jersey*

3.3.4.2 Methods of field investigation
Level 4: 12–15 marks

- Detailed use of a range of appropriate observational, recording and other data collection approaches including sampling.

- Thorough and well-reasoned justification of data collection approaches.

- Detailed demonstration of practical knowledge and understanding of field methodologies appropriate to the investigation of human and physical processes.

- Detailed implementation of chosen methodologies to collect data/information of good quality and relevant to the topic under investigation.

⬥ **Figure 12** *Glasgow waterfront*

3.3.4.3 Methods of critical analysis
Level 4: 15–20 marks

- Effective demonstration of knowledge and understanding of the techniques appropriate for analysing field data and information and for representing results.

- Thorough ability to select suitable quantitative or qualitative approaches and to apply them.

- Thorough ability to interrogate and critically examine field data in order to comment on its accuracy and/or the extent to which it is representative.

- Complete use of the experience to extend geographical understanding.

- Effective application of existing knowledge, theory and concepts to order and understand field observations.

⬥ **Figure 13** *The Cotswolds, Broadway, Worcestershire*

3.3.4.4 Conclusions, evaluation and presentation
Level 4: 12–15 marks

- Thorough ability to write up field results clearly and logically, using a range of presentation methods.

- Effective evaluation and reflection on the fieldwork investigation.

- Complete explanation of how the results relate to the wider context(s).

- Thorough understanding of the ethical dimensions of field research.

- Thorough ability to write a coherent analysis of fieldwork findings in order to answer a specific geographical question.

- Draws effectively on evidence and theory to make a well argued case.

ACTIVITIES

Study Figures **11–13**. Pick one location and discuss the following with a partner.

a The risks of conducting fieldwork in this environment and how you would minimise these risks.

b Two possible hypotheses or key questions to investigate.

c Any ethical dilemmas that might arise as part of your proposed investigation in this place.

AQA material is reproduced by permission of AQA

Your AS or A Level course is a lively and interesting one, full of contemporary topics, such as Changing places as well as old favourites like Hazards. It is important to prepare for your assessment well, and you should take the time to understand the structure and assessment criteria of your course.

What does the **AS** specification include?

Remember that the AS Geography qualification does not count towards an A Level. The topic content is designed to provide you with a real knowledge of geographical themes and skills, along with a real enthusiasm for the subject.

The AS specification consists of six topics, of which you must study three plus a fieldwork investigation. They are grouped into two components: Physical geography and people and the environment; Human geography and fieldwork.

Component 1: Physical geography and people and the environment	Component 2: Human geography and geography fieldwork investigation
• Section A: you must study one of either Water and carbon cycles (Chapter 1 in *AQA Geography A Level & AS Physical Geography*) or Coastal systems and landscapes (Chapter 3 in *AQA Geography A Level & AS Physical Geography*) or Glacial systems and landscapes (Chapter 4 in *AQA Geography A Level & AS Physical Geography*) • Section B: you must study either Hazards (Chapter 5 in *AQA Geography A Level & AS Physical Geography*) or Contemporary urban environments (Chapter 3 in this textbook)	• Section A: Changing places (Chapter 2 in this textbook). This topic is compulsory. • Section B: Geography fieldwork investigation (Chapter 6 in this textbook) and geographical skills. You will have to answer one question on each of these topics.

Remember, there is no coursework at AS, but questions will be included in the exam about the fieldwork you complete as part of your AS course.

What does the **A Level** specification include?

The A Level specification consists of eleven topics of which you must study six plus a fieldwork investigation. They are grouped into three components: Physical geography; Human geography; Geography fieldwork investigation.

Component 1: Physical geography	Component 2: Human geography	Component 3: Geography fieldwork investigation
• Section A: Water and carbon cycles (Chapter 1 in *AQA Geography A Level & AS Physical Geography*). This topic is compulsory. • Section B: you must study one of either Hot desert systems and landscapes (Chapter 2 in *AQA Geography A Level & AS Physical Geography*) or Coastal systems and landscapes (Chapter 3 in *AQA Geography A Level & AS Physical Geography*) or Glacial systems and landscapes (Chapter 4 in *AQA Geography A Level & AS Physical Geography*) • Section C: you must study either Hazards (Chapter 5 in *AQA Geography A Level & AS Physical Geography*) or Ecosystems under stress (Chapter 6 in *AQA Geography A Level & AS Physical Geography*)	• Section A: Global systems and global governance (Chapter 1 in this textbook). This topic is compulsory. • Section B: Changing places (Chapter 2 in this textbook). This topic is compulsory. • Section C: you must study either Contemporary urban environments (Chapter 3 in this textbook) or Population and the environment (Chapter 4 in this textbook) or Resource security (Chapter 5 in this textbook)	• You must complete an individual investigation which must include data collected in the field. • Your investigation must be based on a question or issue defined and developed by the student relating to any part of the specification content

How will you be assessed?

AS Level

There are two exams for AS Geography. Unlike A Level there is no coursework – but you must complete two days of AS fieldwork. The AS exams will involve:

Component 1: Physical geography and people and the environment

There is a total of 80 marks available, worth 50% of the AS qualification. The paper consists of multiple-choice, short answer, levels of response and includes 9- and 20-mark extended prose questions. You may use a calculator.

You will have 1 hour 30 minutes.

- Section A: you must answer either question 1 (Water and carbon cycles) or question 2 (Coastal systems and landscapes) or question 3 (Glacial systems and landscapes) – **40 marks**

- Section B: you must answer either question 4 (Hazards) or question 5 (Contemporary urban environments) – **40 marks**

Component 2: Human geography and geography fieldwork investigation

There is a total of 80 marks available, worth 50% of the AS qualification. The paper consists of multiple-choice, short answer, levels of response and includes 9- and 20-mark extended prose questions. You may use a calculator.

You will have 1 hour 30 minutes.

- Section A: you must answer all questions (Changing places) – **40 marks**

- Section B: this will test fieldwork skills. You will be asked questions on your own fieldwork and the use of fieldwork skills in different situations – **40 marks**

A Level

There are two exams for A Level Geography and one individual fieldwork investigation – you must complete at least four days of A Level fieldwork. The A Level assessment will involve:

Component 1: Physical geography

There is a total of 120 marks available, worth 40% of the A Level qualification. The paper consists of multiple-choice, short answer, levels of response and includes 20-mark extended prose questions. You may use a calculator.

You will have 2 hours 30 minutes.

- Section A: you must answer all questions (Water and carbon cycles) – **36 marks**

- Section B: you must answer either question 2 (Hot desert systems and landscapes) or question 3 (Coastal systems and landscapes) or question 4 (Glacial systems and landscapes) – **36 marks**

- Section C: you must answer either question 5 (Hazards) or question 6 (Ecosystems under stress) – **48 marks**

Component 2: Human geography

There is a total of 120 marks available, worth 40% of the A Level qualification. The paper consists of multiple-choice, short answer, levels of response and includes 20-mark extended prose questions. You may use a calculator.

You will have 2 hour 30 minutes.

- Section A: you must answer all questions (Global systems and global governance) – **36 marks**

- Section B: you must answer all questions (Changing places) – **36 marks**

- Section C: you must answer either question 3 (Contemporary urban environments) or question 4 (Population and the environment) or question 5 (Resource security) – **48 marks**

Component 3: Geography fieldwork investigation

There is a total of 60 marks available for this individual investigation, worth 20% of the A Level qualification. You must complete an individual investigation based on a question or issue defined and developed by you relating to any part of the specification content.

You are advised to write between 3000 and 4000 words. Your investigation will be marked by your teachers and moderated by AQA.

How are the exam papers marked?

Examiners have to know what it is that they are assessing you on, so they use Assessment Objectives, or AOs for short. There are three AOs for AS and A Level.

- **AO1: Demonstrate knowledge and understanding** of places, environments, concepts, processes, interactions and change, at a variety of scales (30–40%).

- **AO2: Apply knowledge and understanding** in different contexts to interpret, analyse and evaluate geographical information and issues (30–40%).

- **AO3: Use a variety of relevant quantitative, qualitative and fieldwork skills** to:
  - investigate geographical questions and issues
  - interpret, analyse and evaluate data and evidence
  - construct arguments and draw conclusions (20–30%).

Understanding the mark schemes

There are two types of mark scheme, which depend on the number of marks allocated for shorter questions (up to 4 marks) and those for extended written answers (6 marks or more).

Shorter answers (worth up to 4 marks)

Questions carrying up to 4 marks are point marked. For every correct point that you make, you earn a mark. Sometimes these are single marks for 1-mark question. Others require the development of a point for a second mark, for example, if you are asked to describe one feature of something for 3 marks. *Outline the role of wind in affecting coastal energy. [3 marks]*

There are three marks for this question and you have to outline the role of wind. You would receive one mark per valid point (e.g. 'wind is responsible for the generation of waves as friction occurs at the surface of the water') with additional credit for development (e.g. 'this has a direct bearing upon the potential for longshore drift depending upon the angle that the waves hit the coastline').

Extended answers (worth 6 marks or more)

Questions carrying 6 marks or more are *level-marked*. The examiner reads your whole answer and then uses a set of criteria – known as levels – to judge its qualities. There are two levels for questions carrying 6 marks, three levels for 9 marks, and four levels for 20 marks.

Consider this question:

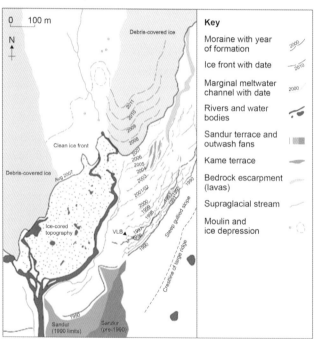

▲ *Figure 1* A geomorphological map of an Icelandic glacier, Virkisjökull-Falljökull

With reference to Figure 1, interpret the evidence that this glacier is changing. [6 marks]

Examiners would mark this question using the mark scheme on page 349, which consists of two levels. The mark scheme is the same for all 6-mark questions. This type of question tests interpretation of the quantitative evidence provided, so it is an AO3 question.

Level	Mark	Descriptor
Level 2	4–6	• Clear interpretation of the quantitative evidence provided, which makes appropriate use of data in support.
		• Clear connection(s) between different aspects of the data and evidence.
Level 1	1–3	• Basic interpretation of the quantitative evidence provided, which makes limited use of data and evidence in support.
		• Basic connection(s) between different aspects of the data and evidence.

Interpretation makes clear links between glacial retreat and evidence provided in Figure **1**. Use and understanding of the map evidence are clear and accurate. Your answer must fit the Level **2** criteria.

9- and 20-mark answers

9-mark questions have three levels in their mark schemes. These are identical for each of the exam papers. The balance of marks is split between 4 marks for AO1 and 5 marks for AO2. You should aim to:

- demonstrate accurate knowledge and understanding throughout (AO1)

- apply your knowledge and understanding (AO2)

- produce a full interpretation that is relevant and supported by evidence (AO2)

- make supported judgements in a balanced and coherent argument (AO2).

20-mark questions have four levels in their mark schemes. The balance of marks is split evenly with 10 marks for AO1 and 10 marks for AO2. In addition to the qualities for Level 3 you should aim to:

- reach a detailed evaluative conclusion that is rational and firmly based on knowledge and understanding which is applied to the context of the question (AO2)

- show a detailed, coherent and relevant analysis and evaluation in the application of knowledge and understanding throughout (AO2)

- show full evidence of links between knowledge and understanding and the application of knowledge and understanding in different contexts (AO2).

Command words

Command words – that is, those words that tell you what you must do – are listed here.

Analyse Break down concepts, information and/or issues to convey an understanding of them by finding connections and causes and/or effects. This assesses AO2.

Annotate Add to a diagram, image or graphic a number of words that describe and/or explain features, rather than just identify them (which is labelling). This assesses AO3.

Assess Consider several options or arguments and weigh them up so as to come to a conclusion about their effectiveness or validity. This assesses AO1 (as you have to know something) but mainly AO2 (as you are asked to apply your understanding and make a judgement).

Compare Describe the similarities and differences of at least two phenomena. This generally assesses AO1 or AO3.

Contrast Point out the differences between at least two phenomena. This generally assesses AO1 or AO3.

Critically Often occurs before 'Assess' or 'Evaluate' inviting an examination of an issue from the point of view of a critic with a particular focus on the strengths and weaknesses of the points of view being expressed. This assesses AO1 (as you have to know something) but mainly AO2 (as you are asked to apply your understanding and make a judgement).

Define…, What is meant by… State the precise meaning of an idea or concept. This assesses AO1.

Describe Give an account in words of a phenomenon which may be an entity, an event, a feature, a pattern, a distribution or a process. For example, if describing a landform say what it looks like, give some indication of size or scale, what it is made of, and where it is in relation to something else (field relationship). This assesses AO1.

Distinguish between Give the meaning of two (or more) phenomena and make it clear how they are different from each other. This assesses AO3.

Evaluate Consider several options, ideas or arguments and form a view based on evidence about their importance/validity/merit/utility. This assesses AO1 (as you have to know something) but mainly AO2 (as you are asked to apply your understanding and make a judgement).

Examine Consider carefully and provide a detailed account of the indicated topic. This assesses AO1.

Explain... Why... Suggest reasons for... Set out the causes of a phenomenon and/or the factors which influence its form/nature. This usually requires an understanding of processes. This assesses AO1 and occasionally AO2.

Interpret Ascribe meaning to geographical information and issues. This assesses AO3.

Justify Give reasons for the validity of a view or idea or why some action should be undertaken. This might reasonably involve discussing and discounting alternative views or actions. This assesses AO1

(as you have to know something) but mainly AO2 (as you are asked to apply your understanding and make a judgement).

Outline..., Summarise... Provide a brief account of relevant information. This assesses AO1.

To what extent... Form and express a view as to the merit or validity of a view or statement after examining the evidence available and/or different sides of an argument. This assesses AO1 (as you have to know something) but mainly AO2 (as you are asked to apply your understanding and make a judgement).

Handy hints

It can be tough to keep cool under pressure. However, these hints should help you perform much better.

> Note the command word – 'outline'! So provide a brief account of relevant info.

Outline the impact of temperature variation on weathering processes in hot deserts. [4 marks]

> The focus of the question – the impact of temperature variation.

> Relate everything you say to weathering processes – and why they're happening in the way they do.

Dissect the question

Look at the example to the right. For this question you should aim to make two well-developed points.

Choose examples and case studies

Case studies are in-depth examples of particular places, used to illustrate big ideas at localised scales. There are plenty in these books; some examples are a paragraph, while others run to several pages. You should use these to answer questions.

Few questions actually ask for examples but including examples can help you better demonstrate a point. The following is a question about Hazards (Chapter 5 in *AQA Geography A Level & AS Physical Geography*), where good examples could produce a better answer: *To what extent do you agree that seismic events will always generate more widespread and severe impacts than volcanic events?* [9 marks]

In this example question, specific examples of seismic and volcanic hazard would enhance your answer – they could be events that you have studied in this book (see sections 5.8 and 5.13 in *AQA Geography A Level & AS Physical Geography*). The important thing is that named examples make your answer precise – and you could find that three explained examples could help you better demonstrate and apply knowledge and understanding of the relative impacts of seismic and volcanic events.

But be selective – no question at this level will ever ask you to write everything you know about a case study.

Plan answers

Students who plan their answers before writing them usually score higher marks than those who don't. This is because planning:

- stops the student from going off-track

- prevents 'memory blanks'.

Your plans need not to be lengthy – allow 5–10% of total exam time for planning. Also, your plans should not be elaborate – brief notes will keep your answer on track.

Keep to time

Many students lose track of time in exams. The exam questions are designed to be completed in the time allowed, so the following suggestion may help you to achieve this. BEFORE the exam:

- work out how long each section should take to complete

- work out 5–10% total planning time, 80–85% writing time, and 10% checking time

- practise timed answers – including shorter answers, as well as longer ones.

Glossary

actors Players on the world stage – the governments and other global institutions, both public and private, that participate in global governance, exercise power, make decisions, solve problems and improve lives

agglomeration economy The economies of scale derived by firms clustering together and sharing ancillary services and public utilities; they also include industrial linkages, the development of a specialised labour force, bulk buying and shared marketing

agribusinesses Large farms, plantations or estates owned and managed by major companies that organise the purchase of all inputs, labour, processing and marketing

agrotechnologies The application of modern technologies, such as irrigation technologies to agriculture

albedo The reflectivity of a surface which varies according to its colour and texture; for example, fresh snow reflects 85 per cent of solar radiation, grass 25 per cent, but tarmac less than 10 per cent

allergen A substance that causes an abnormally vigorous immune response or allergy in the sufferer

alluvial plains Sedimentary deposits ranging from coarse gravel to fine silt spread across floodplains

Anthropocene A term used to convey the scale and importance of human influence in Earth's most recent time period, given widespread evidence of anthropogenic global warming etc. Scientists suggest that this new period of time began approximately 10 000 years ago with the end of the last glacial period.

anthropologists Scientists who study human beings in relation to their physical characteristics, social relations and culture, and the origins and distribution of races

aquifer A porous and permeable rock that acts as a groundwater store

archipelago A closely grouped cluster of islands

artesian basin Low-lying region where groundwater is confined under hydrostatic pressure from surrounding layers of rock; often found where an aquifer lies trapped in a syncline

asylum seeker A person who is seeking international protection but whose claim for refugee status has yet to be determined (see refugee)

bandwidth throttling The deliberate slowing of internet service by an internet service provider (ISP)

big data Contemporary term used to describe extremely large datasets from which we can learn a great deal with effective analysis

bilateral Between two parties

bioprospecting The process of discovery and commercialisation of new products based on biological resources

carbon capture and storage (CCS) Technology designed to replace the pumping of industrial and power-station fossil fuel CO_2 into the atmosphere

carrying capacity The idea of a population ceiling beyond which an environment cannot support people at a high standard of living (or at a subsistence level), for a sustained period of time without environmental degradation

channelisation Straightening and lining of a river channel to improve its rate of flow and navigability

chronic A term used to describe an illness that lasts for a long time or is constantly recurring

COBRA Acronymn for Cabinet Office Briefing Room A; the government emergency response committee, including government ministers, police and intelligence officers

code Title or abbreviation of a title used by a researcher to label and sort sections of text, according to theme to aid analysis; different codes and sub-codes may be applied to individual paragraphs or lines of text within a qualitative data source (e.g. an interview transcript)

colonialism The policy or practice of a power in extending control over weaker nations or peoples

community-focused regeneration Regeneration focusing on the social needs of communities; for example, affordable housing to rent or buy, shops and schools

continuous data Data that can take any value (within a given range), such as life expectancy

counterurbanisation Population movement from large urban areas to smaller urban settlements and rural areas

cratons Large, ancient sections of the Earth's crust that have remained relatively stable for considerable periods of geological time; they are associated with the 'drifted' fragments of Alfred Wegener's 'supercontinent' Pangaea

crude rates A measure of the basic statistics of any population, such as birth or death rates per 1000

decentralisation Process of redistributing people, functions or power away from the centre to the periphery

decile One of ten equal subsections that a population may be divided up into, according to the distribution of the ranked values of a particular variable

DEFRA The UK government's Department for Environment, Food and Rural Affairs

deindustrialised Long-term decline of industry leading to significant social and economic changes

demographers People who study population

demographic dividend A falling birth rate results in a smaller population of young, dependent ages and relatively more people within the economically active adult age groups; this improves the ratio of productive workers to child dependents which can encourage economic growth

deprivation A meaningful measure of poverty, defined in terms of people's lack of access to social and economic necessities. The Townsend Index of material deprivation (1988) incorporated four variables: unemployment; non-car ownership; non-home ownership; and household overcrowding.

dew point The critical temperature at which air, on cooling, becomes saturated with water vapour; as temperature falls further the vapour condenses into cloud droplets; consequently, the dew point is defined by the cloud base

dialect A particular form of a language which is peculiar to a place, region or social group

disability-adjusted life year (DALY) A measure of the overall burden of a disease that combines mortality (early death) and morbidity (ill-health, disability) in a single measure

discrete data Data that can only take certain values, such as number of babies born to a woman

diseases of affluence Degenerative diseases such as cancer, heart disease and dementia most associated with lifestyles in more economically developed countries

drosscape A landscape of industrial dereliction; rapidly urbanised regions that become the waste products of historic economic and industrial processes

Dutch disease The negative consequences as a result of large increases in a country's income; it is usually associated with the discovery of natural resources, particularly oil reserves, but can also result from any large increase in foreign currency such as FDI

ecological footprint A measure of the demands we humans place on ecosystems which support us; the amount of biologically productive land that is used to produce the resources we consume and to absorb the waste we generate

economic migrant A person who has left their own country to seek employment in another country in order to improve their living conditions

edge cities Modern suburban areas that act as an alternative central business district, including shops, offices and entertainment; they are characteristic of low density suburbanisation such as in the USA

emancipated Free from domestic or cultural restraints and so allowed to pursue independent lifestyles and careers

emissions (carbon) trading Effectively trading a permit to pollute. These permits are distributed to polluting organisations that may only emit as much carbon as it has allowances for – hence having to buy the right to pollute from more efficient businesses if it cannot become more environmentally friendly.

empowerment To give power or authority to someone; female empowerment involves the fuller participation of women in a nation's economy and society

endemic Disease that is always present in a population

endogenous factor A key aspect of a place's local geography (physical or human) that helps to shape its unique character, for example, geology

energy dependent The higher the proportion of energy imported, the more energy dependent the country is on others. In such cases, a diversification of both energy sources and suppliers is crucial.

energy poverty Having less energy than is required to meet demand

energy security The uninterrupted availability of energy sources at an affordable price. For example, Russia is very energy secure because of its huge energy surplus. The UK is energy insecure because of its energy deficit and has to import much of its supplies.

epidemic Widespread occurrence of an infectious disease in a community at a particular time

epidemiological transition The theory that populations, typically, shift from being defined by high rates of infant mortality (and low life expectancy) to a state in which average life expectancy is much higher (50+ years)

exogenous factor A relationship with another place/s that help to shape the unique character of a place, for example, membership of the European Union; such relationships can be seen in the movement or flow of people, resources, money and ideas across space

exponential growth Increasingly more rapid growth at a constant rate

exurbs Residential areas that are planned and built beyond the suburbs to create large extended suburbs, typically associated with urban sprawl

Fairtrade A value-based organisation and trademark that aims to tackle injustices of the globalised economy. Fairtrade aims to pay farmers a guaranteed minimum price, offer fair terms of trade and make payment of an additional development premium for reinvestment.

far (distant) place Somewhere that an individual or society perceives as being physically distant, generally inaccessible. Beyond actual spatial distance, such a perception may be shaped by networks of infrastructure (transport, communication) or access to them. Moreover, such a place may be viewed by some people as being different, even alien or exotic.

flexibility of production A method of production that is sufficiently flexible to be able to respond to both planned and unplanned changes, such as strikes or natural disaster

forces of change The individuals, community groups, companies, governments, national and international institutions that influence the place-making process

Fordism System of mass production of goods that involved assembly lines and 'living wages' for employees, pioneered by Henry Ford in the manufacture of cars in the early twentieth century

foreign direct investment (FDI) An investment made by a company, usually a TNC, based in one country into a company based in another country; the investment is usually made to acquire control or to have significant influence over the foreign company

fracking More correctly known as hydraulic fracturing. Oil- and gas-bearing shale is drilled and fractured by high-pressure injection of water, sand and (toxic) chemicals. Cracks are created in the shale through which the oil or gas will flow more freely.

fuel poverty The condition of being unable to afford to heat your home sufficiently, given your level of income and all other outgoing costs, such as food, rent, transport, etc.

gatekeeper Term used by social scientists to describe individuals within a community or institution who can grant a researcher access to people or sources of data by virtue of their economic role or social standing

generic medication A drug that is comparable to a well-known, branded drug in form, strength and performance but that is produced by a different manufacturer, and is cheaper for the patient or health service to purchase

genius loci A term used by planners to describe the key characteristics of a place, with which any new developments must concur

gentrification The improvement of urban areas by individual property owners, which usually leads to increased commercial activity in local retail areas

geopolitics The study of the ways in which political decisions and processes affect the use of space and resources; it is the relationship between geography, economics and politics

geospatial data A type of data that has a spatial or geographic component, meaning it can be mapped; pieces of digital data may have explicit geographic positioning (e.g. latitude and longitude) information linked to them, such as georeferenced satellite images

ghetto An area of a city in which large numbers of people of a particular minority ethnic group live

global product A product that is marketed and branded throughout the world. Many TNCs produce global products, for example, Coca-Cola, Nike and (Jaguar) Land Rover.

heritage tourism Travel to experience places, artefacts and activities that represent the stories and people of the past; the degree to which such an experience may be 'authentic' is a matter of debate

horizontal integration Improving links between different firms in the same stage of production

identity Who a person is, both in terms of how others view them, and how they see themselves. A person's identity is shaped, in part, by where they live and/or their place of birth (their homeland).

index case First identified case of a disease

indigenous Originating in a particular region or environment

infrastructure Basic facilities and installations that allow a city/country to function

insider Someone who feels safe, secure and 'at home' in a place; they understand the social norms of the society and feel included. They can play an active social and economic role in society.

knowledge economy An economy based on creating, evaluating and trading knowledge and high level skills

leachates Toxic waste water containing arsenic, lead, solvents and other contaminants leached from illegal dumps and landfill

leaching The process by which heavy rainfall infiltrates through a soil, removing humus and nutrients in solution

liveability The combination of factors that determine a community's quality of life; they include the built and natural environments, its accessibility, economic prosperity, social stability, educational opportunities and sustainability, and culture, entertainment and recreation

locale A setting where everyday life activities take place, for example an office, park or cruise ship; people behave in a certain way in a locale, according to social norms or rules

location A physical position that can be plotted on a map

malnutrition Inadequately balanced diet whether through undernutrition or overnutrition (obesity)

marine protected areas (MPAs) Marine areas where certain activities are limited or prohibited in order to meet specific conservation, habitat protection, or fisheries management objectives

marine reserves Marine reserves are fully protected areas that are off-limits to all extractive uses, including fishing. They provide the highest level of protection to all elements of the ocean ecosystem.

market economy An economy in which economic decisions, such as those regarding investment and production, are based solely on supply and demand with little government involvement

megaprojects Very large investment projects that typically cost more than (US)$1billion

metropolitan Spatial area that is greater than the limits of the city it relates to; it includes both the densely-populated urban core and its surrounding suburbs (plus, in some cases, other smaller urban areas) that are bound to the city by employment, commerce and/or infrastructure

microclimate Climate within a relatively small area that is distinctively different from the climate of the surroundings; for example, the climate of an urban area contrasting markedly with that of the surrounding countryside

Millennium Development Goals Eight international development goals that were established following the Millennium Summit of the United Nations in 2000; each goal had a target to meet by 2015. Replaced by the Sustainable Development Goals in the same year.

modes Particular forms or means of something, for example, transport

monopoly When a single company or group owns all or nearly all of the market for a given type of product or service – there is little choice and little competition

morbidity The incidence or prevalence rate of a disease or all diseases within a population

mortality Being subject to death, being mortal; the most common indicator of mortality within a population is the death rate

myth Socially-constructed versions of reality that may, as a result of its long history or because it is widely-held, be thought of as common sense

natural population change The pattern of change in population over time that does not take into account the impact of migration

near place Somewhere that an individual/society perceives as being physically close either by virtue of being easily accessible and/or spatially close. Such a place may be viewed as being similar, perhaps, even inextricably linked to the place where an individual/society is located.

neo-Malthusianism Views or attitudes that are in common with Thomas Malthus who believed that there are environmental limits to population growth

NIMBYism 'Not in My Back Yard'. An attitude shared by those who do not want a development, such as a wind farm, in their near locality

non-utilitarian Not practical, such as a beautiful landscape

North Atlantic Treaty Organisation (NATO) Based upon the North Atlantic Treaty signed in April 1949. It is an intergovernmental military alliance whereby its member states agree to mutual defense in response to an attack by any external party.

obese For adults of both sexes, aged 18 and over, obesity is a body mass index (BMI) greater than or equal to 30 (overweight is a BMI between 25 and 29.9)

Office for National Statistics (ONS) largest producer of official statistics in the UK

Organisation of Petroleum Exporting Countries (OPEC) An organisation or cartel that follows a common approach to the sale of oil

optimum population A theoretical concept, and ideal; the number of people that can make the best use of all available resources within a country or region, ensuring that everyone has an adequate standard of living

ores Rocks where the mineral content (usually metal) is of sufficient economic value to justify exploitation

Orientalism Edward Said's theory that, historically, Europeans viewed people in the Orient (a region that included the Middle East, North Africa and Asia), as being not only exotic, but decadent and corrupt. Such a view of the Orient was used, at the time, to justify the actions of imperial powers.

other Someone or something that is different, alien or exotic. A person living in a distant place (or key characteristics of their way of life) may be defined as 'other' by individuals or a society, as a result of the perceived contrast between 'them' and 'us'.

out of place A feeling of not being 'normal', not fitting in, in the context of a particular society or locale

outsider Someone who feels homesick, alienated or excluded from society in a specific place; they may not be able to take an active role, for example, in work or study as a result of socially-constructed barriers

overpopulation Too many people for the resources or technology available in a given area, or country, to support at an adequate standard of living

pandemic An epidemic occurring worldwide, or over a very wide area, crossing international boundaries and usually affecting a large number of people

particulates Tiny particles, such as dust or soot, given off when fossil fuels such as coal or oil are burned

place More than its physical location, a place is a space given meaning(s) by people

place study An investigation of a place and its developing character, involving the collection, analysis and interpretation of both quantitative and qualitative data, including representations of it in the media

placelessness The idea that a particular landscape 'could be anywhere' because it lacks unique features. Some UK high streets have been criticised for being dominated by identically-branded chain stores.

planning blight The reduction of economic activity and/or property values in a particular place resulting from expected or potential future development or restrictions on development

postmodernism A philosophical movement that applies a particular viewpoint to the world in which we live

primary health care Health care that is provided at the point of contact to all in the community

primary products Goods made up of a natural raw material and that have not been through any manufacturing process (such as oil, timber or fish)

privatisation Transfer of ownership from the government to private companies and businesses

property-led regeneration Regeneration led by property developers and financiers with little input from government except the creation of opportunities or tax breaks

provenance The context in which a source or text is produced, which may give clues to its purpose (and its reliability)

qualitative data Data that can only be organised into descriptive categories that are not numerical; such data may include oral sources such as interviews, reminiscences and songs, and visual media include artistic representations

QUANGO An acronym for quasi-autonomous local government organisation; an administrative body funded, but not run, by the government

quantitative data Numerical data to which different techniques of statistical analysis may be applied to test a hypothesis

rebranding A process to rename and reimage a location to boost footfall, investment and the wider social and economic prospects of a place

refugee A person who has been granted leave to stay in a foreign country, having been forced to leave the country of his or her nationality, 'owing to well-founded fear of being persecuted for reasons of race, religion, nationality, membership of a particular social group or political opinion'(UNHCR)

regeneration The process of urban or rural improvement , which may be economic, social or environmental in nature

reliable A social scientist must try to evaluate how accurate or useful a particular source is. This can be done by comparing it to other sources and also finding out about the author or artist, their wider views and those of their patrons (its provenance).

remote sensing The scientific collection of (mass) data from objects or areas without being in physical contact with them; the data is gathered by electronic scanning devices carried in high-flying aircraft or satellites

residential differentiation Demarcation of socio-economic, cultural and ethnic groups within specific housing areas

retrofit To adapt something or add to it, after it has been installed for use, for example, the addition of double glazing to a house once built and single-glazed

reverse osmosis Involves pushing salt or brackish water through a porous membrane that filters out salts and other impurities to produce freshwater

royalties Proportion of profits paid to whoever grants a mining lease

Rust Belt A region that stretches across the north-east of the USA, the Great Lakes and the Midwest states that has suffered from economic decline of key industries; it was once called the Steel Belt

sedentary Too much sitting and insufficient physical exercise

sense of place (place meaning) An individual's subjective and emotional attachment to a place, its place meaning

single-product economy A country which relies on one, or a very small number, of products (usually raw materials) for its export earnings

socially constructed Social processes produce and reproduce the social and economic relations between different groups of people in society, in different locations.

One aspect of this is place meaning. The dominant place meaning of a location may benefit dominant classes and the status quo.

space The three-dimensional surface of the Earth; further defined by human geographers as a container in which objects are located and human behaviour is played out

spatial imbalance An unevenness in any geographical distribution, for example, precipitation

sports-led regeneration Regeneration which focuses on the benefits brought by sports events

spreading of locational risks Spreading of risk by investing in different geographical locations – 'Not putting all the eggs in one basket'

stakeholders Individuals, groups or organisations that are affected by an issue

stewardship The responsible use and protection of the natural environment through conservation and sustainable practices

suburbanisation The outward growth of people, services and employment towards the edges of an urban area

syncline Where rocks have been folded downwards into a basin shape

teleworking A work arrangement that allows an employed person to work from home; this is made possible by the use of technologies such as the internet, satellite technologies and mobile phones

terms of trade The value of a country's exports relative to that of its imports

text A source for analysis; texts used by cultural geographers include works of art and films, as well as written sources like novels, poetry and travelogues

thermal distillation Uses heat to create a vapour from salt or brackish water which is then converted into freshwater

tourist gaze What a visitor sees or experiences of a place of interest (e.g. a historic site). To some extent, it is organised or edited by professionals in the tourist industry.

trade bloc Where a set of countries trade freely with each other with few, if any, barriers such as tariffs. Countries outside this area that wish to trade anywhere within the trade bloc have to pay an agreed tariff.

trade protectionism The use of barriers such as import duties (tariffs) or customs in order to increase the price of imports and so protect domestic production

trailhead attraction A primary attraction to which visitors flock, from which they can be redirected to secondary attractions benefitting the wider district, city or region

trans-boundary pollution Pollution which crosses national boundaries

transhumance Animals are tended on upland pastures during the summer, while fodder crops are grown in the valleys; the animals are then moved down for winter stall feeding

transnational companies (TNCs) Corporations or companies that operate in at least two countries

trickle-down effects The diffusion of the benefits of urbanisation, such as economic growth and prosperity, to poorer districts and people

undernourishment Measure of hunger referring to not having enough food to develop or function normally

underpopulation A population that is too small to fully utilise its resources; or a situation in which the resources could support a larger population without any reduction in living standards

United Nations Environment Programme (UNEP) The leading global environmental authority and advocate. It arguably sets the global environmental agenda, and seeks to promote the coherent implementation of the environmental dimension of sustainable development (within the United Nations system).

urban resurgence Population movement from rural back to urban areas, such as university students and upwardly mobile young people, reviving inner city and CBD areas urban sprawl

vectors Living organisms that can transmit infectious diseases between humans, or from animals to humans; they cause more than 1 million deaths annually

vertical integration An industry where one company either owns or controls multiple stages in the production and distribution chain

virtual water The volume of freshwater used to produce a product, measured at the place where the product was actually made

waste stream The complete flow of waste from its source through to recovery, recycling or disposal

water catchment An area of land through which water from any form of precipitation drains into a body of water (including groundwater supplies)

water scarcity Severe water stress; it is largely accepted that this occurs when annual water supplies fall below $1000\,m^3$ per person

water security The ability of a country to protect access to safe water resources for all the population

water stress When the demand for water exceeds available water resources, or when poor quality restricts its use

world cities Interconnected global economic centres that have significant influence on the world economy

World Trade Organisation (WTO) A global organisation that deals with the rules of trade between nations

zone in transition An area immediately surrounding the city centre of mixed land use – including older industries, terraced housing and areas of improvement and redevelopment

Index